This item must be returned or renewed by the last date shown below. The loan period may be shortened if it is reserved by another reader. A fine will be due if it is not returned on time.

DATE OF RETURN

UNIVERSITY COLLEGE LONDON
Gower Street London WC1E 6BT

Model Organisms
in Drug Discovery

Model Organisms in Drug Discovery

Edited by

Pamela M. Carroll and **Kevin Fitzgerald**

Applied Genomics
Pharmaceutical Research Institute
Bristol-Myers Squibb
Princeton, New Jersey
USA

John Wiley & Sons, Ltd

Copyright © 2003   John Wiley & Sons Ltd, The Atrium, Southern Gate, Chichester,
West Sussex PO19 8SQ, England

Telephone (+44) 1243 779777

E-mail (for orders and customer service enquiries): cs-books@wiley.co.uk
Visit our Home Page on www.wileyeurope.com or www.wiley.com

Cover photograph by Charlotte Raymond of Bristol-Myers Squibb, taken at Exelixis, Inc

Other Wiley Editorial Offices

John Wiley & Sons Inc., 111 River Street, Hoboken, NJ 07030, USA

Jossey-Bass, 989 Market Street, San Francisco, CA 94103-1741, USA

Wiley-VCH Verlag GmbH, Boschstr. 12, D-69469 Weinheim, Germany

John Wiley & Sons Australia Ltd, 33 Park Road, Milton, Queensland 4064, Australia

John Wiley & Sons (Asia) Pte Ltd, 2 Clementi Loop #02-01, Jin Xing Distripark, Singapore 129809

John Wiley & Sons Canada Ltd, 22 Worcester Road, Etobicoke, Ontario, Canada M9W 1L1

Wiley also publishes its books in a variety of electronic formats. Some content that appears in print
may not be available in electronic books.

Library of Congress Cataloging-in-Publication Data

Model organisms in drug discovery / edited by Pamela M. Carroll and
Kevin Fitzgerald.
        p. cm.
Includes bibliographical references and index.
    ISBN 0-470-84893-6 (Cloth : alk. paper)
   1. Pharmacogenomics. 2.   Pharmacogenetics -- Animal models. 3.
Drugs -- Research.   I. Carroll, Pamela M.   II. Fitzgerald, Kevin J., 1967-
    RM301.3.G45M635 2003
    615'.19 -- dc21

                                            2003013114

British Library Cataloguing in Publication Data

A catalogue record for this book is available from the British Library

ISBN 0 470 84893 6

Typeset by Dobbie Typesetting Ltd, Tavistock, Devon
Printed and bound in Great Britain by TJ International, Padstow, Cornwall
This book is printed on acid-free paper responsibly manufactured from sustainable forestry in which at
least two trees are planted for each one used for paper production..

Contents

Model Organisms in Drug Discovery. Edited by Pamela M. Carroll and Kevin Fitzgerald.
© 2003 John Wiley & Sons, Ltd. ISBN 0 470 84893 6

List of Contributors

Hector Beltrandelrio Lexicon Genetics Incorporated, 8800 Technology Forest Place, The Woodlands, TX 77381, USA

Thierry Bogaert Devgen NV, Technologiepark 9, B-9052 Ghent-Zwijnaarde, Belgium

Lynn Butler Devgen NV, Technologiepark 9, B-9052, Ghent-Zwijnaarde, Belgium

Pamela M. Carroll Department of Applied Genomics, Pharmaceutical Research Institute, Bristol-Myers Squibb Company, PO Box 5400, Princeton, NJ 08543, USA

Neil Dear Ingenium Pharmaceuticals AG, Fraunhoferstrasse 13, D-82152 Martinsried, Germany

Steven A. Farber Department of Microbiology & Immunology, Kimmel Cancer Center, Thomas Jefferson University, Philadelphia, PA, USA

Kevin Fitzgerald Department of Applied Genomics, Pharmaceutical Research Institute, Bristol-Myers Squibb Company, PO Box 5400, Princeton, NJ 08543, USA

Barbara Froesch The Genetics Company, Inc., Wagistrasse 27, CH-8952 Zürich-Schlieren, Switzerland

Dan Garza Department of Functional Genomics, Novartis Pharmaceuticals Corporation, 556 Morrive Avenue, SEF2026, Summit, NJ 07901, USA

Johannes Grosse Ingenium Pharmaceuticals AG, Fraunhoferstrasse 13, D-82152 Martinsried, Germany

Ernst Hafen The Genetics Company, Inc., Wagistrasse 27, CH-8952 Zürich-Schlieren, Switzerland

Shiu-Ying Ho Department of Microbiology & Immunology, Kimmel Cancer Center, Thomas Jefferson University, Philadelphia, PA, USA

Model Organisms in Drug Discovery. Edited by Pamela M. Carroll and Kevin Fitzgerald.
© 2003 John Wiley & Sons, Ltd. ISBN 0 470 84893 6

Titus Kaletta   Devgen NV, Technologiepark 9, B-9052 Ghent-Zwijnaarde, Belgium

Francis Kern   Lexicon Genetics Incorporated, 8800 Technology Forest Place, The Woodlands, TX 77381, USA

Rachel Kindt   Exelixis, Inc., 170 Harbor Way, South San Francisco, CA 94083-0511, USA

Thomas Lanthorn   Lexicon Genetics Incorporated, 8800 Technology Forest Place, The Woodlands, TX 77381, USA

Hao Li   Department of Functional Genomics, Novartis Pharmaceuticals Corporation, 556 Morrive Avenue, SEF2026, Summit, NJ 07901, USA

Geert Mudde   Ingenium Pharmaceuticals AG, Fraunhoferstrasse 13, D-82152 Martinsried, Germany

Michael Nehls   Ingenium Pharmaceuticals AG, Fraunhoferstrasse 13, D-82152 Martinsried, Germany

Tamas Oravecz   Lexicon Genetics Incorporated, 8800 Technology Forest Place, The Woodlands, TX 77381, USA

Michael Pack   Departments of Medicine and Cell and Developmental Biology, University of Pennsylvania School of Medicine, Room 1212 BRB 2/3, 421 Curie Blvd., Philadelphia, PA 19104, USA

James Piggott   Lexicon Genetics Incorporated, 8800 Technology Forest Place, The Woodlands, TX 77381, USA

David Powell   Lexicon Genetics Incorporated, 8800 Technology Forest Place, The Woodlands, TX 77381, USA

Ramiro Ramirez-Solis   Lexicon Genetics Incorporated, 8800 Technology Forest Place, The Woodlands, TX 77381, USA

Petra Ross-Macdonald   Department of Applied Genomics, Pharmaceutical Research Institute, Bristol-Myers Squibb Company, PO Box 5400, Princeton, NJ 08543, USA

Andreas Russ   Ingenium Pharmaceuticals AG, Fraunhoferstrasse 13, D-82152 Martinsried, Germany

Arthur T. Sands   Lexicon Genetics Incorporated, 8800 Technology Forest Place, The Woodlands, TX 77381, USA

Andreas Schröder   Ingenium Pharmaceuticals AG, Fraunhoferstrasse 13, D-82152 Martinsried, Germany

Stefan Schulte-Merker Exelixis, Deutschland GmbH, Spemannstrasse 35, D-72076, Tübingen, Germany

Corina Schütt The Genetics Company, Inc., Wagistrasse 27, CH-8952 Zürich-Schlieren, Switzerland

Reinhard Sedlmeier Ingenium Pharmaceuticals AG, Fraunhoferstrasse 13, D-82152 Martinsried, Germany

Gabriele Stumm Ingenium Pharmaceuticals AG, Fraunhoferstrasse 13, D-82152 Martinsried, Germany

Sigrid Wattler Ingenium Pharmaceuticals AG, Fraunhoferstrasse 13, D-82152 Martinsried, Germany

Brian Zambrowicz Lexicon Genetics Incorporated, 8800 Technology Forest Place, The Woodlands, TX 77381, USA

Acknowledgments

It has been an exciting time to translate model systems and the new age of genomics into relevant technologies for drug discovery. We thank all of the excellent authors for their creative contributions. We also thank our colleagues at Bristol-Myers Squibb for their continued support of our work and for allowing us to pursue model organism approaches. In particular, we would like to acknowledge the visionary leadership of Drs Elliott Sigal and Mark Cockett at Bristol-Myers Squibb and Drs Geoff Duyk and Greg Plowman at Exelixis. We thank our excellent editor Joan Marsh and editorial assistant Layla Paggetti at John Wiley & Sons, Ltd for their guidance, suggestions, expertise and especially their patience.

Model Organisms in Drug Discovery. Edited by Pamela M. Carroll and Kevin Fitzgerald.
© 2003 John Wiley & Sons, Ltd. ISBN 0 470 84893 6

1

Introduction to Model Systems in Drug Discovery

Kevin Fitzgerald and **Pamela M. Carroll**

A major challenge in the 'post-genomic' world is to rapidly uncover the proteins that may become the high-quality therapeutic targets of the future. This book will focus on the utility of model organisms as a systematic approach to a broad array of disease-based questions. The recent publication of the human genome revealed the most complete set of human genes to date, yet most of these genes have not been assigned a biological function and an even smaller number have been linked to a human disease process. Comparative genomic analysis of simple model systems with that of the human has revealed the evolutionary conservation of gene and protein structure as well as 'gene networks'. This evolutionary conservation is now being exploited with model systems as critical 'functional genomics' linchpins, in associating conserved genes with therapeutic utilities. Genes of unknown function can now be studied in the more tractable model systems and inferences can be drawn about their roles in complex biological processes.

1.1 Integrating model organism research with drug discovery

Pharmaceutical drugs in the modern era are something we all take for granted. We swallow a pill if we have a headache and magically the pain abates. Infections that in the past caused limb amputations, paralysis, lung damage or death are treated by antibiotic tablets and the infection and symptoms abate.

Model Organisms in Drug Discovery. Edited by Pamela M. Carroll and Kevin Fitzgerald.
© 2003 John Wiley & Sons, Ltd. ISBN 0 470 84893 6

Diseases such as diabetes, AIDS, high blood pressure and cholesterol that often resulted in a host of serious and medical issues are now controlled with medications. Life expectancy has increased and the quality of life in old age continues to improve. Drug discovery and development have a remarkable history of success considering that the quest for new pharmaceuticals traditionally has encompassed searching for a needle in a chaotic and disorganized haystack of complex human biology and disease. It was not until the release of a complete draft of the human genome sequence in 2001 that scientists were provided with a list of all possible drug targets for pharmaceutical intervention. The current and future challenges are to identify those genes implicated in disease and to leverage the genome information into an understanding of complex biological systems, efficiently paving the way for drug discovery.

The genome information provides the rudimentary gene list for all possible drug targets but still leaves scientific research a great distance from understanding the role of each of these protein targets in normal biology and disease processes. Years from now the sequencing of not only the human genome but the genomes of *Saccharomyces cerevisae* (yeast), *Caenorhabditis elegans* (nematode), *Drosophila melanogaster* (fruit fly), *Danio rerio* (zebrafish) and *Mus musculus* (mouse), as well as a large number of unpleasant pathogenic bacteria and viruses, will be looked upon as watershed events in the development of novel medicines. Parallel to the sequencing of the genome are advances in chemistry, engineering, microscopy and genetics that are having a major impact on the drug discovery process. The purpose of this book is to update and forecast how these technological advances are being combined with model organisms in biology to have an impact on modern drug discovery.

A useful analogy of model organism studies is the hobby of constructing 'model' cars or planes. Such model kits arrive with a parts list, a large number of pieces and an assembly manual that describes the function of each part and how the various parts fit together into a three-dimensional working object. Models can be manipulated by removing a part and determining the overall structure and function of the model without that part. The same is true of model organisms in drug discovery. The genome sequences of 'model' systems described in this book are the list of parts. Of course, we are not handed the assembly manual (therein lies both the challenge and the promise) but biologists are arduously writing this very complex manual in small bits at a time. Organisms arrive whole and functioning, and scientists strive to deconstruct the functioning end product into its various parts and then hypothesize about the functions of individual parts and the connections between them. This is actually more akin to someone handing you a functioning F-16 fighter jet along with a parts list and requiring you, without any instruction manual, to assemble a new fighter jet or, in an analogy to a

Table 1.1 Genome comparisons of model organisms

Organism	Transcriptome size	% Genes[1] similar to a human gene	Cellular complexity	Generation time
Yeast	6200 genes	46%	1 cell	2 h
Nematode	18 300 genes	43%	~959 cells	3 days
Drosophila	14 400 genes	61%	$>10^6$ cells	10 days
Zebrafish	30 000–80 000 genes	>80%	$>10^8$	6–8 weeks
Mouse	30 000–80 000 genes	95–97%	$>10^9$ cells	6 weeks

[1]From Lander, E. S., *et al.* (2001) *Nature* **409**, 860–921.

human disease state, to diagnose and fix a malfunctioning jet. The progress in genetic and molecular tools has allowed us to begin the process of deconstructing normal and disease biology, but the process remains daunting and in reality will most likely take decades to complete. Because we cannot dismantle the human organism, we rely upon the fact that biology has evolved in a similar fashion from the single cell yeast to the system complexity of the mouse. We utilize organisms such as *C. elegans* and *Drosophila* because scientists have the tools to deconstruct these organisms and ask questions about the functions of every gene. Scientists can leverage the fact that evolution, for the most part, did not reinvent the same processes many times. For instance, the process by which one cell divides to make a second cell is a conserved function and biological pathway in yeast and humans. Throughout this book you should begin to gain an appreciation for how few biological differences there are between animal models and humans, and how to exploit this similarity to uncover the causes of and find new treatments for human disease.

This book will review the technical and innovative advantages that are specific for each model organism, as well as provide detailed accounts of 'disease models' in simple organisms that have had an impact on the understanding of human biology. The model organisms of focus are yeast, nematodes, fruitflies, zebrafish and mice. Many of these organisms have the advantage of a complete genome sequence and recent sophisticated advances in 'forward' (going from a phenotype *in vivo* to the causative gene mutation) and 'reverse' (going from a gene to the phenotype of a mutation in that gene *in vivo*) genetic tools that allow for genome-wide functional discoveries.

Table 1.1 offers a glance at comparisons of the systems in terms of the number of genes, similarity to humans and life cycle length (personal communication with Ethan Bier). When embarking on research projects it is not always clear which organism to choose for human relevance and speed of discovery. With increasing biological complexity comes greater similarities to humans; therefore, the mouse would be the clear system of choice if it were not for its long generation time and cumbersome technologies. For example, when carrying out mutation studies, embryonic lethal mutations are often more easily characterized in the zebrafish than the mouse. In the last decade, we have seen experimental models such as *Xenopus laevis* (the frog) lose favor. In the case of *X. laevis* this is due to a large and polyploid genome making genomics and genetic undertakings unreasonable. On the horizon are new model systems that have not entered the subject of this book but may soon be on all our research radar screens. Sometimes a new system needs the commitment of powerful scientists to lead the research community. Would zebrafish have seen the massive worldwide undertaking of genetic screens and technologies without the commitment of *Drosophila* geneticist and Nobel Laureate Christian Nusslein-Volhard? Will Sydney Brenner, the founding father of *C. elegans* as a model organism and Nobel Laureate, leverage his interest in the Japanese pufferfish (*Fugu*) and its complete genome into an important experimental model?

Specific model organisms were chosen as this book's focus because they are widely accepted as valuable experimental models in genomics and genetics. Many biotechnology and pharmaceutical companies have programs centered on model organisms for an array of drug discovery and development platforms. Applications covered herein range from target identification, target validation, compound discovery and toxicology screening. Important models in drug development, such as rat and monkey, were not included largely due to less developed genetic tools. Each model system has a set of unique advantages and disadvantages offered by that particular genetic model. The biological problems that are chosen for study in each system depend on how likely a model system is to yield insights into human biology. For example, zebrafish offers an unparalleled visualization of a multi-organ vertebrate system and many of the organ systems (such as the circulatory system) are good models for human organs, but the technologies available for forward and reverse genetics are still relatively costly and time-consuming. Conversely, yeast offers rapid, efficient genetic approaches, but only about 50% of the gene networks are functionally conserved with humans and they lack the complex nature of human organ tissue systems. *Drosophila* in many cases represents a good 'happy medium' in that they integrate multiple complex organ systems yet have the rapid genetic tools used to deconvolute complex biology.

The chapters of this book are ordered along increases in evolutionary complexity towards humans, starting with yeast, nematodes and fruitflies and then proceeding into chapters centered around zebrafish and mice. One could also view this as a progression of technology development with an abundance of powerful genetic tools available in yeast, fruitflies and nematodes and the quest of zebrafish and mice researchers to develop similar technologies. The book will detail the incorporation of advances in the application of bio-informatics, proteomics, genomics, biochemical and automation technologies to simple organisms and how these advances constitute an integrated drug discovery platform. Detailed accounts of the application of model organism technology to specific therapeutic areas will be covered. The authors include leading experts in each field who will examine state-of-the-art applications of individual model systems, describe real-life applications of these systems and speculate on the impact of model organisms in the future. The first of these authors will delve into the relatively simple model organism, yeast.

Chapter 2 by Ross-Macdonald of Bristol-Myers Squibb describes the history of *Saccharomyces cerevisae* (yeast) research in drug discovery and how this simple eukaryote historically has been utilized mainly as a production vehicle due to its ability to produce compounds and proteins but also as a valuable tool in understanding biology. Yeast researchers have an unparalleled breadth of reagents to probe the genome, making it a natural choice for studying conserved targets and mechanisms of basic biological processes. With the sequencing of the yeast genome and the advent of such tools as transcriptional profiling, protein–protein interaction assays and genetic tools such as deficiency, overexpression and haploinsufficiency strain sets, yeast is now a workhorse in uncovering hidden links among genes and defining cell signaling circuits. Many of the genomics tools that are being applied to the other model systems were developed in yeast and the yeast model system continues to be an invaluable source of innovation and technology development. For this review, Ross-Macdonald has chosen to highlight the contributions of biotechnology and pharmaceutical researchers in order to focus this broad field.

Caenorhabditis elegans is a tiny worm composed of just around 900 cells and a life cycle of about three days, yet it contains many of the cell types and genes found in humans. It was the first multicellular organism to have its complete genome sequenced. It is in *C. elegans* where we begin to see the development of rudimentary tissues, organs and the beginnings of a more sophisticated nervous system. The level of complexity (complex but not so complex as to have little chance of ever understanding all of the various neuronal connections) is one of the attributes of *C. elegans* that first attracted Sydney Brenner to *C. elegans* as a model system. Research into *C. elegans* has played an essential role in our general understanding of more complex human diseases such as cancer (i.e. Ras oncogene), depression (i.e. neuronal signaling and drug mechanism of action), Alzheimer's disease (i.e. presenilin genes) and cell death. In Chapter 3, Kaletta, Butler and Bogaert from DevGen review the short but impactful career of *C. elegans* in drug discovery. They also take us through the detailed process of applying *C. elegans* technologies of 'high-throughput' target identification and compound screening. Clearly, there is a great future for *C. elegans* in drug discovery.

For nearly 100 years *Drosophila* genetics has been a central contributor of research on inheritance, genome organization and the development of an organism. *Drosophila* represents a 'happy medium' in that terrific genetic tools are available and yet there is a level of complexity to the organism that more closely resembles vertebrates. In *Drosophila* there is the emergence of a complex nervous system and visual and digestive organs. Chapter 4, authored by Li and Garza from Novartis, describes the *Drosophila* technologies that have evolved over this long history, and in Chapter 5 Ernst Hafen and colleagues at the Genetics Company and the University of Zurich show how these technologies have been implemented to decipher several important disease pathways. For example, recent genetic studies have revealed the *Drosophila* insulin-mediated signaling pathway and its astounding similarity to mammals, suggesting that *Drosophila* research deserves a place in the studies of metabolic diseases such as diabetes. Any discussion of drug discovery would be incomplete without a clear discussion of compounds that lie at the very heart of and are the ultimate goal of the process. It is clear that one of the emerging areas of model systems will be 'chemical genetics'.

Chemical genetics consists of combining the genetic tools of model organisms with novel compounds in order to get a better understanding of their mode of action. It also encompasses screening for compounds that interfere with biological processes and then using those compounds as tools, which, when combined with genetics, allow you to unravel pathways of gene interaction. Every chapter of the book touches upon this new emerging field and Chapter 6, authored by the editors and Rachel Kindt at Exelixis, is dedicated to this concept. Perhaps the most striking revelation contained in these pages is that compounds work on conserved targets across species and, although ultimately the compound affinities may differ, the mechanisms of action are similar. Chapter 6 highlights the utility and benefits of having multiple genetic systems to unravel a problem. Examples of relevance in understanding the mode of action of gamma secretase inhibitors in Alzheimer's disease and natural products in inflammation are discussed, and these examples explore the integration of compounds with genetics.

The emerging power of the zebrafish system is captured in Chapter 7 by Schulte-Merker at Exelixis and in Chapter 8 by Ho, Farber and Pack at Thomas Jefferson University and the University of Pennsylvania. Zebrafish are a vertebrate model that develop externally and transparently; thus the formation of many structures and biological processes can be easily monitored. The progress of genome mapping, mutagenesis screens and new 'knock-out' and overexpression technologies will provide significant insights into these biological processes (Chapter 7). Chapter 8 discusses a specific model where zebrafish are being utilized to study lipid metabolism with strong parallels to those found in humans.

Finally, Chapters 9 and 10 explore the advances in one of the workhorses of modern drug discovery, the mouse. Mice have been involved in drug discovery for some time as models of human disease but the adaptation of higher throughput technologies is just beginning to have an impact on the search for novel targets. In addition, the mouse model is coming into its own as a tool to 'de-orphan' the biology of novel targets and allow compounds to be tested in mouse models lacking any gene. In some areas such as neuroscience, a phenotype in a mouse model is the gold standard (besides active compounds or human genetics) that associates a given gene with a disease. The mouse-focused chapters are divided into forward genetic approaches contributed by Ingenium AG (Chapter 9) and the reverse genetics approaches based on work at Lexicon Genetics (Chapter 10). In forward genetics a phenotype is identified first and then the molecular basis of a given trait is identified. Historically, the process of phenotype to mutation has been laborious and time-consuming, but new genomics technology is rendering the process more robust. Chapter 9 reveals new approaches for novel, rapid, chemical genetic screens and mutation identification that allow for *in vivo* target discovery in unprecedented ways. Conversely, Lexicon Genetics (Chapter 10) describes its undertaking of systematic large-scale gene knock-outs of the 'druggable genome' in mice and the process in place to associate a gene's functions with disease. Because most drugs act as antagonists, knock-out phenotypes should mimic drug action.

An exciting paradigm for drug discovery is evolving. The current processes by which drugs are discovered are long and expensive. Many compounds still fall out of the discovery pipeline due to lack of efficacy and mechanism-based toxicity. Central to these reasons is a failure to understand properly all of the biological roles of potential drug targets in normal and disease processes (also referred to as 'target validation'). This knowledge failure results in ignorance of the many potential unpleasant consequences that could be rendered by compound modulation of the target's activity *in vivo*. The integration of model systems into the drug discovery process, the speed of the tools and the amount of *in vivo* validation data that these models can provide will clearly help to define better the disease biology and thereby result in better validated targets. Better targets will lead to high efficacy and less toxic therapeutic compounds. The future will see a merging of the genetics of model systems with proteomics, bioinformatics, structural biology and compound screening, creating the exciting new framework of drug discovery for the 21st century.

2
Growing Yeast for Fun and Profit: Use of *Saccharomyces cerevisiae* as a Model System in Drug Discovery

Petra Ross-Macdonald

Yeast has great utility as a surrogate system to study aspects of mammalian biology. This utility extends to the drug discovery process, where yeast has been used to reveal the mechanism of action of compounds, to discover and characterize components of signaling pathways and to dissect protein function. These applications of yeast are illustrated by examples of research published by major pharmaceutical companies.

2.1   Introduction

This chapter is intended to illustrate the use of yeast (*Saccharomyces cerevisiae*) as a model organism in drug discovery research. Yeast has had a long utility as the workhorse of pharmaceutical discovery research, whether as a representative of its pathogenic cousins or as a living eukaryotic vessel for bringing together reagents such as the two-hybrid system components or carrying reporter constructs for screening. However, I will confine this review to applications where yeast has been used as a true 'model' for vertebrate biology in the area of disease. To demonstrate the value of yeast in applied

Model Organisms in Drug Discovery. Edited by Pamela M. Carroll and Kevin Fitzgerald.
© 2003 John Wiley & Sons, Ltd. ISBN 0 470 84893 6

Pathway biology

- mechanism of drug action
- pathway dissection
- pathway construction

Protein biology

- inference by homology
- heterologous expression
- structure/function analysis
- biochemical assays

Figure 2.1 Outline of the areas in which yeast has been used as a model system for the biology of higher eukaryotes. Pharmaceutical research in these areas is described in the text

pharmaceutical research, my examples and citations are almost completely culled from publications by research scientists at major pharmaceutical companies (i.e. roughly the top 20 companies based on the market share). This approach results in the omission of many fine pieces of academic work that may have had publication priority, but the aim of this chapter is to demonstrate the type of yeast research that drug discovery organizations historically have regarded as worthwhile, informative and likely to affect their bottom line. Unfortunately this approach also unwittingly leads to the omission of much excellent biomedical research using the fission yeast *Schizosaccharomyces pombe*, because few examples of its application have been published by pharmaceutical companies. The uses of yeast described in this chapter are laid out in Figure 2.1; they include sections on the use of yeast in elucidating pathways and their components, including pathways that are not native to yeast and pathways involved in the mechanism of action of compounds. I will also describe more targeted experiments to characterize the functions of specific proteins. Finally, I will review the 'post-genome' tools, technologies and information resource advances that now enable yeast research.

2.2 *Saccharomyces cerevisiae* and its genome: a brief primer

Commonly known as baker's, brewer's or budding yeast, *S. cerevisiae* has been a standard laboratory microorganism since the 1950s. It has many endearing attributes, including the ability to fill a laboratory with a pleasant 'warm-bread' odor, yet also to survive years of abandonment in a fridge or freezer or even on a desiccated piece of agar in a forgotten petri dish. (Almost every yeast biologist has had the need to test this last assertion.) It is cheap to feed, non-pathogenic and divides every 2 h. It can grow either aerobically or anaerobically, depending on the nutrients provided, and in solid or liquid media. It can exist stably as a haploid or a diploid, and haploids can be mated and put through meiosis to recover haploid progeny in a matter of days. Although a unicellular organism, it can on occasion display such group characteristics as pseudohyphal growth, intercellular signaling and programmed cell death. Finally, a highly versatile transformation (transfection) system has been available for several decades. You can choose a vector that is linear, circular or integrating, high or low copy number, with a positive or negative selection system, and you can express your favorite gene from several types of regulated promoters. In addition, homologous recombination occurs with high efficiency, allowing the integration of transformed DNA into chromosomes at precise locations, replacing and deleting host DNA as desired.

The *S. cerevisiae* genome sequence was completed and almost entirely annotated for genes in 1996 (Goffeau *et al.*, 1996) but it has not remained static (Kumar *et al.*, 2002c). By comparison to most eukaryotes, coding regions are enviably simple to identify in yeast: about 70% of the genome encodes protein, and only about 4% of yeast genes contain introns, usually as a small insertion very near the 5' end of the coding region. For expedience, the primary annotaters of the genome set the ability to encode a 100-amino-acid protein as the cutoff for a gene (unless other evidence existed). Each resulting open reading frame (ORF) was given a unique and informative seven-character identifier, e.g. YOR107w. This name immediately tells a yeast biologist that the gene lies on the Watson strand of the right arm of chromosome XV, 107 genes distal from the centromere. Unlike the Dewey decimal system, this left no room for additions; fortunately there have been relatively few subsequent modifications of genes because these have had to be dealt with by inelegant suffixes (A, -B, etc.; inconsistencies in their syntax are a common source of error in data handling). This systematic name complements and conforms to the yeast genetic nomenclature adopted by consensus in the 1960s, in which the upper case notation informs us that a wild-type gene is being discussed, a lower case notation would indicate a mutant and Yor107w is the name of the encoded protein. All yeast genes thus have a systematic ORF name; about half of them also have one or more traditional three-letter gene names that are intended to reflect some property of interest, e.g. *RAD1* to identify the first gene identified from a mutant screen for radiation sensitivity. Yeast biologists have concluded that clarity in the literature is more important than their egos and nowadays they commonly agree on a single, rational primary gene name maintained in a central registry. These names are certainly duller than those for *Drosophila* – yeast never had *ether-a-gogo* but for over a

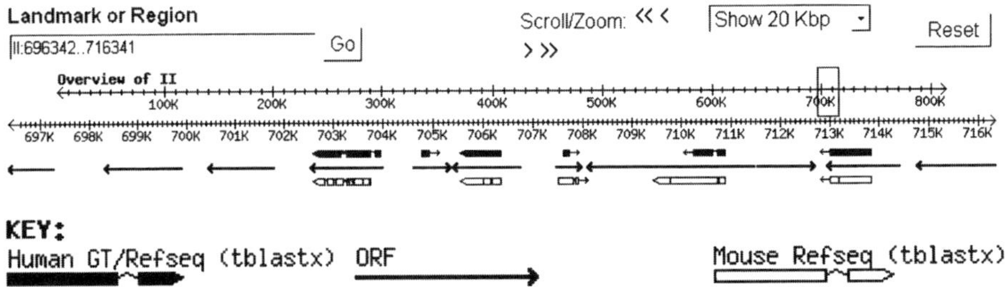

Figure 2.2   A graphical view of a 20-kilobase region of yeast chromosome II, showing 11 open reading frames (ORFs) encoding proteins. The NCBI Reference Sequence project (RefSeq) clones from human and mouse that show significant homology at the protein level are overlaid on their yeast homologs. The view was generated using the browser created by the Generic Model Organism Database Project (www.gmod.org) following customization by Dr N. Siemers

decade it did have *WHI1* (whiskey1, named in a pub in Scotland) until the title's overturn by the more prosaic name *CLN3* (cyclin3).

The number of recognized genes in yeast hovers just above 6000, remaining in flux due to continued research on which of these are spurious and what additions should be made (see Kumar *et al.*, 2002c). Figure 2.2 provides a visual snapshot of a region of the yeast genome and illustrates the significant homology between some of the proteins coded therein and proteins from the mouse and human genomes. Unfortunately, no quantitative cross-comparison between yeast and human genomes has been published since 'completion' of the human genome sequence. An analysis performed in 1997 found that about one-third of yeast proteins had significant homology to a mammalian GenBank sequence (Botstein *et al.*, 1997); by 1997 the results from Bassett *et al.* (1996) had been updated to suggest the existence of yeast homologs for 34% of the 84 disease-related human genes that were positionally cloned at the time. In 1998 a very stringent comparison between yeast and the newly finished *Caenorhabditis elegans* genome (Chervitz *et al.*, 1998) predicted that about 40% of yeast proteins were orthologous to about 20% of those encoded in worm. Many of the remaining 80% of worm proteins contained domains also present in yeast, but their arrangement within proteins was not identical. Because 80% of *C. elegans* proteins apparently lack a close relative in yeast, it might seem that there is a low probability of a given gene from a multicellular organism having a yeast homolog that can be studied productively. However, these numbers are skewed by the 'bulking out' of the *C. elegans* proteome by gene duplication events that lead to huge multigene families such as that for the nuclear hormone receptors. Within core metabolic and structural functions there is virtually complete conservation across eukaryotes. A recent comparison between the predicted proteins of the *S. pombe* and *S. cerevisiae* genomes and 289 human disease proteins found 182 *S. cerevisiae* proteins with significant similarity with about 50 probable orthologs (Wood *et al.*, 2002). The shared proteins covered a range of human disease areas from neurological to metabolic, the largest group being those implicated in cancer. Also, in many situations where a more intensive analysis has been brought to bear, proteins previously cited as absent from yeast have been found. A recent example is the identification of a caspase-type protein in yeast (Uren *et al.*, 2000) and demonstration of its orthology to metazoan caspases (Madeo *et al.*, 2002).

2.3 Yeast in pathway and mechanism elucidation

Selection of appropriate targets remains a major hurdle in drug discovery. When a biological pathway is of interest for therapeutic intervention, a broad understanding of its components is essential to allow the design of assays that can address both desired and undesired effects of that intervention. Knowledge of pathway biology is at its most advanced in yeast, owing to the ground cleared by decades of academic yeast research. Observations of cell cycle mutants of *S. pombe* and *S. cerevisiae* in the 1970s led directly to identification of the same pathway in humans and to the first generation of cyclin-dependent kinase (CDK) inhibitors currently in the clinic (Senderowicz, 2000). Yet examples of the use of yeast by pharmaceutical companies in further dissection of this pathway are rare, although Novartis has reported a yeast system to screen for Cdk4-specific antagonists (Moorthamer *et al.*, 1998). It seems that translation of observations in yeast to the relevance in mammalian systems and into pharmaceutical application continues to be underutilized. Mammalian biologists often feel that yeast is too simple to be of relevance to the process they study, or they point to incongruities in data to insist that yeast 'does it differently'.

Such reservations are partly justified: there are many examples of mammalian target proteins or drug effector mechanisms that are simply not present in yeast. For example, components of the cholesterol biosynthesis pathway, including the target for basic biochemical inhibitory action of the statin drugs, are largely conserved from yeast to humans. Yeast was used extensively by companies such as Bristol-Myers Squibb (Robinson *et al.*, 1993) and Zeneca (Summers *et al.*, 1993) in the identification and characterization of targets within this pathway. However, statins exert the majority of their cholesterol-lowering effect in humans by a feedback mechanism that leads to upregulation of the hepatic low-density lipoprotein (LDL) receptor, and this protein is not conserved in yeast (although feedback mechanisms responding to lowered sterol level do exist). Yeast also has no nuclear hormone receptors and thus lacks a form of regulation that overlays many conserved metabolic pathways in higher eukaryotes. Conversely, examples also exist of cases where yeast has proved to contain the target for a drug, even though that drug has its therapeutic effect in a process such as immunity, which has no apparent parallel in yeast. There are also cases where a very clear conservation exists and yet the published work is almost exclusively academic, e.g. the use of yeast in the determination of the mechanism of action of the topoisomerase inhibitors (reviewed by Bjornsti *et al.*, 1994). A search of the literature on camptothecin produces only one example of the use of yeast by industry: Takeda laboratories used *S. pombe* to demonstrate that the mechanism of a novel topoisomerase I inhibitor differs from that of camptothecin (Horiguchi and Tanida, 1995).

2.4   An example of mechanism elucidation: immunosuppressive agents

Three sterling examples of how yeast can contribute to the identification of a drug target and characterization of the responding pathway are provided by the immunosuppressive agents cyclosporin A, FK506 and rapamycin. The story of this research is also the story of what would have been an overwhelmingly difficult mechanism of action study without yeast, because it is a case where compounds interact with structurally unrelated binding partners to affect the same target and, conversely, compounds interact with the same binding partner to affect different targets (see Figure 2.3). The mechanism runs contrary to established wisdom on the feasibility of modulating protein–protein interactions. Finally, the binding partners are not the therapeutic target but, to throw in a couple of red herrings, they do have a common enzymatic activity that is inhibited by the compound! Without academic and industry groups striving neck and neck for the answer, and without yeast to identify additional components and provide genetic dissection and stringent hypothesis-testing, determination of their mechanisms within a decade of research is extremely unlikely to have occurred. Ironically, the ultimate targets are a kinase and a phosphatase, and today no right-thinking pharmaceutical company would put any money into a compound that took such a convoluted path to reach these targets. But these compounds were clinical successes before their mechanisms were established, and their efficacy has yet to be matched by small molecules from a rational development process. Cyclosporin A was identified in the 1970s at Sandoz (now Novartis) and approved for use as a transplant rejection therapeutic in 1983. As an interesting footnote, Novartis's own web page states that the initial observations on the natural product indicated a very weak compound that was regarded as being of little practical value. Fortunately an intellectual curiosity prevailed and allowed work to continue until Dr Jean Francois

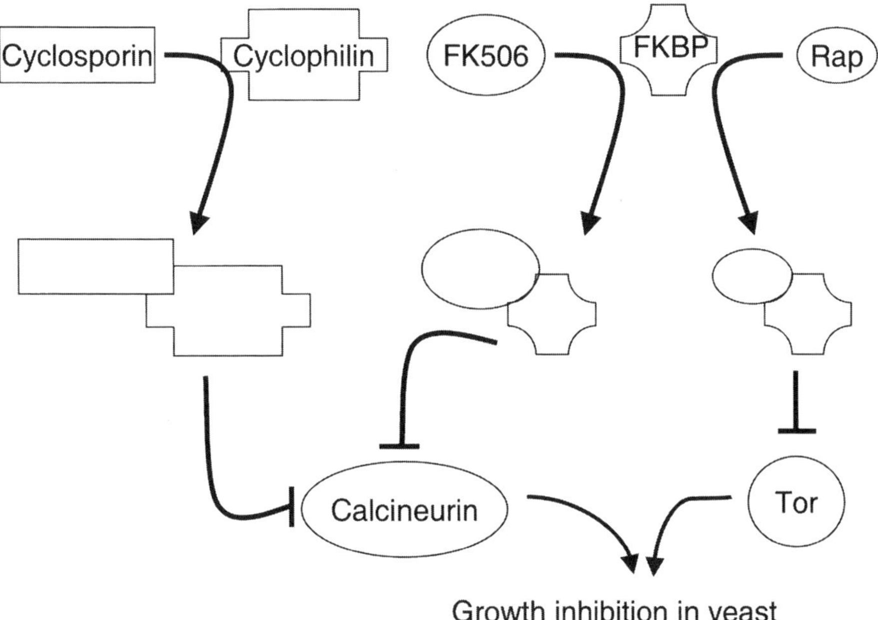

Figure 2.3   Binding partners and mechanism of action of cyclosporin A, FK506 and rapamycin (Rap). Cyclosporin A binds the cyclophilins, which are members of a family of proteins with peptidyl–prolyl isomerase activity. Both FK506 and rapamycin bind the same targets – a family of FK506-binding proteins (FKBPs). The FKBPs are members of a class of peptidyl–prolyl isomerases that are structurally unrelated to the cyclophilins. The cyclosporin A–cyclophilin and FK506–FKBP complexes both inhibit the protein phosphatase calcineurin. The rapamycin–FKBP complex inhibits the Tor kinases

Borel's team discovered the selective T-cell effects and purified the compound that became Sandimmune and Neoral, both long-running blockbusters for Novartis. Tacrolimus (FK506), marketed by Fujisawa as Prograf, was discovered in 1984 and gained FDA approval in 1994, whereas the related macrolide rapamycin (sirolimus), marketed by Wyeth as Rapamune, was discovered in 1975 and approved in 1999.

Far flung in origin, produced by fungi in the soil of Norway or bacteria from the shores of Easter Island or the Tsukuba region of Japan, all these immunosuppressive agents selectively block T-cell activation, with FK506 and cyclosporin A acting to block the transcription of early activation genes, and rapamycin blocking downstream events. However, they had begun their pharmaceutical careers as antibiotics, and scientists in academia (most notably the groups of Michael Hall and Joseph Heitman) and in industry applied yeast to understanding their mechanism and a search for the molecular target. The contributions of many scientists to this work are covered in a comprehensive review by Cardenas *et al.* (1994).

By 1990, cyclosporin A had been determined biochemically to bind and inhibit a target protein named cyclophilin that had been purified also as a peptidyl–prolyl *cis–trans* isomerase (PPIase). Academic work had shown that cyclophilin existed in yeast, and that CsA resistance in yeast correlated with the loss of cyclophylin interaction (Tropschug *et al.*, 1989). Yeast contributed to extensive structure/activity investigations of cyclosporin A at Sandoz (Baumann *et al.*, 1992). In 1990, Merck scientists reported that FK506 also bound and inhibited a protein that had PPIase activity. This protein (FKBP12) was from a novel class of PPIases. It was not lymphoid-specific and it was conserved from yeast to humans (Siekierka *et al.*, 1990). Because FK506 and CsA each inhibited the PPI activity of their binding partner, these 'immunophilins' were obvious candidates for the biological effector in mammals and for the lethality observed in yeast. Yet cloning and disruption of the yeast gene for the protein FKBP, *FKB1* (now *FPR1*), revealed that it was non-essential (Wiederrecht *et al.*, 1991). Scientists from SmithKline Beecham identified a second yeast cyclosporin-A-binding protein, Cyp2 (Koser *et al.*, 1990), and then a third (McLaughlin *et al.*, 1992), suggesting that a protein family was also targeted in humans. Although both Cyp1 and Cyp2 had PPI activity that could be inhibited by cyclosporin A, a triple deletion (*cyp1 cyp2 fpr1*) was viable. Although the existence of further PPI proteins giving functional redundancy was possible, these strains had very little PPIase activity, thus separating PPIase inhibition from lethality. Intriguingly, a genomic disruption of the *CYP1* gene gave cyclosporin A resistance in yeast, providing some of the first unequivocal evidence that the drug–immunophilin complex was a toxic agent (Koser *et al.*, 1991).

Following research from Stuart Schreiber's laboratory suggesting that the target of that toxicity was the protein phosphatase calcineurin (Liu *et al.*, 1992), Merck scientists showed that, like human FKBP12, yeast Fpr1 complexed with FK506 had the ability to inhibit this enzyme (Rotonda *et al.*, 1993). They also observed that the compound L-685,818, which acted as an FK506 antagonist in an immunosuppression assay and failed to inhibit calcineurin when complexed with human FKBP12, nonetheless proved to be an active inhibitor in complex with yeast Fpr1. Despite such differences in the behavior of drug–protein complexes, the crystal structure of yeast Fpr1 with FK506 was very similar to that of human FKBP12 with FK506, and pointed to structural modifications that could be made to improve potency (Rotonda *et al.*, 1993). Structure/function relationships between FKBP and its ligands were also explored by a group at SmithKline Beecham, who correlated the effects of an amino acid alteration with catalytic and ligand-binding properties and with protein function in yeast (Bossard *et al.*, 1994).

Work from Merck had been among the first to suggest that FK506 and rapamycin had different biological effects, indicating different targets (Dumont *et al.*, 1990), and yet the compounds acted as reciprocal antagonists and appeared to compete for binding to FKBP12. Scientists at SmithKline Beecham attempted to resolve this paradox by identifying rapamycin target proteins *in vivo* using yeast. The gene that they cloned by virtue of the rapamycin resistance of a mutant, *RBP1*, proved identical to that for the FK506 binding protein Fpr1. They showed that both rapamycin and FK506 inhibited the PPIase activity of Fpr1, and that heterologous expression of human FKBP12 restored rapamycin sensitivity to the rapamycin-resistant *fpr1* mutant, indicating a true functional equivalence (Koltin *et al.*, 1991). They identified mutations in two further genes, *DRR1* and *DRR2*, that showed a dominant phenotype of rapamycin resistance. Both *DRR1* and *DRR2* were proved to encode proteins of the phosphatidylinositol 3-kinase family (Cafferkey *et al.*, 1993), and are now called Tor1 and Tor2. Further characterization revealed that for both proteins it was a point mutation of a conserved serine residue that had been responsible for the resistance to the FK506–Fpr1 complex (Cafferkey *et al.*, 1994). The Tor proteins are now known to be part of a conserved signaling pathway that activates eIF-4E-dependent protein synthesis (reviewed by Schmelzle and Hall, 2000).

Although publications from industry have waned, academic research using yeast continues to illuminate the processes affected by these immunosuppressants and to indicate new targets in the pathway. Some of this work illustrates the application of genomic tools that will be described in the second part of this chapter, e.g. genome deletion collections (Chan *et al.*, 2000) and microarrays (Shamji *et al.*, 2000).

2.5 Application in pathway elucidation: G-protein-coupled receptor/mitogen-activated protein kinase signaling

The area of G-protein signaling pathways is one where the relevance and utility of yeast biology was not appreciated for many years. G-protein-coupled receptors (GPCRs) represent the most fertile area of therapeutic intervention, with GPCR agonists and antagonists accounting for over 50% of marketed drugs (cited in Gutkind, 2000). Targeting the receptor itself usually provides the requisite specificity and yet an understanding of the biology around the coupled heterotrimeric G protein and downstream signal transduction events is essential to address issues such as desensitization. Yeast possesses two GPCR-coupled pathways, and the biology of the mitogen-activated protein kinase (MAPK) cascade coupled to the mating receptor via Gβ/Gγ is unparalleled in the degree to which it has been dissected into molecular components (Dohlman, 2002). However, for many years mammalian GPCR effects were considered to be mediated solely via the Gα subunit, and yeast was regarded as an oddity for signaling via the Gβ/Gγ subunits. It was not until the mid 1990s that mammalian $G\beta/G\gamma/\text{MAPK}$ interactions were characterized and the direct analogy between yeast and metazoan pathways became obvious (reviewed by Gutkind, 1998). Components of GPCRs such as regulator of G-protein signaling (RGS) proteins, and MAPK pathway components such as scaffold proteins, were first identified in yeast but continue to find metazoan counterparts (see review by Gutkind, 2000). The genetic tractability of yeast allows for intelligent investigation of their function: see the use of scaffold/pathway fusion proteins to dissect control and specificity in MAPK signaling (Harris *et al.*, 2001).

As an adjunct to their extensive use of the yeast mating signal transduction pathway as a reporter system for GPCR ligand screening, two companies have published further characterizations of its components. Scientists at Glaxo Wellcome have characterized interactions between the $G\alpha$ subunit and the pathway scaffold protein Ste5 (Dowell *et al.*, 1998) that may have relevance to the recent identification of scaffold proteins in mammalian pathways (reviewed by Gutkind, 2000). Wyeth-Ayerst researchers collaborated in a study of the interplay between $G\alpha$ and the RGS protein Sst2, succeeding in uncoupling the regulation (DiBello *et al.*, 1998). Such observations raise the possibility that small molecules could modulate RGS function and thus GPCR signaling (Zhong and Neubig, 2001).

2.6   Applications in pathway deconstruction/reconstruction

An alternative use of yeast in the study of pathway biology has been to select a pathway where yeast lacks (or appears to lack) components, and to add these back. For example, a group at Glaxo used *S. pombe* as a host to reconstitute signaling through platelet-derived growth factor β to phospholipase $C\gamma2$ (Arkinstall *et al.*, 1995) and to investigate the structure/function behavior of the SHP-2 phosphatase (Arkinstall *et al.*, 1998). A more widely applied example is the use of yeast to study apoptosis. Until recently components of programmed cell death had seemed lacking in yeast, and observations suggesting that apoptosis did exist (reviewed by Frohlich and Madeo, 2000) were largely ignored. Thus, yeast seemed an ideal vessel in which to investigate determinants of the process. Researchers from Novartis were among those to observe that the apoptosis effector Bax can induce cell death in yeast, and that this effect was overcome by mammalian apoptosis inhibitors such as Bcl-2 and Bcl-x(L) (Greenhalf *et al.*, 1996). Novartis used the yeast system to identify two novel inhibitors of apoptosis – BASS1 and BASS2 (Greenhalf *et al.*, 1999) – and to characterize the structure/function behavior of Bax (Clow *et al.*, 1998) and Bfl-1 (Zhang *et al.*, 2000). Glaxo Wellcome used a Bak-mediated lethality screen in *S. pombe* to characterize host proteins involved in mediating that lethality, identifying calnexin 1 as a necessary component (see Torgler *et al.*, 2000). Researchers at Merck recently used homology to a yeast protein to clone sphingosine-1-phosphate phosphatase (SPP1), a human enzyme with a key role in the interconversion of metabolites that regulate apoptosis. Human SPP1 partially complements the loss of the yeast gene function, and overexpression induces apoptosis in mammalian cell culture (Mandala *et al.*, 2000). It remains to be seen whether interpretation of such data will be modified by the recent demonstration of a yeast caspase-related protease that regulates a genuine apoptotic effect (Madeo *et al.*, 2002), and the identification of molecules that induce the process (Narasimhan *et al.*, 2001).

2.7   Applications to the study of protein function

The conservation of protein structure and function among eukaryotes, and the ease of genetic and molecular manipulation make yeast a natural choice for studies of protein function. These range from inferring a human protein's function based on that of its yeast homolog, to detailed dissection of structural dependencies.

Inference of function

It is perhaps a measure of the acceptance of conserved roles that nowadays researchers seeking a role for a mammalian gene may cite the involvement of the yeast in a particular process as a powerful reason for examining that same role in mammalian biology. For example, scientists studying the REDK kinase at SmithKline Beecham note that the homologous yeast protein is a negative regulator of cell division. The function of the yeast homolog is presented as evidence in support of their hypothesis that REDK acts as a brake upon erythropoiesis (Lord *et al.*, 2000). Where the existing academic literature on a homolog is not sufficient to the needs of industry, researchers have performed studies to validate the function of a yeast protein. Thus, functional studies by Glaxo on the yeast Duk1 (Tok1) protein, which proved to be the founder member of a new structural class of potassium channels (Reid *et al.*, 1996), were only narrowly preceded by the same work from an academic group (Ketchum *et al.*, 1995).

Heterologous expression

The ability of proteins from multicellular eukaryotes to substitute for the yeast function has long been recognized. Back in 1996, the XREFdb project (Ploger *et al.*, 2000) had already reported the existence of 71 examples of human/yeast complementation. Use of a cloned mammalian gene to substitute functionally for a yeast protein has been widely used in industry, both as a means to isolate proteins and to prove their equivalence. Several examples were presented above in the research on mechanism of action of immunosuppressives, and in the characterization of cell cycle control components and apoptosis-regulating proteins. An additional example is the demonstration by researchers from Roche that three human RNA polymerase subunits could correctly assemble into multiprotein complexes and functionally substitute for the essential role of their yeast homologs (McKune *et al.*, 1995).

In some cases, human genes have been isolated deliberately based on their homology to a yeast protein. Examples include mSPP1, discussed in the section on apoptosis (Mandala *et al.*, 2000), or Chk2, the mammalian homolog of the *S. cerevisiae* Rad53 and the *S. pombe* Cds1 kinases. The latter was cloned by scientists at SmithKline Beecham and subsequently shown to complement partially the Cds1 function and to act as a downstream effector in the DNA damage checkpoint pathway (Chaturvedi *et al.*, 1999). Alternatively, novel proteins identified from a mammalian screen may be analyzed subsequently in yeast. For example, research at Eli Lilly identified a novel kinase, pancreatic eukaryotic kinase (PEK), from rat pancreatic islet cells and noted primary and structural homology to elongation initiation factor 2 kinases (eIF-2α kinases) but also a substantial and distinctive amino-terminal region. Despite this difference, they were able subsequently to demonstrate functional substitution by PEK for the yeast eIF-2α kinase GCN2, including use of the correct phosphorylation target site on eIF-2α (Shi *et al.*, 1998). These examples of functional complementation underscore the remarkable conservation of cellular machinery in eukaryotes.

Structure/function and structure/activity

Going one step beyond functional complementation are examples where heterologously expressed proteins are altered, or mutant forms of medical significance are used, in an attempt to correlate their structure with their properties. One example of an attempt to correlate the effects of a mutation with the role of a protein in disease is the use of yeast as a model to study the conductance regulator that is mutated in cystic fibrosis. Although academic research is still active in this area, pharmaceutical industry interest in this approach (as measured by publication) seems to have waned after research at Glaxo found that an early yeast model did not correctly mimic the mammalian disease biology (Paddon *et al.*, 1996). However, the general concept of using yeast for such analyses is undoubtedly of merit. There are several recent examples from academia where yeast has proved successful, e.g. in providing a model for the cellular defect (Pearce *et al.*, 1999b) and even suggesting a therapeutic route (Pearce *et al.*, 1999a) in Batten Disease, a progressive neurodegenerative disorder of a class that affects one in 12 500 births.

There is a more successful example of simple structure/function analysis from the pharmaceutical industry: after identifying SAG as a novel human protein involved in apoptosis that had a yeast homolog (Duan *et al.*, 1999), scientists from Warner Lambert (now Pfizer) demonstrated complementation of the yeast *hrt1* mutant function with SAG and showed a requirement for the RING protein (Swaroop *et al.*, 2000); SAG proved to be a novel homolog of the ROC1/Rbx1/Hrt1 protein, which interacts with the Skp–cullin–F-box protein complex to generate an active E3 ubiquitin ligase. This ligase promotes degradation of CDK inhibitory proteins, and mutants that had lost E3 ligase activity were unable to complement the yeast *hrt1* mutant. Upon withdrawal of SAG expression, an *hrt1* mutant arrests with a very heterogeneous DNA content, and transcription profiling identified responsive genes from both the G1/S and G2/M checkpoints (Swaroop *et al.*, 2000). This group also used complementation of *hrt1* in yeast to test whether a SAG splicing variant encoded a functional protein (Swaroop *et al.*, 2001).

Another example of the use of yeast in structure/function analysis is provided by the human phosphoacetylglucosamine mutase genes HsAGM1 and HsAGX1, which were cloned using yeast by scientists at the Nippon Roche Research Center. Gene HsAGX1 encodes a UDP-*N*-acetylglucosamine pyrophosphorylase that may be involved in antibody-mediated male infertility. After cloning based on homology, it was shown to substitute functionally for the loss of yeast Qri1 (Uap1), and key catalytic residues were investigated by site-directed mutagenesis (Mio *et al.*, 1998). Gene HsAGM1 was cloned by functional complementation in yeast and, after sequence comparisons with other family members identified as likely key residues, site-specific mutagenesis was successfully combined with *in vitro* and *in vivo* yeast assays to identify residues essential for catalytic activity (Mio *et al.*, 2000). In both cases, identification of likely catalytic residues to target for mutagenesis was facilitated by extensive characterization of the hexose phosphate mutase family in yeast (Boles *et al.*, 1994).

Mitogen-activated protein kinases and their associated pathways are currently a hot area of pharmaceutical research. The p38α kinase is an active target of several major anti-inflammatory programs (Drosos, 2002). As always with kinases, issues of specificity are at the forefront (Scapin, 2002). The potential utility of yeast in this field is shown by research at SmithKline Beecham directed at dissecting functional differences between p38/CSBP1 and an uncharacterized splice variant that they called CSBP2. They were able to demonstrate complementation of yeast *hog1* mutants by human CSBP1 and by mutants of CSBP2, but not native CSBP2, and to obtain structure/function information for kinase activity and the salt-responsiveness of the enzymes

(Kumar *et al.*, 1995). Hog1 is a yeast MAPK that responds to osmotic stress; the mutant phenotype also can be rescued partially by stress-activated protein kinase/Jun N-terminal kinase (SAPK/JNK) 1. Kinase p38α and SAPK/JNK activation by hyperosmolarity also seems to be conserved in some mammalian cell lines (reviewed by Kultz and Burg, 1998).

Several cases where the yeast protein structure could be correlated with compound activity (structure/activity relationship studies) in drug discovery research have been cited above in the work on cyclosporin, FK506 and rapamycin. Another published example is the inclusion of yeast farnesyl–protein transferase and geranyl–geranyl–protein transferase in structure/ activity evaluations of several chaetomellic acid chemotypes under study at Merck for inhibition of prenyl transferase activity (Singh *et al.*, 2000). Finally, yeast may also act as an *in silico* surrogate for mammalian proteins in structure/activity work: scientists at Novartis have described the use of the yeast crystal structure for the 20S proteasome to guide analog design for mammalian proteasome inhibitors that have therapeutic potential as antitumor agents (Furet *et al.*, 2001).

Biochemical assays

Many studies published by pharmaceutical companies have used yeast as a source of biochemical data; however, the majority most likely represent enzymes that are targets for antifungal drug discovery rather than those that model a vertebrate protein. Respiratory uncoupling proteins, which are implicated in the regulation of energy expenditure and the development of obesity, represent an area where *in vivo* biochemical studies in yeast have been used to characterize function. For example, when Merck scientists identified a novel member of the uncoupling protein family, they used expression in yeast to show that it caused a loss of mitochondrial membrane potential (Liu *et al.*, 1998). Novartis has also described heterologous expression of human uncoupling protein 1 (UCP1) and UCP3, measurement of their effects on mitochondrial polarization and modulation of their effects with purine nucleotides (Hinz *et al.*, 1999). The type of detailed kinetic data that can be obtained on a yeast enzyme *in vitro* is illustrated by an analysis of the steady-state mechanism of decarboxylation by orotidine-5′-phosphate decarboxylase, published by scientists from Glaxo Wellcome (Porter and Short, 2000).

2.8 Reagents and resources available in yeast

Ironically, from this fairly complete survey of publications from major pharmaceutical companies, it seems that utilization of yeast as a model system in drug discovery has fallen even as the ease of working with it has increased. Perhaps this is because mammalian systems have also become tractable enough to tip the balance, or perhaps exploratory biologists working in industry are publishing less of their basic research. Financial considerations have driven yeast researchers in academia to seize upon the information and tools generated by the complete genome, and direct their research in ever-more disease relevant and commercially applicable directions. However, the gap between an interesting observation and a drug in humans is still huge, and even the best academic research cannot substitute for the applied use of model systems within 'big pharma'.

Yeast is often cited for its ease and rapidity of use and for the range and sophistication of techniques available for genetic manipulation. Common techniques for manipulating yeast are covered in numerous texts; an excellent basic laboratory manual is provided by Guthrie and Fink's *Guide to Yeast Genetics and Molecular Biology* (Guthrie and Fink, 1991). Below I shall describe the range of 'genomic tools' now broadly available and the information resources they have generated, wherever possible giving examples of their use in disease-related research. The topics to be covered are laid out in Figure 2.4.

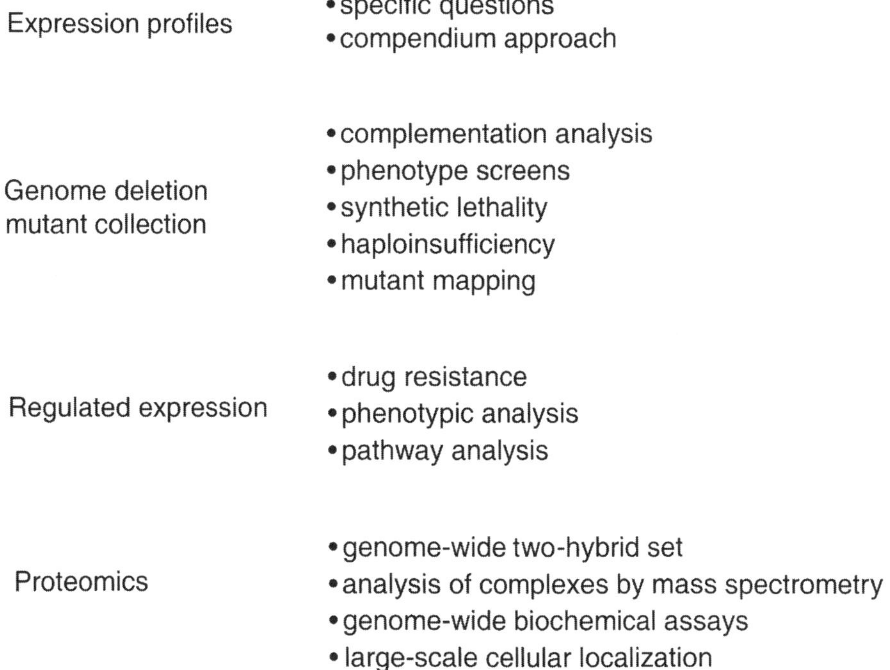

Figure 2.4 Reagents and techniques available in yeast, and their applications as discussed in the text

2.9 Gene expression profiling using microarrays

From the seminal publication in 1996 (Shalon *et al.*, 1996), yeast served as the test bed for academic and commercial development of microarrayed DNA probes and was the first organism for which whole genome arrays were available. Incyte's Yeast Proteome Database (YPD; Costanzo *et al.*, 2001) currently contains data culled from nearly 50 yeast genome microarray publications, including some covering as many as 119 experimental conditions (Roberts *et al.*, 2000). There is currently a variety of sources for yeast genome arrays, including primer sets for polymerase chain reaction (PCR) amplification of each ORF (Invitrogen), sets of 50-mer (MGW Biotech) or 70-mer (Invitrogen) oligonucleotides that each probe one ORF, ready-arrayed 50-mer oligonucleotides (MGW Biotech) and Affymetrix gene chips that use sets of 25-mer oligonucleotides to provide a readout for each gene. These reagents are listed in ascending order of price; they are also listed in descending order of effort to implement. The relative merits in this tradeoff can be harder to determine for yeast than other systems. In the use of mammalian-based assays for a transcriptional profiling experiment, the cost of generating samples is often much higher than that of the chip. Yeast, however, allows many hundreds of assays to be run in relatively short times with low reagent cost. This allows the creation of a database of profiles to which new entries can be compared in a manner analogous to BLAST searches on GenBank. The creation of an agglomerated public database of yeast transcriptional profiles still lies in the future, but the usefulness of such a resource was demonstrated by work at Acacia Biosciences. After creating a large dataset of response profiles generated by mutations and compounds of known mechanism, they were able to use clustering algorithms to categorize new compounds or compounds presented 'blind' into functional groups. Complexities of blocking a pathway such as isoprene synthesis at different steps could be revealed (Dimster-Denk *et al.*, 1999). This approach was extended by scientists at Rosetta Inpharmatics who compared transcription profiles of cyclosporin-A- and FK506-treated cells with those from null mutants for the immunophilins CPH1 and FPR1 and for the ultimate target protein calcineurin. They demonstrated that the ability of a compound to inhibit pathways other than its intended target can be quantified by such experiments, providing a means to group and rationally select desirable chemotypes (Marton *et al.*, 1998). By 2000 Rosetta had constructed a reference database of transcription profiles for 300 diverse mutations and chemical treatments in *S. cerevisiae*. 'Homology' of profiles within this database suggested functions for eight uncharacterized ORFs; such functions were then confirmed by more detailed individual analysis. As an example of the utility of such a database in compound classification, the observation that the profile for yeast treated with dyclonine resembled those of yeast mutants with blocks in sterol synthesis, and specifically that of an *erg2* mutant, suggests that this compound might mediate effects in humans by binding the sigma receptor, the closest homolog of Erg2 (Hughes *et al.*, 2000). The sigma receptor binds a number of neuroactive drugs, including the antipsychotic haloperidol (Haldol; Ortho-McNeil), which also likely inhibits yeast Erg2 (Acacia Biosciences, unpublished data). Dyclonine is a widely used topical anesthetic (Dyclone; Astra) that is reportedly longer lasting than benzocaine; if you want to test the effect on your own sigma receptors, there are a couple of milligrams of dyclonine in every Sucret throat lozenge (except original mint flavor)!

2.10   Deletion collections: reinventing traditional screens

Saccharomyces cerevisiae was early to benefit from reagents that allowed analysis of gene function on a genome-wide scale (Ross-Macdonald *et al.*, 1999). It now has a resource that will remain unparalleled in any system: a set of strains that comprise a start-to-stop-codon deletion for nearly every annotated ORF in the genome (Winzeler *et al.*, 1999; Giaever *et al.*, 2002). One immediate and obvious utility of the Yeast Genome Deletion Collection for the use of yeast as a model organism is that complementation of a particular gene defect by a putative human ortholog can be tested very rapidly because the necessary mutant strain already exists. Where the function of the ORF proved essential for viability under normal growth conditions, a heterozygous diploid containing one wild-type copy was created; for all other genes, both haploid and diploid homozygous disruptants were made in addition to the heterozygous diploid (Winzeler *et al.*, 1999). There are thus four collections: the heterozygous diploids (ca. 6000 strains); the haploid disruptants of each mating type; and the homozygous diploids (ca. 4800 each). These strains can be obtained from either ResGen (Invitrogen, Carlsbad, CA) or the American Type Culture Collection (Manassas, VA) as individual tubes, in microtiter format, or as pools.

An additional feature of the collection is that for each ORF the deletion construct was individually designed and constructed, allowing the insertion of two unique 20-base elements into the genome at the site of the deletion (Figure 2.5). The existence of these 'tags' allows the identity of a particular strain to be confirmed rapidly by sequencing of a PCR product. It also allows the presence or absence of a particular strain among a group of strains to be measured by various PCR strategies. One strategy is to generate a labeled PCR product containing the tag sequence as shown in Figure 2.5 and then hybridize to microarrays of oligonucleotides that are complementary to the tags (Shoemaker *et al.*, 1996). Figure 2.6 shows an example from my laboratory of such a pool of PCR products hybridized to the 'Tag3' array, custom produced by Affymetrix. Use of these 'tag arrays' represents a very new area

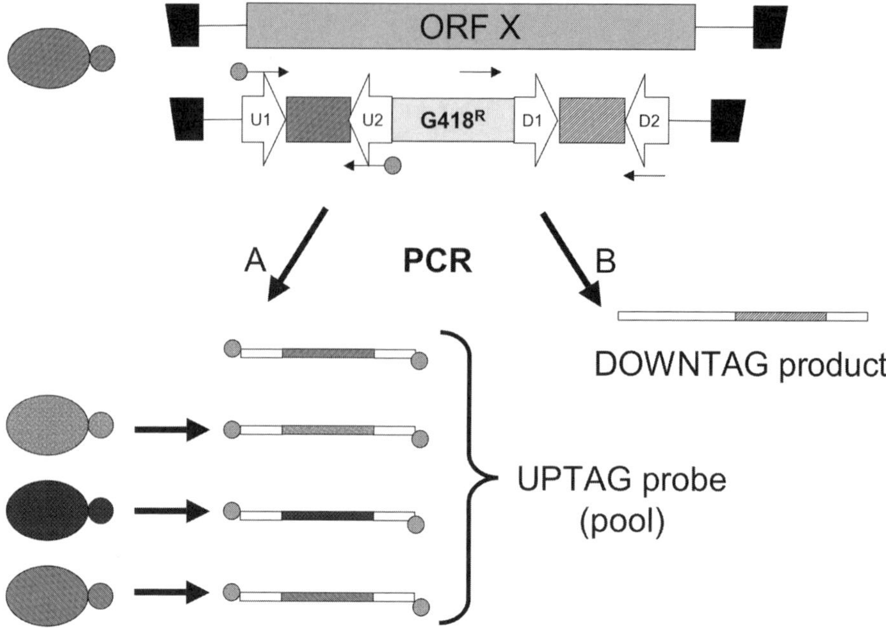

Figure 2.5 Utility of features incorporated into the Yeast Genome Deletion Collection. For each ORF in the *S. cerevisiae* genome (ORF X in the example), a specific deletion cassette was constructed. This cassette contained flanking sequences that targeted it to replace the ORF from the start codon to the stop codon. The DNA substituted for the ORF contained a gene conferring resistance to the antibiotic G418 (G418^R). At each end, it also contained unique 20-base-pair sequences not found in the yeast genome. Called the 'uptag' and the 'downtag', these 20-mer tags are flanked by short sequences that are common to each construct, indicated as U1, U2, D1 and D2. These common sequences can be used as priming sites for polymerase chain reaction (PCR), allowing every tag present in a pool to be amplified in a single reaction. In this example, (A) shows an amplification using primers U1 and U2, where U1 carries a molecular probe such as biotin or a fluorophore. The resulting pool of PCR products could be hybridized to an oligonucleotide array to determine its composition. Alternatively, PCR (B) uses a primer complementary to the G418^R marker region in combination with primer D2 to generate a longer PCR product suitable for sequencing. Because the downtag is unique to the strain carrying the deletion of ORF X, this sequencing reaction immediately reveals the identity of the strain

with few publications, and the availability of arrays is a limiting factor. However, it is anticipated that the ability to pool hundreds or indeed all 6000 strains, perform a selection and then identify all the changes that have occurred in the population in a rapid, multiplex fashion will enable new types of screens that were too onerous to perform by traditional methods. Although they have been available for less than 2 years, the collections are already finding wide use as detailed below.

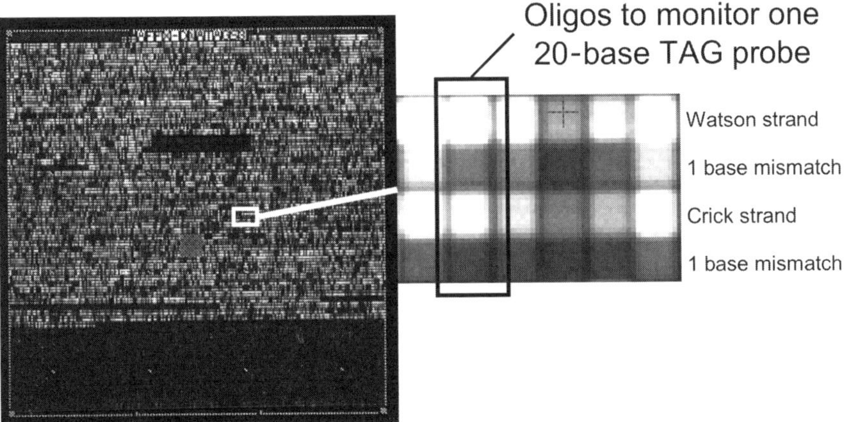

Figure 2.6 Use of 'Tag array' chips to assay the Yeast Genome Deletion Collection. Using genomic DNA from the collection of ca. 6000 heterozygous deletion mutant strains as a template, a pool of biotinylated polymerase chain reaction (PCR) products was generated as shown in example (A) of Figure 2.5. Both the 'uptag' and the 'downtag' regions were amplified in separate reactions. The pool of ca. 12 000 PCR products were hybridized to a 'Tag3' custom oligonucleotide array provided by Affymetrix (Santa Clara, CA). This array allows probing with 16 000 distinct sequences and because only ca. 12 000 of these sequences were used in generating the Yeast Genome Deletion Collection the bottom quarter of the chip does not show hybridization. Each 'tag' that was used in the Yeast Genome Deletion Collection is represented by four features on this array: two features that represent a perfect match to either strand of the tag region and two features that contain a single base mismatch substitution in the center of the perfect match sequence. As shown to the right, the PCR product containing a particular tag sequence should hybridize strongly to the two 'perfect match' features and with reduced efficiency to the 'mismatch' features

Phenotypic screens

The most immediate application of the deletion collection is to direct phenotypic screening. Publications on screens for genes involved in rapamycin sensitivity (Chan *et al.*, 2000), autophagy (Barth and Thumm, 2001), glycogen storage (Wilson *et al.*, 2002), mitochondrial function (Dimmer *et al.*, 2002), DNA repair (Ooi *et al.*, 2001), proteasome inhibition (Fleming *et al.*, 2002) and intracellular transport (Muren *et al.*, 2001) have already appeared. These screens have used analysis of individual strains. Only one published example to date has employed a chip-based assay on the pooled collection: Ooi *et al.* reported the identification of known and novel yeast gene deletion mutants that are incapable of circularizing a plasmid introduced in linear form. Instead of performing over 4800 individual assays, one pool was used (Ooi *et al.*, 2001).

Synthetic lethality

Synthetic lethality is another established genetic technique to receive a new twist from the deletion collections. In this approach, mutations that are individually of little or no effect are revealed to synergize and cause cell death. This approach is particularly useful for characterizing pathways because it can identify redundant functions or synergistic effects. Historically, this was a laborious technique that involved random mutagenesis followed by close visual inspection of thousands of colonies, followed by a cloning attempt that might prove unsuccessful. Now, Tong *et al.* (2001) have described a system for the use of the haploid deletion collection to perform systematically a version of synthetic lethality analysis that is scalable and amenable to automation. A haploid containing the mutant of interest is mated to the ca. 4800 viable haploid deletion strains. Following meiosis, the viable haploid progeny are examined for co-segregation of the mutation of interest and the deletion mutation. This sounds complex, but in fact is achieved by a series of simple steps using selective media. Lack of growth occurs when there is an inability to carry the two mutations in the same haploid cell, indicating that a synthetic lethality is occurring.

Haploinsufficiency analysis

Another utility for the deletion collection that has published proof-of-principle is 'haploinsufficiency' screening for drug targets (Giaever *et al.*, 1999). Conceptually, this approach relies on the increased drug sensitivity of a strain that has reduced gene dosage for the target protein. Such hypersensitive strains can be identified by individual assays on each strain from the heterozygous mutant collection; alternatively, the inclusion of the individual 'tags' in each deletion construct allows analysis of pooled strains with a chip-based readout. The ability to use pooled strains is a major advantage for drug discovery applications, where limited amounts of compound are available. Haploinsufficiency is a sound theory, but in practice many mutant strains will be hypersensitive to a given compound, and robust statistical analysis of a large body of compound and control data is essential to identify strains that respond specifically to a test compound (D. Shoemaker, personal communication; Rachel Kindt, personal communication).

Mutant gene mapping

Use of a complete collection of gene deletion mutants eliminates much of the uncertainty involved in phenotypic screens: issues of mutagenesis of multiple genes, screen saturation and gene recovery are eliminated. What is lost is the ability to generate specific changes such as conditional alleles of essential genes; such point mutants historically have been the richest source of information. To illustrate this assertion: although complete deletion of the gene for yeast immunophilin Fpr1 results in rapamycin resistance, isolation of point mutations in the essential Tor proteins by Cafferkey *et al.* (1994) was required for a complete understanding of the mechanism of action. However, another way to look at the deletion collection is as a comprehensive set of mapped markers. Thus, a point mutation of interest can be mapped by mating a haploid mutant to the ca. 4800 viable haploid deletion strains and then examining the haploid meiotic progeny of each diploid for linkage between the mutation and the G418 resistance marker. The same system applied by Tong *et al.* to high-throughput synthetic lethal screening can be applied here to render this process rapid and automated (Tong *et al.*, 2001).

2.11   Overexpression analysis: enough is enough

As yet, no genome-wide reagent for systematic overexpression of yeast genes exists, although several are in construction. Such a collection will have broad utility. Historically, several drug targets have been identified in yeast by virtue of the resistance caused by introduction of a genomic fragment containing the gene on a high-copy plasmid (Rine *et al.*, 1983) and 'high-copy suppressors' of mutant phenotypes are a standard tool in analysis of gene function. More recently, overexpression analysis has been used to examine effects on MAPK signaling, identifying new kinases that can modulate a well-characterized pathway (Burchett *et al.*, 2001). Overexpression was also used by Stevenson *et al.* to identify new proteins implicated in cell cycle control (Stevenson *et al.*, 2001). Kroll *et al.* (1996) described synthetic lethality when a protein of interest is overexpressed in the background of an otherwise benign mutation as a method of detecting specific genetic interactions. This technique was applied in a screen for genes whose overexpression is lethal in a proteasome-impaired mutant, and revealed six novel genes capable of inducing apoptotic death in yeast (Ligr *et al.*, 2001). A standardized regulated genome-wide collection of expression constructs is arguably the next great yeast genomic reagent.

2.12   Proteomics: would you like chips with that?

Although drug discovery is in a period where protein targets are screened in splendid isolation, that is not how they exist in the cell and ultimately some information about interactions and modifications is likely to prove necessary. Our ability to study such characteristics of a protein has greatly increased, and yeast has served as both the test bed for many techniques and as a surrogate for mammalian target proteins. Several complete-genome reagents and their use have been described.

Two-hybrid analysis of interactions

Two-hybrid analysis originated in yeast, and the ease of high-throughput assays in this system has made it the host of choice for most commercial and academic analysis of mammalian protein interactions (Uetz, 2002), although this is likely to change as mammalian systems become more tractable. The assay requires two fusion constructs to be expressed in the same cell: if an interaction occurs between the proteins under test, it reconstitutes their attached domains into a protein that can generate a measurable output (e.g. a transcription factor). Performing a comprehensive analysis involves mating of a strain with one fusion construct (the 'bait') to an array of strains carrying possible interactors (the 'prey'). Although a genome-wide analysis of every protein in yeast is theoretically possible, it would require over 38 million matings. However, if you wish to perform your own screen on a protein of interest, the Fields' laboratory makes the complete set of yeast fusion 'prey' constructs (Uetz *et al.*, 2000) available to all interested researchers.

Several large-scale two-hybrid studies have been reported to date: each tested only a subset of the genome and/or used pooling strategies (Fromont-Racine *et al.*, 1997; Flores *et al.*, 1999; Ito *et al.*, 2000; Uetz *et al.*, 2000). Such data can be synthesized to provide an interaction map for a eukaryote proteome and to suggest a function for uncharacterized proteins (Schwikowski *et al.*, 2000). Integration of the data into yeast information resources such as YPD and MIPS mean that results for orthologs of human proteins are readily accessible. An example of yeast as a model for a target of therapeutic relevance is a recent dissection of interactions within the 26S proteasome. Thirty-one proteasome components were screened against the entire proteome, and novel interacting components could be validated further by mutant analysis and reporter assays (Cagney *et al.*, 2001).

Analysis of complexes by mass spectrometry

This relatively recent addition to the set of techniques available is fast proving valuable. For various reasons discussed in the publications below, it usually produces quite different answers than two-hybrid analysis, and the datasets that are obtained complement each other. To achieve sufficient specificity, mass spectrometry must be applied to protein complexes that can be purified physically. Usually this means epitope tagging of the protein of interest and then passing through multiple rounds of affinity purification (TAP) followed by gel purification, although one report on the 40s ribosomal subunit directly analyzed complexes physically separated by other means (Link *et al.*, 1999). Honey *et al.* reported the use of TAP and mass spectrometry to characterize components of the active yeast CDK complex tagged on its cyclin subunit (Honey *et al.*, 2001). More recently two commercial entities reported far larger scale projects: Cellzome's work included 1143 yeast orthologs of relevance to human biology (Gavin *et al.*, 2002), whereas MDS Proteomics tagged 725 proteins, including a large number implicated in DNA damage responses (Ho *et al.*, 2002). These commercial projects represent pilots for mammalian work, as well as providing a large body of data for many yeast proteins that have mammalian homologs.

Biochemical analysis

A surprising pursuit, in this day and age, is to click around the links in various metabolic pathway websites and to discover how many of the described biochemical activities do not have a yeast gene linked to them. To eliminate the onerous task of purifying such activities to identify the responsible protein, Martzen *et al.* (1999) created expression constructs for all yeast ORFs in which the yeast protein was fused to glutathione-*s*-transferase (GST). These GST-fusion proteins can be purified and screened for enhancement of a particular activity; they are also a useful resource for hypothesis testing with cross-linkable ligands. As a further refinement, Zhu *et al.* report attachment of such tagged yeast proteins to microarrays and their screening for kinase activity (Zhu *et al.*, 2000) and for affinity to calmodulin and phospholipids (Zhu *et al.*, 2001). It is easy to conceive of future use of such arrays to identify molecular targets for labeled compounds.

Localization data

In addition to the most complete protein interaction data resources, yeast has a large volume of information on subcellular protein localization. Greatly extending data provided by individual studies and by an earlier large-scale project (Ross-Macdonald *et al.*, 1999), Kumar *et al.* conducted a genome-wide epitope-tagging and immunocytochemistry project resulting in annotation of nearly half the proteins in yeast to one of six subcellular localization sites (Kumar *et al.*, 2002a,b).

2.13   Web-accessible databases: bringing it all back home

- Saccharomyces   Genome   Database: http://genome-www.stanford.edu/ Saccharomyces/

- Comprehensive Yeast Genome Database: http://mips.gsf.de/proj/yeast/CYGD/db/index.html

The community of yeast researchers numbers in the tens of thousands and, coupled with the tools described above, the capacity to generate 'omic' scale information is almost overwhelming (Zhu and Snyder, 2002). In addition, the ability to measure and modulate so many parameters in yeast means that it is a natural test bed for systems biology (Ideker *et al.*, 2001). Gene-centric information for yeast is compiled into several databases that have made commendable efforts to cross-reference each other. Principal among these are the Saccharomyces Genome Database (SGD) based at Stanford, USA (Dwight *et al.*, 2002), the Comprehensive Yeast Genome Database (CYGD) at MIPS-GSF (Germany) (Mewes *et al.*, 2000) and Incyte's YPD (Costanzo *et al.*, 2001). The latter is a commercial subscription database historically provided free to academic researchers and has served as a template for Incyte's Human-PSD and GPCR-PSD databases. Many other databases exist to collate specialized information in greater detail; these are indexed off the sources listed above.

2.14 Conclusion

Analysis of the genomic sequences of both humans and yeast has led to a renewed appreciation of the shared biology of these long-separated eukaryotes. Although the understanding of this relationship is broader in the academic community, this review illustrates the wide range of uses that yeast has served in the pharmaceutical industry. As the technologies available become more powerful every year, it is to be hoped that we do not lose our appreciation of the insight that this small organism can continue to provide.

2.15 References

Arkinstall, S., Payton, M. and Maundrell, K. (1995). Activation of phospholipase C gamma in *Schizosaccharomyces pombe* by coexpression of receptor or nonreceptor tyrosine kinases. *Mol. Cell. Biol.* **15**, 1431–1438.

Arkinstall, S., Gillieron, C., Vial-Knecht, E. and Maundrell, K. (1998). A negative regulatory function for the protein tyrosine phosphatase PTP2C revealed by reconstruction of platelet-derived growth factor receptor signalling in *Schizosaccharomyces pombe*. *FEBS Lett.* **422**, 321–327.

Barth, H. and Thumm, M. (2001). A genomic screen identifies AUT8 as a novel gene essential for autophagy in the yeast *Saccharomyces cerevisiae*. *Gene* **274**, 151–156.

Bassett, D. E., Jr., Boguski, M. S. and Hieter, P. (1996). Yeast genes and human disease. *Nature* **379**, 589–590.

Baumann, G., Zenke, G., Wenger, R., Hiestand, P., Quesniaux, V., Andersen, E. and Schreier, M. H. (1992). Molecular mechanisms of immunosuppression. *J. Autoimmun.* **5**, 67–72.

Bjornsti, M. A., Knab, A. M. and Benedetti, P. (1994). Yeast *Saccharomyces cerevisiae* as a model system to study the cytotoxic activity of the antitumor drug camptothecin. *Cancer Chemother. Pharmacol.* **34**, S1–S5.

Boles, E., Liebetrau, W., Hofmann, M. and Zimmermann, F. K. (1994). A family of hexosephosphate mutases in *Saccharomyces cerevisiae. Eur. J. Biochem.* **220**, 83–96.

Bossard, M. J., Bergsma, D. J., Brandt, M., Livi, G. P., Eng, W. K., Johnson, R. K. and Levy, M. A. (1994). Catalytic and ligand binding properties of the FK506 binding protein FKBP12: effects of the single amino acid substitution of Tyr82 to Leu. *Biochem. J.* **297**, 365–372.

Botstein, D., Chervitz, S. A. and Cherry, J. M. (1997). Yeast as a model organism. *Science* **277**, 1259–1260.

Burchett, S. A., Scott, A., Errede, B. and Dohlman, H. G. (2001). Identification of novel pheromone-response regulators through systematic overexpression of 120 protein kinases in yeast. *J. Biol. Chem.* **276**, 26472–26478.

Cafferkey, R., Young, P. R., McLaughlin, M. M., Bergsma, D. J., Koltin, Y., Sathe, G. M., Faucette, L., *et al.* (1993). Dominant missense mutations in a novel yeast protein related to mammalian phosphatidylinositol 3-kinase and VPS34 abrogate rapamycin cytotoxicity. *Mol. Cell. Biol.* **13**, 6012–6023.

Cafferkey, R., McLaughlin, M. M., Young, P. R., Johnson, R. K. and Livi, G. P. (1994). Yeast TOR (DRR) proteins: amino-acid sequence alignment and identification of structural motifs. *Gene* **141**, 133–136.

Cagney, G., Uetz, P. and Fields, S. (2001). Two-hybrid analysis of the *Saccharomyces cerevisiae* 26S proteasome. *Physiol. Genom.* **7**, 27–34.

Cardenas, M. E., Lorenz, M., Hemenway, C. and Heitman, J. (1994). Yeast as model T cells. *Perspect. Drug Discov. Design* **2**, 103–126.

Chan, T. F., Carvalho, J., Riles, L. and Zheng, X. F. (2000). A chemical genomics approach toward understanding the global functions of the target of rapamycin protein (TOR). *Proc. Natl. Acad. Sci. USA* **97**, 13227–13232.

Chaturvedi, P., Eng, W. K., Zhu, Y., Mattern, M. R., Mishra, R., Hurle, M. R., Zhang, X., *et al.* (1999). Mammalian Chk2 is a downstream effector of the ATM-dependent DNA damage checkpoint pathway. *Oncogene* **18**, 4047–4054.

Chervitz, S. A., Aravind, L., Sherlock, G., Ball, C. A., Koonin, E. V., Dwight, S. S., Harris, M. A., *et al.* (1998). Comparison of the complete protein sets of worm and yeast: orthology and divergence. *Science* **282**, 2022–2028.

Clow, A., Greenhalf, W. and Chaudhuri, B. (1998). Under respiratory growth conditions, Bcl-x(L) and Bcl-2 are unable to overcome yeast cell death triggered by a mutant Bax protein lacking the membrane anchor. *Eur. J. Biochem.* **258**, 19–28.

Costanzo, M. C., Crawford, M. E., Hirschman, J. E., Kranz, J. E., Olsen, P., Robertson, L. S., Skrzypek, M. S., *et al.* (2001). YPD, PombePD and WormPD: model organism volumes of the BioKnowledge library, an integrated resource for protein information. *Nucleic Acids Res.* **29**, 75–79.

DiBello, P. R., Garrison, T. R., Apanovitch, D. M., Hoffman, G., Shuey, D. J., Mason, K., Cockett, M. I., *et al.* (1998). Selective uncoupling of RGS action by a single point mutation in the G protein alpha-subunit. *J. Biol. Chem.* **273**, 5780–5784.

Dimmer, K. S., Fritz, S., Fuchs, F., Messerschmitt, M., Weinbach, N., Neupert, W. and Westermann, B. (2002). Genetic basis of mitochondrial function and morphology in *Saccharomyces cerevisiae. Mol. Biol. Cell.* **13**, 847–853.

Dimster-Denk, D., Rine, J., Phillips, J., Scherer, S., Cundiff, P., DeBord, K., Gilliland, D., *et al.* (1999). Comprehensive evaluation of isoprenoid biosynthesis regulation in *Saccharomyces cerevisiae* utilizing the Genome Reporter Matrix. *J. Lipid Res.* **40**, 850–860.

Dohlman, H. G. (2002). G proteins and pheromone signaling. *Annu. Rev. Physiol.* **64**, 129–152.

Dowell, S. J., Bishop, A. L., Dyos, S. L., Brown, A. J. and Whiteway, M. S. (1998). Mapping of a yeast G protein betagamma signaling interaction. *Genetics* **150**, 1407–1417.

Drosos, A. A. (2002). Newer immunosuppressive drugs: their potential role in rheumatoid arthritis therapy. *Drugs* **62**, 891–907.

Duan, H., Wang, Y., Aviram, M., Swaroop, M., Loo, J. A., Bian, J., Tian, Y., *et al.* (1999). SAG, a novel zinc RING finger protein that protects cells from apoptosis induced by redox agents. *Mol. Cell. Biol.* **19**, 3145–3155.

Dumont, F. J., Staruch, M. J., Koprak, S. L., Melino, M. R. and Sigal, N. H. (1990). Distinct mechanisms of suppression of murine T cell activation by the related macrolides FK-506 and rapamycin. *J. Immunol.* **144**, 251–258.

Dwight, S. S., Harris, M. A., Dolinski, K., Ball, C. A., Binkley, G., Christie, K. R., Fisk, D. G., *et al.* (2002). Saccharomyces Genome Database (SGD) provides secondary gene annotation using the Gene Ontology (GO). *Nucleic Acids Res.* **30**, 69–72.

Fleming, J. A., Lightcap, E. S., Sadis, S., Thoroddsen, V., Bulawa, C. E. and Blackman, R. K. (2002). Complementary whole-genome technologies reveal the cellular response to proteasome inhibition by PS-341. *Proc. Natl. Acad. Sci. USA* **99**, 1461–1466.

Flores, A., Briand, J. F., Gadal, O., Andrau, J. C., Rubbi, L., Van Mullem, V., Boschiero, C., *et al.* (1999). A protein–protein interaction map of yeast RNA polymerase III. *Proc. Natl. Acad. Sci. USA* **96**, 7815–7820.

Frohlich, K. U. and Madeo, F. (2000). Apoptosis in yeast – a monocellular organism exhibits altruistic behaviour. *FEBS Lett.* **473**, 6–9.

Fromont-Racine, M., Rain, J. C. and Legrain, P. (1997). Toward a functional analysis of the yeast genome through exhaustive two-hybrid screens [see comments]. *Nat. Genet.* **16**, 277–282.

Furet, P., Imbach, P., Furst, P., Lang, M., Noorani, M., Zimmermann, J. and Garcia-Echeverria, C. (2001). Modeling of the binding mode of a non-covalent inhibitor of the 20S proteasome. Application to structure-based analogue design. *Bioorg. Med. Chem. Lett.* **11**, 1321–1324.

Gavin, A. C., Bosche, M., Krause, R., Grandi, P., Marzioch, M., Bauer, A., Schultz, J., *et al.* (2002). Functional organization of the yeast proteome by systematic analysis of protein complexes. *Nature* **415**, 141–147.

Giaever, G., Shoemaker, D. D., Jones, T. W., Liang, H., Winzeler, E. A., Astromoff, A. and Davis, R. W. (1999). Genomic profiling of drug sensitivities via induced haploinsufficiency. *Nat. Genet.* **21**, 278–283.

Giaever, G., Chu, A. M., Ni, L., Connelly, C., Riles, L., Veronneau, S., Dow, S., *et al.* (2002). Functional profiling of the *Saccharomyces cerevisiae* genome. *Nature* **418**, 387–391.

Goffeau, A., Barrell, B. G., Bussey, H., Davis, R. W., Dujon, B., Feldmann, H., Galibert, F., *et al.* (1996). Life with 6000 genes [see comments]. *Science* **274**, 546, 563–547.

Greenhalf, W., Stephan, C. and Chaudhuri, B. (1996). Role of mitochondria and C-terminal membrane anchor of Bcl-2 in Bax induced growth arrest and mortality in *Saccharomyces cerevisiae. FEBS Lett.* **380**, 169–175.

Greenhalf, W., Lee, J. and Chaudhuri, B. (1999). A selection system for human apoptosis inhibitors using yeast. *Yeast* **15**, 1307–1321.

Guthrie, C. and Fink, G. R., eds. (1991). *Guide to Yeast Genetics and Molecular Biology*. San Diego: Academic Press.

Gutkind, J. S. (1998). The pathways connecting G protein-coupled receptors to the nucleus through divergent mitogen-activated protein kinase cascades. *J. Biol. Chem.* **273**, 1839–1842.

Gutkind, J. S. (2000). Regulation of mitogen-activated protein kinase signaling networks by G protein-coupled receptors. *Sci. STKE* **2000**, RE1.

Harris, K., Lamson, R. E., Nelson, B., Hughes, T. R., Marton, M. J., Roberts, C. J., Boone, C., *et al.* (2001). Role of scaffolds in MAP kinase pathway specificity revealed by custom design of pathway-dedicated signaling proteins. *Curr. Biol.* **11**, 1815–1824.

Hinz, W., Gruninger, S., De Pover, A. and Chiesi, M. (1999). Properties of the human long and short isoforms of the uncoupling protein-3 expressed in yeast cells. *FEBS Lett.* **462**, 411–415.

Ho, Y., Gruhler, A., Heilbut, A., Bader, G. D., Moore, L., Adams, S. L., Millar, A., *et al.* (2002). Systematic identification of protein complexes in *Saccharomyces cerevisiae* by mass spectrometry. *Nature* **415**, 180–183.

Honey, S., Schneider, B. L., Schieltz, D. M., Yates, J. R. and Futcher, B. (2001). A novel multiple affinity purification tag and its use in identification of proteins associated with a cyclin-CDK complex. *Nucleic Acids Res.* **29**, E24.

Horiguchi, T. and Tanida, S. (1995). Rescue of *Schizosaccharomyces pombe* from camptothecin-mediated death by a DNA topoisomerase I inhibitor, TAN-1518 A. *Biochem. Pharmacol.* **49**, 1395–1401.

Hughes, T. R., Marton, M. J., Jones, A. R., Roberts, C. J., Stoughton, R., Armour, C. D., Bennett, H. A., *et al.* (2000). Functional discovery via a compendium of expression profiles. *Cell* **102**, 109–126.

Ideker, T., Thorsson, V., Ranish, J. A., Christmas, R., Buhler, J., Eng, J. K., Bumgarner, R., *et al.* (2001). Integrated genomic and proteomic analyses of a systematically perturbed metabolic network. *Science* **292**, 929–934.

Ito, T., Tashiro, K., Muta, S., Ozawa, R., Chiba, T., Nishizawa, M., Yamamoto, K., *et al.* (2000). Toward a protein–protein interaction map of the budding yeast: a comprehensive system to examine two-hybrid interactions in all possible combinations between the yeast proteins. *Proc. Natl. Acad. Sci. USA* **97**, 1143–1147.

Ketchum, K. A., Joiner, W. J., Sellers, A. J., Kaczmarek, L. K. and Goldstein, S. A. (1995). A new family of outwardly rectifying potassium channel proteins with two pore domains in tandem. *Nature* **376**, 690–695.

Koltin, Y., Faucette, L., Bergsma, D. J., Levy, M. A., Cafferkey, R., Koser, P. L., Johnson, R. K., *et al.* (1991). Rapamycin sensitivity in *Saccharomyces cerevisiae* is mediated by a peptidyl–prolyl *cis–trans* isomerase related to human FK506-binding protein. *Mol. Cell. Biol.* **11**, 1718–1723.

Koser, P. L., Sylvester, D., Livi, G. P. and Bergsma, D. J. (1990). A second cyclophilin-related gene in *Saccharomyces cerevisiae*. *Nucleic Acids Res.* **18**, 1643.

Koser, P. L., Bergsma, D. J., Cafferkey, R., Eng, W. K., McLaughlin, M. M., Ferrara, A., Silverman, C., *et al.* (1991). The CYP2 gene of *Saccharomyces cerevisiae* encodes a cyclosporin A-sensitive peptidyl–prolyl *cis–trans* isomerase with an N-terminal signal sequence. *Gene* **108**, 73–80.

Kroll, E. S., Hyland, K. M., Hieter, P. and Li, J. J. (1996). Establishing genetic interactions by a synthetic dosage lethality phenotype. *Genetics* **143**, 95–102.

Kultz, D. and Burg, M. (1998). Evolution of osmotic stress signaling via MAP kinase cascades. *J. Exp. Biol.* **201**, 3015–3021.

Kumar, A., Agarwal, S., Heyman, J. A., Matson, S., Heidtman, M., Piccirillo, S., Umansky, L., *et al.* (2002a). Subcellular localization of the yeast proteome. *Genes Dev.* **16**, 707–719.

Kumar, A., Cheung, K. H., Tosches, N., Masiar, P., Liu, Y., Miller, P. and Snyder, M. (2002b). The TRIPLES database: a community resource for yeast molecular biology. *Nucleic Acids Res.* **30**, 73–75.

Kumar, A., Harrison, P. M., Cheung, K. H., Lan, N., Echols, N., Bertone, P., Miller, P., *et al.* (2002c). An integrated approach for finding overlooked genes in yeast. *Nat. Biotechnol.* **20**, 58–63.

Kumar, S., McLaughlin, M. M., McDonnell, P. C., Lee, J. C., Livi, G. P. and Young, P. R. (1995). Human mitogen-activated protein kinase CSBP1, but not CSBP2, complements a hog1 deletion in yeast. *J. Biol. Chem.* **270**, 29043–29046.

Ligr, M., Velten, I., Frohlich, E., Madeo, F., Ledig, M., Frohlich, K. U., Wolf, D. H., *et al.* (2001). The proteasomal substrate Stm1 participates in apoptosis-like cell death in yeast. *Mol. Biol. Cell* **12**, 2422–2432.

Link, A. J., Eng, J., Schieltz, D. M., Carmack, E., Mize, G. J., Morris, D. R., Garvik, B. M., *et al.* (1999). Direct analysis of protein complexes using mass spectrometry. *Nat. Biotechnol.* **17**, 676–682.

Liu, J., Albers, M. W., Wandless, T. J., Luan, S., Alberg, D. G., Belshaw, P. J., Cohen, P., *et al.* (1992). Inhibition of T cell signaling by immunophilin–ligand complexes correlates with loss of calcineurin phosphatase activity. *Biochemistry* **31**, 3896–3901.

Liu, Q., Bai, C., Chen, F., Wang, R., MacDonald, T., Gu, M., Zhang, Q., *et al.* (1998). Uncoupling protein-3: a muscle-specific gene upregulated by leptin in ob/ob mice. *Gene* **207**, 1–7.

Lord, K. A., Creasy, C. L., King, A. G., King, C., Burns, B. M., Lee, J. C. and Dillon, S. B. (2000). REDK, a novel human regulatory erythroid kinase. *Blood* **95**, 2838–2846.

Madeo, F., Herker, E., Maldener, C., Wissing, S., Lachelt, S., Herlan, M., Fehr, M., *et al.* (2002). A caspase-related protease regulates apoptosis in yeast. *Mol. Cell* **9**, 911–917.

Mandala, S. M., Thornton, R., Galve-Roperh, I., Poulton, S., Peterson, C., Olivera, A., Bergstrom, J., *et al.* (2000). Molecular cloning and characterization of a lipid phosphohydrolase that degrades sphingosine-1- phosphate and induces cell death. *Proc. Natl. Acad. Sci. USA* **97**, 7859–7864.

Marton, M. J., DeRisi, J. L., Bennett, H. A., Iyer, V. R., Meyer, M. R., Roberts, C. J., Stoughton, R., *et al.* (1998). Drug target validation and identification of secondary drug target effects using DNA microarrays [see comments]. *Nat. Med.* **4**, 1293–1301.

Martzen, M. R., McCraith, S. M., Spinelli, S. L., Torres, F. M., Fields, S., Grayhack, E. J. and Phizicky, E. M. (1999). A biochemical genomics approach for identifying genes by the activity of their products. *Science* **286**, 1153–1155.

McKune, K., Moore, P. A., Hull, M. W. and Woychik, N. A. (1995). Six human RNA polymerase subunits functionally substitute for their yeast counterparts. *Mol. Cell. Biol.* **15**, 6895–6900.

McLaughlin, M. M., Bossard, M. J., Koser, P. L., Cafferkey, R., Morris, R. A., Miles, L. M., Strickler, J., *et al.* (1992). The yeast cyclophilin multigene family: purification, cloning and characterization of a new isoform. *Gene* **111**, 85–92.

Mewes, H. W., Frishman, D., Gruber, C., Geier, B., Haase, D., Kaps, A., Lemcke, K., *et al.* (2000). MIPS: a database for genomes and protein sequences. *Nucleic Acids Res.* **28**, 37–40.

Mio, T., Yabe, T., Arisawa, M. and Yamada-Okabe, H. (1998). The eukaryotic UDP-N-acetylglucosamine pyrophosphorylases. Gene cloning, protein expression, and catalytic mechanism. *J. Biol. Chem.* **273**, 14392–14397.

Mio, T., Yamada-Okabe, T., Arisawa, M. and Yamada-Okabe, H. (2000). Functional cloning and mutational analysis of the human cDNA for phosphoacetylglucosamine mutase: identification of the amino acid residues essential for the catalysis. *Biochim. Biophys. Acta* **1492**, 369–376.

Moorthamer, M., Panchal, M., Greenhalf, W. and Chaudhuri, B. (1998). The p16(INK4A) protein and flavopiridol restore yeast cell growth inhibited by Cdk4. *Biochem. Biophys. Res. Commun.* **250**, 791–797.

Muren, E., Oyen, M., Barmark, G. and Ronne, H. (2001). Identification of yeast deletion strains that are hypersensitive to brefeldin A or monensin, two drugs that affect intracellular transport. *Yeast* **18**, 163–172.

Narasimhan, M. L., Damsz, B., Coca, M. A., Ibeas, J. I., Yun, D. J., Pardo, J. M., Hasegawa, P. M., *et al.* (2001). A plant defense response effector induces microbial apoptosis. *Mol. Cell* **8**, 921–930.

Ooi, S. L., Shoemaker, D. D. and Boeke, J. D. (2001). A DNA microarray-based genetic screen for nonhomologous end-joining mutants in *Saccharomyces cerevisiae*. *Science* **294**, 2552–2556.

Paddon, C., Loayza, D., Vangelista, L., Solari, R. and Michaelis, S. (1996). Analysis of the localization of STE6/CFTR chimeras in a *Saccharomyces cerevisiae* model for the cystic fibrosis defect CFTR delta F508. *Mol. Microbiol.* **19**, 1007–1017.

Pearce, D. A., Carr, C. J., Das, B. and Sherman, F. (1999a). Phenotypic reversal of the btn1 defects in yeast by chloroquine: a yeast model for Batten disease. *Proc. Natl. Acad. Sci. USA* **96**, 11341–11345.

Pearce, D. A., Ferea, T., Nosel, S. A., Das, B. and Sherman, F. (1999b). Action of BTN1, the yeast orthologue of the gene mutated in Batten disease. *Nat. Genet.* **22**, 55–58.

Ploger, R., Zhang, J., Bassett, D., Reeves, R., Hieter, P., Boguski, M. and Spencer, F. (2000). XREFdb: cross-referencing the genetics and genes of mammals and model organisms. *Nucleic Acids Res.* **28**, 120–122.

Porter, D. J. and Short, S. A. (2000). Yeast orotidine-5′-phosphate decarboxylase: steady-state and pre-steady-state analysis of the kinetic mechanism of substrate decarboxylation. *Biochemistry* **39**, 11788–11800.

Reid, J. D., Lukas, W., Shafaatian, R., Bertl, A., Scheurmann-Kettner, C., Guy, H. R. and North, R. A. (1996). The *S. cerevisiae* outwardly-rectifying potassium channel (DUK1) identifies a new family of channels with duplicated pore domains. *Receptors Channels* **4**, 51–62.

Rine, J., Hansen, W., Hardeman, E. and Davis, R. W. (1983). Targeted selection of recombinant clones through gene dosage effects. *Proc. Natl. Acad. Sci. USA* **80**, 6750–6754.

Roberts, C. J., Nelson, B., Marton, M. J., Stoughton, R., Meyer, M. R., Bennett, H. A., He, Y. D., *et al.* (2000). Signaling and circuitry of multiple MAPK pathways revealed by a matrix of global gene expression profiles. *Science* **287**, 873–880.

Robinson, G. W., Tsay, Y. H., Kienzle, B. K., Smith-Monroy, C. A. and Bishop, R. W. (1993). Conservation between human and fungal squalene synthetases: similarities in structure, function, and regulation. *Mol. Cell. Biol.* **13**, 2706–2717.

Ross-Macdonald, P., Coelho, P. S., Roemer, T., Agarwal, S., Kumar, A., Jansen, R., Cheung, K. H., *et al.* (1999). Large-scale analysis of the yeast genome by transposon tagging and gene disruption [see comments]. *Nature* **402**, 413–418.

Rotonda, J., Burbaum, J. J., Chan, H. K., Marcy, A. I. and Becker, J. W. (1993). Improved calcineurin inhibition by yeast FKBP12-drug complexes. Crystallographic and functional analysis. *J. Biol. Chem.* **268**, 7607–7609.

Scapin, G. (2002). Structural biology in drug design: selective protein kinase inhibitors. *Drug Discov. Today* **7**, 601–611.

Schmelzle, T. and Hall, M. N. (2000). TOR, a central controller of cell growth. *Cell* **103**, 253–262.

Schwikowski, B., Uetz, P. and Fields, S. (2000). A network of protein–protein interactions in yeast. *Nat. Biotechnol.* **18**, 1257–1261.

Senderowicz, A. M. (2000). Small molecule modulators of cyclin-dependent kinases for cancer therapy. *Oncogene* **19**, 6600–6606.

Shalon, D., Smith, S. J. and Brown, P. O. (1996). A DNA microarray system for analyzing complex DNA samples using two-color fluorescent probe hybridization. *Genome Res.* **6**, 639–645.

Shamji, A. F., Kuruvilla, F. G. and Schreiber, S. L. (2000). Partitioning the transcriptional program induced by rapamycin among the effectors of the Tor proteins. *Curr. Biol.* **10**, 1574–1581.

Shi, Y., Vattem, K. M., Sood, R., An, J., Liang, J., Stramm, L. and Wek, R. C. (1998). Identification and characterization of pancreatic eukaryotic initiation factor 2 alpha-subunit kinase, PEK, involved in translational control. *Mol. Cell. Biol.* **18**, 7499–7509.

Shoemaker, D. D., Lashkari, D. A., Morris, D., Mittmann, M. and Davis, R. W. (1996). Quantitative phenotypic analysis of yeast deletion mutants using a highly parallel molecular bar-coding strategy. *Nat. Genet.* **14**, 450–456.

Siekierka, J. J., Wiederrecht, G., Greulich, H., Boulton, D., Hung, S. H., Cryan, J., Hodges, P. J., *et al.* (1990). The cytosolic-binding protein for the immunosuppressant FK-506 is both a ubiquitous and highly conserved peptidyl–prolyl *cis–trans* isomerase. *J. Biol. Chem.* **265**, 21011–21015.

Singh, S. B., Jayasuriya, H., Silverman, K. C., Bonfiglio, C. A., Williamson, J. M. and Lingham, R. B. (2000). Efficient syntheses, human and yeast farnesyl-protein transferase inhibitory activities of chaetomellic acids and analogues. *Bioorg. Med. Chem.* **8**, 571–580.

Stevenson, L. F., Kennedy, B. K. and Harlow, E. (2001). A large-scale overexpression screen in *Saccharomyces cerevisiae* identifies previously uncharacterized cell cycle genes. *Proc. Natl. Acad. Sci. USA* **98**, 3946–3951.

Summers, C., Karst, F. and Charles, A. D. (1993). Cloning, expression and characterisation of the cDNA encoding human hepatic squalene synthase, and its relationship to phytoene synthase. *Gene* **136**, 185–192.

Swaroop, M., Wang, Y., Miller, P., Duan, H., Jatkoe, T., Madore, S. J. and Sun, Y. (2000). Yeast homolog of human SAG/ROC2/Rbx2/Hrt2 is essential for cell growth, but not for germination: chip profiling implicates its role in cell cycle regulation. *Oncogene* **19**, 2855–2866.

Swaroop, M., Gosink, M. and Sun, Y. (2001). SAG/ROC2/Rbx2/Hrt2, a component of SCF E3 ubiquitin ligase: genomic structure, a splicing variant, and two family pseudogenes. *DNA Cell Biol.* **20**, 425–434.

Tong, A. H., Evangelista, M., Parsons, A. B., Xu, H., Bader, G. D., Page, N., Robinson, M., *et al.* (2001). Systematic genetic analysis with ordered arrays of yeast deletion mutants. *Science* **294**, 2364–2368.

Torgler, C. N., Brown, R. and Meldrum, E. (2000). Exploiting the utility of yeast in the context of programmed cell death. *Methods Enzymol.* **322**, 297–322.

Tropschug, M., Barthelmess, I. B. and Neupert, W. (1989). Sensitivity to cyclosporin A is mediated by cyclophilin in *Neurospora crassa* and *Saccharomyces cerevisiae*. *Nature* **342**, 953–955.

Uetz, P. (2002). Two-hybrid arrays. *Curr. Opin. Chem. Biol.* **6**, 57–62.

Uetz, P., Giot, L., Cagney, G., Mansfield, T. A., Judson, R. S., Knight, J. R., Lockshon, D., *et al.* (2000). A comprehensive analysis of protein–protein interactions in *Saccharomyces cerevisiae* [process citation]. *Nature* **403**, 623–627.

Uren, A. G., O'Rourke, K., Aravind, L. A., Pisabarro, M. T., Seshagiri, S., Koonin, E. V. and Dixit, V. M. (2000). Identification of paracaspases and metacaspases: two ancient families of caspase-like proteins, one of which plays a key role in MALT lymphoma. *Mol. Cell* **6**, 961–967.

Wiederrecht, G., Brizuela, L., Elliston, K., Sigal, N. H. and Siekierka, J. J. (1991). FKB1 encodes a nonessential FK 506-binding protein in *Saccharomyces cerevisiae* and contains regions suggesting homology to the cyclophilins. *Proc. Natl. Acad. Sci. USA* **88**, 1029–1033.

Wilson, W. A., Wang, Z. and Roach, P. J. (2002). Systematic identification of the genes affecting glycogen storage in the yeast *Saccharomyces cerevisiae. Mol. Cell. Proteom.* **1**, 232–242.

Winzeler, E. A., Shoemaker, D. D., Astromoff, A., Liang, H., Anderson, K., Andre, B., Bangham, R., *et al.* (1999). Functional characterization of the *S. cerevisiae* genome by gene deletion and parallel analysis. *Science* **285**, 901–906.

Wood, V., Gwilliam, R., Rajandream, M. A., Lyne, M., Lyne, R., Stewart, A., Sgouros, J., *et al.* (2002). The genome sequence of *Schizosaccharomyces pombe. Nature* **415**, 871–880.

Zhang, H., Cowan-Jacob, S. W., Simonen, M., Greenhalf, W., Heim, J. and Meyhack, B. (2000). Structural basis of BFL-1 for its interaction with BAX and its anti-apoptotic action in mammalian and yeast cells. *J. Biol. Chem.* **275**, 11092–11099.

Zhong, H. and Neubig, R. R. (2001). Regulator of G protein signaling proteins: novel multifunctional drug targets. *J. Pharmacol. Exp. Ther.* **297**, 837–845.

Zhu, H. and Snyder, M. (2002). 'Omic' approaches for unraveling signaling networks. *Curr. Opin. Cell Biol.* **14**, 173–179.

Zhu, H., Klemic, J. F., Chang, S., Bertone, P., Casamayor, A., Klemic, K. G., Smith, D., *et al.* (2000). Analysis of yeast protein kinases using protein chips. *Nat. Genet.* **26**, 283–289.

Zhu, H., Bilgin, M., Bangham, R., Hall, D., Casamayor, A., Bertone, P., Lan, N., Jansen, R., Bidlingmaier, S., Houfek, T., *et al.* (2001). Global analysis of protein activities using proteome chips. *Science* **293**, 2101–2105.

3
Caenorhabditis elegans Functional Genomics in Drug Discovery: Expanding Paradigms

Titus Kaletta, Lynn Butler and **Thierry Bogaert**

In the past 10 years genomics has integrated rapidly into the process of drug discovery. Consequently, a wealth of novel targets need to be validated and screened to deliver more drugs in a shorter time. This asks for an animal model that is complex enough to acknowledge the complexity of modern medicine but also simple enough to be used in high-throughput applications. *Caenorhabditis elegans* is the ideal model organism and was identified by Sydney Brenner about 40 years ago. It is a spool-shaped worm ca. 1 mm long with 959 cells that eats bacteria. It is genetically amenable and transparent, so every cell division and differentiation could be followed directly under the microscope. Brenner demonstrated in 1974 that mutations could be introduced into many genes and visualized as distinct changes in organ formation. Through his visionary work Brenner created an important research tool: the nematode had made it into the inner circle of research and its utility for biomedical research has just been awarded a Nobel prize. This chapter describes *C. elegans* as a modern industrial tool for drug discovery. After an introduction into the drug discovery process and into *C. elegans*, various sections cover the design of *C. elegans* disease models, target identification technologies and genome-wide target validation approaches. Subsequent sections cover such topics as *C. elegans* compound assay design, *C. elegans* high-throughput screening and *C. elegans* pharmacology. The reader will be

Model Organisms in Drug Discovery. Edited by Pamela M. Carroll and Kevin Fitzgerald.
© 2003 John Wiley & Sons, Ltd. ISBN 0 470 84893 6

guided through the *C.-elegans*-based drug discovery pipeline by a discovery project for antidepression.

3.1 The drug discovery process

Until the late 20th century, drug discovery was mainly a linear process based on the screening and testing of thousands of chemical substances for therapeutic activity. The drug discovery process could be broken down into the following steps: target selection, assay development, primary screening for chemical hits, hit to lead compound optimization, preclinical and clinical development and, finally, market launch. Early bottlenecks such as the typically limited availability of discovery compounds and the often low-throughput analysis of compounds were reduced significantly during the 1990s with the introduction of combinatorial chemistry and high-throughput screening technologies. Modern ultrahigh-throughput screening allows the analysis of 100 000 compounds per day (Croston, 2002). Surprisingly, despite more than 40 years of research during what is regarded as the modern age of drug discovery, the pool of therapeutic targets used by the pharmaceutical industry remains at less than 500 of the 26 000–40 000 genes that comprise the human genome (Drews, 2000).

With the advent of genome research during the past decade, the traditional concept of drug development started to change and the number of new potential therapeutic targets is rising. The availability of sequence information for the entire human genome makes it possible to browse *in silico* for complete gene families, e.g. kinases and G-protein-coupled receptors (GCPRs). Genomics-based technologies such as high-throughput expression profiling are able to identify targets at a pace that exceeds that of the time required to analyze and prioritize the utility of the targets (Lander *et al.*, 2001; Szymkowski, 2001; Venter *et al.*, 2001). Because it is estimated that more than 10 000 drug targets exist, target identification is no longer a critical obstacle, rather their abundance presents a dilemma. The question of which targets to choose is one of the major challenges faced by the drug discovery industry today because a wealth of poorly characterized potential targets can clog up the discovery pipelines. Thus, rapid and specific tools to validate the *in vivo* functional utility of targets have become an increasingly important component of the drug discovery process. One attractive solution to target validation and prioritization bottlenecks is the use of model organisms.

Caenorhabditis elegans gives a competitive edge

An animal that is complex enough to study behavior or development yet simple enough to be used in the laboratory – this was Sydney Brenner's thought about 40 years ago when, in 1965, he chose the nematode *Caenorhabditis elegans* as a model organism. At first glance, this nematode appears to be nothing more than a transparent tube comprising a mouth and a gut. However, *C. elegans* exhibits sophisticated biology such as organogenesis and displays complex traits such as chemotaxis and mating behavior (Riddle *et al.*, 1997). Brenner published a landmark paper in 1974 in which he described more than 100 genes that are required for *C. elegans* behavior and in which he introduced *C. elegans* as a model organism for biomedical research (Brenner, 1974).

Caenorhabditis elegans is a soil nematode that feeds on bacteria. A wild-type population consists almost exclusively of self-fertile hermaphrodites, a trait that facilitates the growth of genetically homogenous laboratory cultures. The proportion of males in a population can be increased easily under laboratory conditions, which is another advantage for the performance of genetic experiments. The worm is transparent and grows to ca. 1 mm in length, thus anatomy and processes such as embryogenesis can be studied easily in the living animal. *Caenorhabditis elegans* has an invariant development that has allowed for the determination of the complete cell lineage, including the position, fate and tissue type of each cell in the organism (Sulston, 1988). The adult hermaphrodite has 959 somatic cells, which subdivide into many different cell types and tissues, including muscle, hypodermis (skin), intestine, reproductive system, glands and neurons. About 302 (30%) of the cells form the nervous system, and a map of all neurons and all synaptic connectivities has been generated (White, 1988).

Caenorhabditis elegans was initially used as a model for the study of development, neuronal guidance, neurodegeneration and synaptic properties, but it has also provided significant insights into other processes such as programmed cell death. The study of the function and interactions of *C. elegans* cell death genes has greatly enhanced the understanding of the process of apoptosis in vertebrates (Kaufmann and Hengartner, 2001). Experiments using human cell death genes expressed in *C. elegans* have confirmed that the human and *C. elegans* versions of the genes perform the same function in both species. This can be considered a significant step toward the use of *C. elegans* as an important model for biomedical research (Vaux *et al.*, 1992; Miura *et al.*, 1993; Hengartner and Horvitz, 1994). Today, *C. elegans* is widely used for the study of numerous other areas of mammalian biology, such as metabolism, cell–cell signaling, aging and gender determination, because most pathways are significantly conserved between mammals and *C. elegans* (Riddle *et al.*, 1997).

Caenorhabditis elegans has played a crucial role during the genomics era. It was the first multicellular organism to be sequenced fully and its use has pioneered the development of whole-genome mapping, sequencing and bioinformatic tools for 100–1000 Mbp genomes (The *C. elegans* Sequencing

Route A

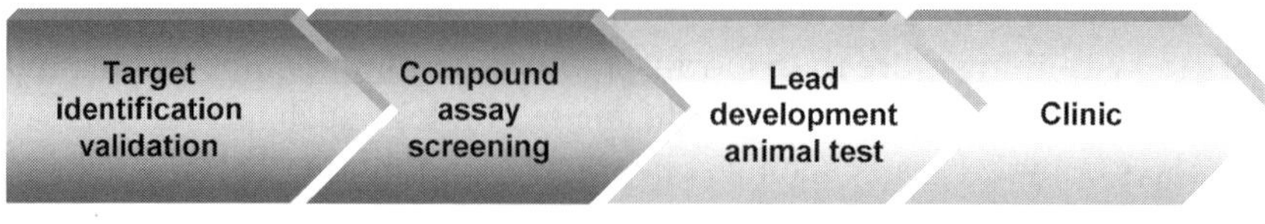

Route B

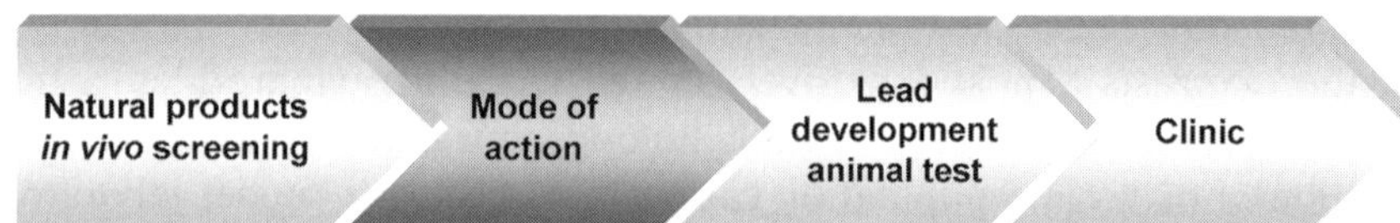

Figure 3.1 *Caenorhabditis elegans* in drug discovery. Target identification and validation are the initial steps in genomics-based drug discovery and *C. elegans* plays an important role during this phase (dark grey). *Caenorhabditis elegans* assay development and *in vivo* high-throughput screening capacities are used for hit generation (dark grey). As soon as vertebrate studies and clinical trials are required, the utility of *C. elegans* diminishes (light grey). However, the drug discovery process is no longer linear and feedback loops are possible between the various phases. For example, an alternative 'route B' is screening in animal and insect models to obtain *in vivo* hits of high quality. The caveat of this approach is that often the molecular target will be unknown. *Caenorhabditis elegans* can be used for mode-of-action studies to identify molecular targets as needed for lead development and registration (dark grey). Obviously, in the case where a novel target is identified, route B would be merged into route A

Consortium, 1998). These methods have become the basis for other high-throughput genome sequencing projects, particularly the human genome project. Comparison of the human and *C. elegans* genomes revealed that many disease genes and disease pathways are present in *C. elegans*. This has stimulated many studies to establish *C. elegans* as a model for the study of a range of human disorders. For example, mutations in the human presenilin-1 gene are associated with early-onset familial Alzheimer's disease. Mutations in the corresponding *C. elegans* ortholog *sel-12* cause defects in neurons as well. Experiments in which these defects have been restored by transgenic expression of human presenilin-1 demonstrated a remarkable functional conservation between *C. elegans* and humans (Levitan *et al.*, 1996; Wittenburg *et al.*, 2000). Another well-conserved pathway, the Ras pathway, provides the possibility to use *C. elegans* in the discovery of anticancer therapeutics. A compelling example is given by the analysis of the effects of farnesyl transferase inhibitors on activated Ras in *C. elegans* mutants. Farnesyl transferase inhibitors inhibit the requisite processing of a number of proteins, including the proto-oncogene Ras, and have been shown to afford good antitumor efficacy (Karp *et al.*, 2001). These inhibitors specifically revert the multivulva phenotype of Ras gain-of-function *C. elegans* mutants (Hara and Han, 1995). The long list of human diseases studied in *C. elegans* also includes metabolic disorders (e.g. diabetes), central nervous system (CNS) disorders (e.g. depression) and several congenital disorders such as Duchenne muscular dystrophy and polycystic kidney disease (Bessou *et al.*, 1998; Barr and Sternberg, 1999; Habeos and Papavassiliou, 2001).

The above attributes have prompted the entry of *C. elegans* into the drug discovery industry in recent years. It is amenable to high-throughput compound screening, mode-of-action analysis and large-scale target validation (Figure 3.1). Millions of animals can be grown daily for screening campaigns, either in liquid or on plates. Conservation of disease pathways, considerable transferability of human drug action into *C. elegans* and drug uptake through the gut membrane allow large-scale *in vivo* pharmacology. A short 3-day life cycle and amenability to molecular, genetic, biochemical and physiological analyses speed up the dissection of entire pathways and target validation programs. Finally, and importantly, the growth and maintenance requirements of *C. elegans* are of relatively low cost.

In the following pages we will describe how to apply *C. elegans* technologies to drug discovery. As an example, we will describe the successful use of *C. elegans* within a CNS disease project. This example will serve as a guide throughout the following chapters.

3.2 From disease to target

Hunt for validated targets

Many diseases are caused by heritable disturbances in gene function whereby the disease is manifested during gestation or shortly after birth. However, the majority of human diseases such as cancer, stroke and diabetes, although also linked to malfunctions in genes, are manifested only later in life. The causes of the malfunctions are case dependent and may involve acquired point mutations, pathogenic mis-expression of genes or may be related to other specific perturbations of cell biology. Importantly, the most common human diseases are often characterized by uncontrolled signaling within several biological pathways. An understanding of the molecular mechanism of diseases opens many opportunities to develop new therapies, including those tailored to the genetic profiles of individual patients.

In this chapter we describe an efficient route leading from the molecular analysis of human disease in the model organism *C. elegans* to the discovery of validated therapeutic targets (Figure 3.2). The process starts with the development of a *C. elegans* disease model, exemplified here via a discussion of a *C. elegans* unipolar depression model. *Caenorhabditis elegans* disease

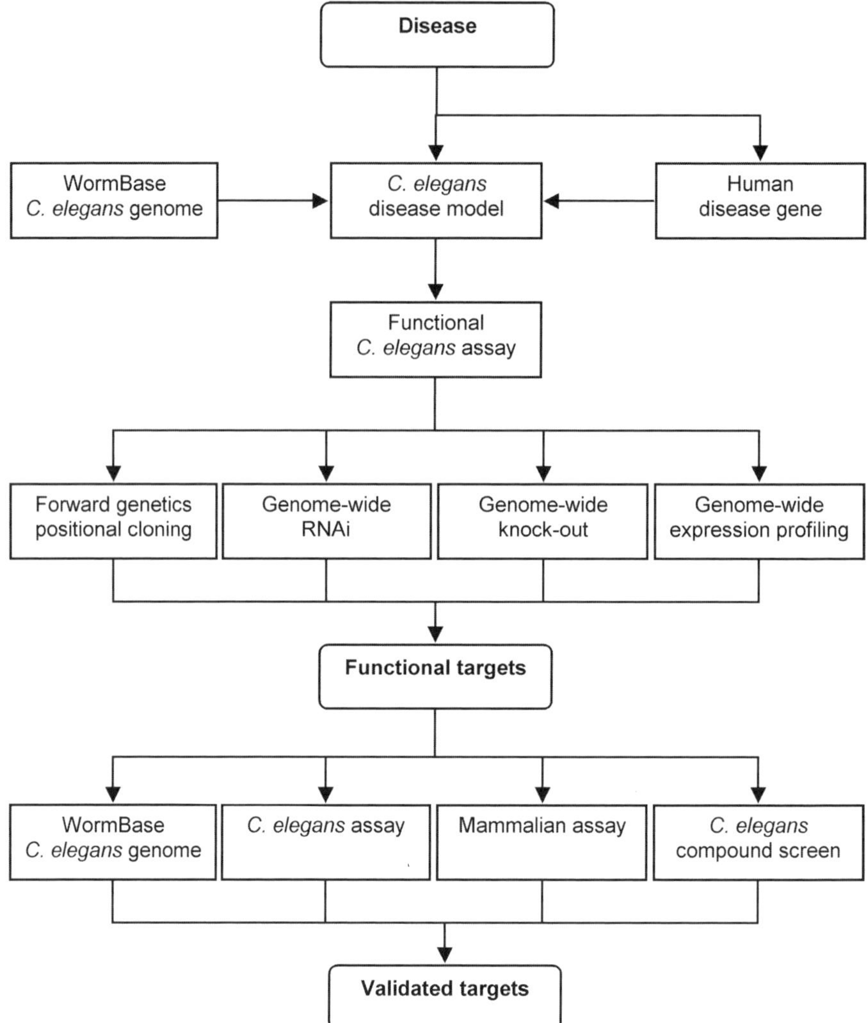

Figure 3.2   Target validation in *C. elegans*

models are designed to mimic the molecular aspects of a disease and to be useful for the conduct of a functional assay suitable for high-throughput technologies. For example, RNA interference technology (RNAi) can be used for selectively knocking down any of the 19 000 genes that make up the *C. elegans* genome. One can select for those genes that, when knocked down, cause a disease-linked and measurable phenotype. In contrast to the use of chip experiments to identify genes whose expression patterns change in response to a given physiological stress, the *C. elegans* target identification technologies described in this chapter are based on functionally validated *C. elegans* models and hence yield targets that have a functional effect in a disease. These molecular targets are then validated further before entry into chemical screening campaigns. For the purposes of our discussion, validated targets are defined as molecular targets that both modify the relevant disease biology and are druggable (e.g. the molecular target codes for a member of a protein family with a history of successful chemistry campaigns). A molecular target may be tested in a battery of *C. elegans* and mammalian assays, and compound screens with *C. elegans* may be conducted to confirm the potential to identify a chemical ligand against the target. The following is a discussion of the utility of *C. elegans* for the identification of validated targets for the treatment of depression.

Depression – a case study

Depression and anxiety are the most frequently occurring mental disorders. These diseases are commonly expressed together rather than as separate syndromes. More than 20% of the adult population suffer from these conditions at some time during their life. The Word Health Organization (WHO) predicts that depression will become the second leading cause of premature death or disability worldwide by the year 2020 (Buller and Legrand, 2001). Surprisingly, depression is still underdiagnosed and under-treated (Hirschfeld *et al.*, 1997; Lepine *et al.*, 1997). Consequently, only 15% of individuals who have recovered from an initial episode of depression do not experience relapse (Thase, 1992).

The study of tryptophan (a serotonin precursor) levels in depressed patients has led to the hypothesis that depression arises from decreased neurological response to, or repressed levels of, serotonin (Coppen, 1967). This implies that increases in the level of, or sensitivity to, serotonin (5-HT) would improve mood. The first-generation antidepressants, such as monoamine oxidase inhibitors (MAOI) and tricyclic antidepressants (TCA), act on neurotransmission by blocking the reuptake of monoamines, inhibiting neurotransmitter degradation or binding directly to specific receptors. Advances in the understanding of these mechanisms have led to the development of drugs with enhanced specificity, such as the selective serotonin reuptake inhibitor (SSRI) fluoxetine (launched in 1988). The clinical relevance of interruptions to serotonin concentrations has been demonstrated with inhibitors of the serotonin transporter. Inhibition of the serotonin transporter increases serotonin concentration at the synaptic cleft and hence increases serotonin activity (de Montigny *et al.*, 1981; Blier and de Montigny, 1994; Czachura and Rasmussen, 2000). Although the role of serotonin concentrations has been demonstrated, the precise molecular mechanism of depression is known to be quite complex because the onset of the therapeutic benefit of SSRIs usually occurs only 2–3 weeks after the onset of therapy. Thus, other mechanisms in addition to signal enhancement may play a role in the treatment of mood disorders. Attempts have been made to modify other pathways that may contribute to mood, such as modulation of the endocrine system to treat depression. Clinical studies have been launched with corticotropin-releasing hormones, but their efficacy is still an open question (Holsboer, 1999; Zobel *et al.*, 2000). Other potential targets include neurokinin receptors, 5-HT$_2$ receptors and *N*-methyl-D-aspartate (NMDA) receptors (Saria, 1999; Petrie *et al.*, 2000; Middlemiss *et al.*, 2002). Despite extensive research, the underlying pathology of depression and anxiety remains poorly understood.

A *C. elegans* model for unipolar depression

How can *C. elegans* help to identify additional molecular mechanisms that influence mood and also to reveal novel targets and drugs for the treatment of depression? The serotonin pathway plays an important role in neuromodulation and metabolism in *C. elegans*. *Caenorhabditis elegans* mutants that lack tryptophan hydroxylase (*trp-1*), the enzyme that initiates serotonin synthesis, show abnormalities in a range of behaviors such as feeding and egg-laying (Sze *et al.*, 2000) (Figure 3.3). They also accumulate large amounts of fat and have a reduced life expectancy, in parallel to the effects of serotonin fluctuations on metabolism and obesity in higher order mammals. How can serotonin influence feeding in *C. elegans* and how can the study of these effects be extrapolated to the treatment of depression and other diseases?

Caenorhabditis elegans feeds on bacteria that are taken up into its mouth and passed into its pharynx where it is ground, processed and pumped into the intestine (Riddle *et al.*, 1997). The pharynx is a tube consisting of muscles, neurons and marginal cells surrounded by a single layer of epithelial cells. The pharynx functions largely as an autonomous unit that pumps rhythmically up to 300 times per minute. The frequency and strength of the contractions are regulated by several neurotransmitter systems. Serotonin functions as the pacemaker for the basal pumping activity and modulates the frequency of pumping in response to food availability and metabolic status. Pumping activity is reduced in the absence of food. Food is sensed by the dopaminergic neurons ADE and CEP in ciliated cells located in the nose of *C. elegans*. Recognition of food by these neurons generates signal transmission, via the serotonergic interneuron RIH and the serotonergic motor neuron NSM, to the pharynx in order to stimulate pumping (Ward, 1973; Perkins *et al.*, 1986; White, 1986; Sawin *et al.*, 2000). The coordination between egg-laying and locomotion in *C. elegans* provides another example of the role of serotonin signaling. *Caenorhabditis elegans* switches regularly between phases of egg-laying activity, with eggs laid in bursts. During such egg-laying bursts, *C. elegans* increases the velocity and direction of movement to enable spatial

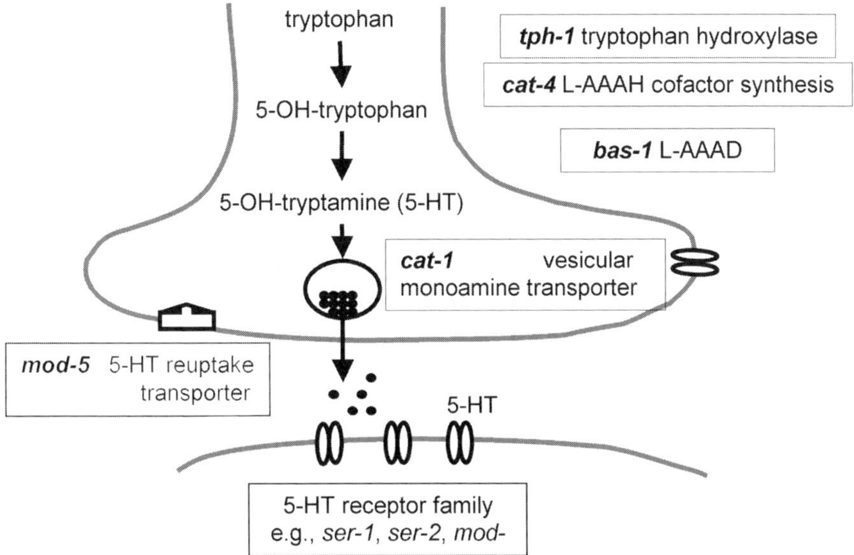

Figure 3.3 *Caenorhabditis elegans* orthologs genes in a serotonergic synapse. *Caenorhabditis elegans* gene names have a three-letter and one-number code and are printed in italic type. Tryptophan is processed with tryptophan hydroxylase (the *C. elegans* ortholog *tph-1*), GTP cyclohydrolase I (*cat-4*), which is an enzyme necessary for synthesis of the cofactor biopterin that is required by (among others) all aromatic amino acid hydroxylases (AAAHs), and the aromatic amino acid decarboxylase (L-AAAD; *bas-1*) (Brownlee and Fairweather, 1999). Serotonin is finally released into the synaptic cleft via the vesicular monoamine transporter *cat-1*. Serotonin activates a range of 5-HT receptors to transmit the signal. *Caenorhabditis elegans* has various 5-HT receptors such as *ser-1*, *ser-2* or *mod-1* (Ranganathan *et al.*, 2000). Serotonin is transported back into the synapse via the 5-HT reuptake transporter *mod-5* (Ranganathan *et al.*, 2001)

distribution of the eggs. This modulation of activity is coordinated by decision-making interneurons that regulate locomotion (AVF) and motor neurons that regulate egg-laying (HSN). Again, serotonin plays a role in modulating the pathway that controls behavioral coordination. Movement itself is controlled by the command interneuron AVB, with the neurotransmitters acetylcholine and GABA acting at the neuromuscular junction.

Although the biological role of serotonin in modulating *C. elegans* behaviour is well understood, it is debatable whether *C. elegans* exhibits a behavior that could be characterized as a mood disorder. Such behaviors could include reduced feeding related to a loss of appetite or a decreased movement linked to suffering. The relevant interpretation of the study of feeding, egg-laying and movement behaviors in *C. elegans* is that the animal has a complex nervous system that operates under the control of the serotonin pathway and other neurotransmitters linked to the manifestation and treatment of numerous human neurological disorders. These behaviors have served as important tools to study human disease-relevant neurological signaling in *C. elegans* (Figure 3.4).

We have mentioned earlier several disease pathways that could be studied in *C. elegans*. It is impossible to discuss all *C. elegans* disease models in sufficient detail but we have outlined in Figure 3.4 three entry points for the development of a *C. elegans* disease model. A certain biological process such as drinking can be chosen as a genetically tractable phenotype to model synapse function. A thorough knowledge of neurotransmitter signaling in *C. elegans* and the availability of drugs has been used to develop this phenotype into a disease-relevant model of serotonergic signaling. A more common approach is the use of gene knock-downs to create disease models. A disease gene such as Ras can be knocked down or overexpressed to create genetically tractable phenotypes (Hara and Han, 1995). It is also possible to express the human gene in *C. elegans* to induce a phenotype. The modeling of a disease in *C. elegans* immediately raises the question: how many genes are actually conserved between humans and *C. elegans*? Depending on the bioinformatics approach, a *C. elegans* homolog has been identified for 65–78% of human genes (Sonnhammer and Durbin, 1997; Kuwabara and O'Neil, 2001). A more rigorous prediction of the number of *C. elegans* homologs that are putative disease gene orthologs has been made based on a comparison of *C. elegans* sequences with genes of the OMIM database (the OMIM database is a catalogue of human genes and genetic disorders, http://www.ncbi.nlm.nih.gov/omim). A *C. elegans* homolog has been found for about 85% of 100 analyzed disease genes. When blasting these *C. elegans* homologs against the human genome, the human disease gene was the closest human gene to the *C. elegans* gene for 42% of the tested genes (Culetto and Sattelle, 2000).

Developing a functional assay

A successful exploitation of model organisms such as *C. elegans* as tools to study human diseases is dependent upon the availability of reliable assays to study gene and pathway function. The primary challenge is to develop an assay that models a disease at the molecular level in a format appropriate for large-scale genetics and compound screening. Regarding the serotonin pathway, the question is how to convert the measurement of a serotonin-related *C. elegans* behavioral phenotype into an assay that can identify genes or compounds that increase activity at the serotonergic synapse. As described previously, pharynx contraction in *C. elegans* is regulated by serotonin. The measurement of pharynx contractions is too laborious for use on a large scale but the eating and drinking behavior of *C. elegans* can be used as an indirect measure of pharynx contraction. Devgen, a drug discovery company based in Belgium, uses a dye that fluoresces only when taken up into the gut of

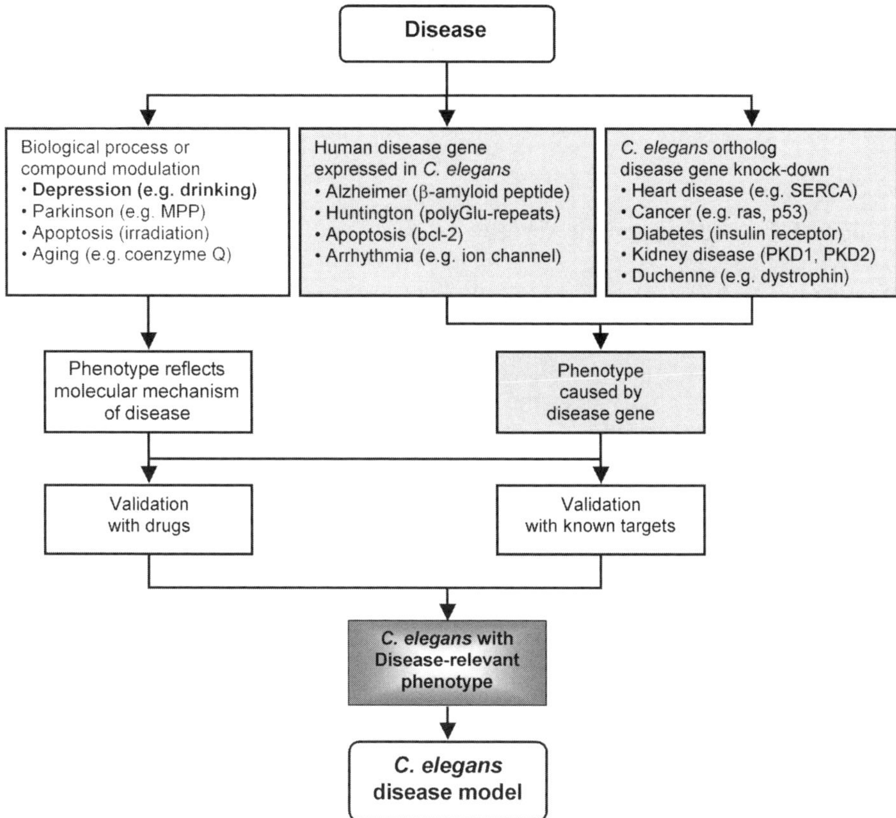

Figure 3.4 Development of *C. elegans* disease models. The development of a *C. elegans* disease model and a functional assay requires a disease-relevant phenotype. There are several ways to engineer a *C. elegans* disease model. A biological process such as synapse transmission reflects the underlying mechanism of a disease and can be used to develop a disease model. We discuss unipolar depression as a case study. Another example is the use of 1-methyl-4-phenylpyridinium (MPP) to induce dopamine neuronal death in *C. elegans* as a model for Parkinson's disease (Nass *et al.*, 2002). Coenzyme Q diets control the lifespan of *C. elegans* and are one of the many ways to model aging in *C. elegans* (Larsen and Clarke, 2002; Tissenbaum and Guarente, 2002). A human gene is expressed in *C. elegans* to cause phenotypes similar to the human disease. For example, expression of human β-amyloid peptide in *C. elegans* causes amyloid deposits. These deposits cause paralysis, which is a genetically tractable phenotype (Link, 1995, 2001). Expression of an NH_2-terminal Huntington fragment in *C. elegans* causes neuronal malfunction and the formation of aggregates (Faber *et al.*, 1999). The configuration of ion channel screens in *C. elegans* is described later in the text. Most *C. elegans* disease models carry knock-outs in *C. elegans* orthologs of disease genes. These models are discussed and referenced in the text. A phenotype identified in *C. elegans* needs to be validated with known reference drugs or by knock-down of disease-related genes to establish a valid link between *C. elegans* and the disease of interest

C. elegans. Fluorescence activity in the gut is proportional to food uptake or drinking, and hence pumping frequency. The assay can be used for both genetic analysis and to screen for compounds that effect pathways involved in pumping. In the following, we demonstrate the use of the assay in the identification of genetic targets in the serotonergic pathway.

Design of *C. elegans* genetic screens

Genetic analysis has been the preferred tool for the study of genes and proteins for nearly a century. In classical or forward genetics, the genome of a model organism is randomly mutagenized. Mutants that exhibit the desired phenotype are used to discover the identity of genes responsible for producing the phenotype. The following simple procedure highlights some specific aspects of a *C. elegans* genetic screen. A typical genetic screen uses the mutagen ethylmethanesulfonate (EMS), which induces G/C to A/T transitions and small deletions in genes. Hermaphrodites are incubated in 50 mM EMS for 4 h in order to accumulate 10–20 mutations per genome. After treatment, the worms are distributed on Petri dishes and left to grow for two generations, resulting in homozygous mutants. The F2 progeny is scored for the desired phenotype and isolated mutants are retested for the phenotype. The strains can be conveniently preserved for long-term storage in liquid nitrogen. *Caenorhabditis elegans* hermaphrodites produce homozygous offspring, thus a simple F2 screen for recessive mutations can be completed within only two weeks. Owing to the high mutation frequency, such a screening campaign requires only 10 to 20 000 haploid genomes to recover a few mutants per gene. Therefore, the mutant strain needs to be out-crossed several times, but this step is also rapid and can be completed in less than a month.

Devgen has used the *C. elegans* 'drinking assay' to screen for mutations in genes that enhance pumping activity and hence they are candidate genes associated with serotonergic signaling. A set of mutant strains has been isolated that exhibit a positive effect in the 'drinking assay', as measured by a significant increase in dye uptake. Before we illustrate the process of the positional cloning of one of these mutants, we shall describe a few examples of more complex genetic screens.

Genetic screens commonly lead to the identification of three categories of genes. The first category contains the genes that contribute directly to the biological process of interest. Pertaining to the 'drinking assay', this category would include the genes that, when mutated, directly increase the serotonergic tonus at the synapse, such as the serotonin reuptake transporter. The second category includes those genes that influence indirectly the process of interest. Taking our example, mutations that constitutively switch on a signal to feed would stimulate drinking and they could be members of the dopaminergic signaling pathway. The third category of genes includes miscellaneous or 'bystander' genes, which would not be of interest to elucidate the biology under study.

Three basic types of genetic screens have been successfully developed and applied using *C. elegans*. The first type of screen, like the screen described above, isolates genetic mutations that induce a measurable phenotype associated with a particular area of biology. The second type of screen is an enhancer/suppressor screen that maps out complete pathways and the third type is a resistance/sensitivity screen that identifies the mode of action of a drug. Enhancer/ suppressor screens have been applied successfully to decipher many *C. elegans* pathways, such as Ras signaling, apoptosis, Alzheimer's disease, transforming growth factor β (TGF-β) and insulin signaling. For example, a model to study the epidermal growth factor (EGF)/Ras pathway in *C. elegans* is the vulva development (Sternberg and Han, 1998; Chang and Sternberg, 1999). The vulva consists of 22 cells and is located in the middle of the hermaphrodite. The eight muscles of the vulva mediate egg-laying. We have already stated that egg-laying is highly regulated by serotonin and acetylcholine. This EGF/Ras signaling cascade induces three out of six candidate vulva precursor cells to adopt vulval fates during vulva development. Mutations in the *C. elegans* homolog of the EGF receptor, LET-23, interrupts this signal and inhibits differentiation of precursors into vulval cells, resulting in a vulva-less phenotype. Gain-of-function mutations in the *C. elegans* Ras kinase homolog, LET-60, lead to overactivation of the pathway whereby all six precursor cells produce vulvae, resulting in a multivulva phenotype. Genetic studies in *C. elegans*, based on mutational outcomes measured via the vulval phenotypes, provided the first indication, in any organism, that Ras proteins have roles in cell specification and differentiation as opposed to cell growth and proliferation (Han and Sternberg, 1990). This work elucidated the cellular function of Ras and established a *C. elegans* model for EGF/Ras-related oncogenesis. A nematode-based enhancer/ suppressor screen for genes within the EGF/Ras pathway identified the *C. elegans* homolog of the proto-oncogen c-cbl, SLI-1 (Yoon *et al.*, 1995). An epistatic analysis of SLI-1 was used to study interactions of the gene with other pathway components to indicate that c-cbl acts as a negative regulator of the EGF/Ras pathway. This hypothesis has been confirmed in c-cbl-deficient mice, leading to an improved understanding of mammalian c-cbl function (Murphy *et al.*, 1998).

Mode-of-action studies

The third type of genetics screen is often referred to as 'chemical genetics' (Alaoui-Ismaili *et al.*, 2002; Zheng and Chan, 2002). The principle is similar to an enhancer/suppressor screen but, instead of using a mutant background, a compound is employed to screen for mutants that are either resistant or hypersensitive to the effect of the compound. The use of chemical ligands in target identification and validation programs also allows for the concurrent analysis of a target's role in a disease process in parallel with an assessment of the target's druggability. In this reverse chemical genetics approach, small, bioavailable and target-specific compounds are used to study biological questions and to expand the pathways around validated drug targets. For example, the acetylcholinesterase inhibitor aldicarb has been used extensively in genetic screens to identify genes involved in synaptic vesicle exocytosis, such as *unc-18*/nSec-1 (Hosono and Kamiya, 1991; Hosono *et al.*, 1992). Mutations in the gene *unc-18* are resistant to the paralyzing effect of aldicarb. *Caenorhabditis elegans*-based studies of this gene provided the first evidence for the role of *unc-18* in synaptic vesicle fusion (Gengyo-Ando *et al.*, 1993; Garcia *et al.*, 1994). Aldicarb has been used to identify presynaptic genes. The acetylcholine receptor agonist levamisol has been used to find postsynaptic targets (Lewis *et al.*, 1980a,b; Kim *et al.*, 2001a).

Genetic screens also can be readily configured as mode-of-action (MOA) assays to identify the molecular targets of drug candidates. This forward chemical genetics approach is extremely useful for natural product molecules or lead compounds arising from *in vivo* screens. Identification of the target allows for the development of assays to enable lead optimization or the identification of further chemical hits and leads. Yet hits and leads generated from 'on-target' screens may, nevertheless, induce clinically relevant effects through interactions with additional targets. Such effects require further MOA analysis. An example is the antidepressant fluoxetine, which inhibits the serotonin reuptake transporter and potentially interacts with other targets. The effect of fluoxetine in *C. elegans* resembles the effect of an SSRI in that it enhances drinking or a particular movement behavior called the 'slow-down response'. Further evidence for fluoxetine action on the *C. elegans* serotonin reuptake transporter, MOD-5, is that *mod-5* mutants are resistant to fluoxetine (Ranganathan *et al.*, 2001). Fluoxetine also induces a 'nose contraction' phenotype in *C. elegans*, suggesting that fluoxetine acts on additional targets. A genetic screen for fluoxetine-resistant mutants identified two novel genes, *nrf-6* and *ndg-4*, that define a novel gene family of multipass transmembrane proteins (Choy and Thomas, 1999). The role of these genes in serotonergic signaling and in depression is currently under investigation.

The suitability of the MOA studies described above depends largely on the conservation of the binding site of the test compound. Although the conservation of genes and pathways between humans and *C. elegans* is remarkably high, a compound's action often depends upon interaction of the compound with only a few amino acid residues. A striking example of the conservation of a compound's binding site is given by the thapsigargin-resistant isoform of the sarcoplasmatic/endoplasmatic reticulum Ca^{2+}-ATPase (SERCA). This SERCA removes Ca^{2+} from the sarcoplasmatic or endoplasmatic reticulum and plays a role in several diseases, such as congestive heart failure. Chinese hamster SERCA is resistant to thapsigargin inhibition due to an F256V mutation (Yu *et al.*, 1999). Introduction of the same mutation at the homolog's position in the *C. elegans* SERCA renders thapsigargin resistance in animals carrying the transgene (Zwaal *et al.*, 2001).

Genetic screens in *C. elegans* are sufficiently fast and effective to permit their incorporation into sophisticated assay formats. As an extreme example, the *C. elegans* homolog of a human potassium channel has been identified in a screen in which each mutated nematode underwent surgery followed by an electrophysiological examination (Davis *et al.*, 1999).

Rapid gene mapping using single-nucleotide polymorphisms

The phenotypic analysis of mutant animals reveals important information about biological processes, but a full elucidation of the molecular basis of the biology of interest requires decoding of the involved genes. Gene identification using positional cloning is a straightforward approach in *C. elegans* that entails two steps: mapping and gene confirmation. The researcher of today can rely on the availability of a detailed genetic and physical map organized in the database ACeDB. Several mapping strategies exist and positional cloning incorporating single-nucleotide polymorphism (SNP) technology has emerged over the last two years (Jakubowski and Kornfeld, 1999; Swan *et al.*, 2002). Single nucleotide polymorphisms are detectable as single base pair changes in the genes of strains or individuals, but small deletions, duplications or insertions are also found. Single-nucleotide polymorphisms occur once every 100–300 bases in the human genome (NCBI, April 2002, http://www.ncbi.nlm.nih.gov/SNP) and can correlate with changes in the amino acid composition of the expressed protein, thereby changing the activity of the protein. The fact, that an SNP can also alter the interaction between the protein and a given drug has received much attention in the pharmaceutical industry. Under the term 'pharmacogenomics', SNP profiles of individual patients are evaluated to tailor drugs and drug regimens to a patient's genetic profile, enabling individualized medicine. The true potential of predicting a patient's response to a drug, based on an SNP haplotype, will be shown in the future (Jazwinska, 2001).

Nevertheless, SNP profiling has and will continue to contribute to the process of target identification. Single-nucleotide polymorphism analysis allows for rapid gene mapping in *C. elegans*, mice and humans (Wang *et al.*, 1998; Lindblad-Toh *et al.*, 2000; Wicks *et al.*, 2001) and can be conducted not only quickly but also cost effectively in *C. elegans*. The *C. elegans* laboratory strain Bristol N2 has little sequence variation from individual to individual but by using this strain in combination with a second strain from Hawaii a high-density polymorphism map has been established. A density of one SNP every 872 bp in these strains has been predicted based on a 5.4 Mbp aligned sequence, with many of the SNPs causing changes in restriction sites such as restriction fragment length polymorphisms (RFLPs) or snip-SNPs (Wicks *et al.*, 2001).

Studies in the Bristol N2 and Hawaii strains have been used to generate a map showing that snip-SNPs exist throughout the *C. elegans* genome at a frequency of one every 91 ± 56 kb. This map can be used to zoom in to a resolution of ca. 0.3 map units. Fine mapping of SNPs then can be conducted easily in the region of interest. Isolation of recombinants is the actual limiting factor rather than the availability of SNP markers. Typically, the mutant strain is crossed with Hawaiian males and the homozygotes of the F2 progeny are submitted to snip-SNP analysis. Next, lysates of either wild-type or mutant animals are pooled and snip-SNPs are then amplified and digested to map the mutation onto a chromosome. Fine mapping is performed via SNP analysis on single animals. In the instance that snip-SNPs are not available for the region of interest, alternative approaches for genotyping SNPs are available (Kwok, 2001).

The *C. elegans* genome has been fully sequenced and genetic map positions can be correlated directly with the physical map. A map unit in *C. elegans* corresponds to ca. 300 genes in the clusters and a dozen genes outside the cluster (Barnes *et al.*, 1995). Databases such as WormBase offer lists of all genes between two given markers and provide efficient tools for the nomination of candidate genes resulting from target hunts (Stein *et al.*, 2001). Candidate genes can be confirmed by rescue of the mutant phenotype, by microinjection or by phenocopy using the RNA interference approach.

Continuing our example of the 'drinking assay', Devgen has positionally cloned several mutants that exhibit high drinking rates. This assay has been confirmed as an 'on-pathway' assay because genes known to be involved in the serotonin pathway have been returned in the screen, such as the *C. elegans* homolog of the human serotonin reuptake transporter MOD-5; MOD-5 has been identified independently in a screen for mutants lacking formaldehyde-induced fluorescence in the NSM neuron after serotonin administration (Ranganathan *et al.*, 2001). These studies have validated the utility of the *C. elegans* 'drinking assay' as a tool for studying genes and drugs acting in human depression.

Genome-wide RNAi

Forward genetics remains a fundamentally important approach for target identification. However, positional cloning can be laborious and time consuming so geneticists have sought alternative technologies for forward genetics. Genome-wide RNA interference technology (RNAi) has become very popular because it supersedes the need for positional cloning. The RNAi phenomenon was first observed in *C. elegans* (Montgomery and Fire, 1998). Double-stranded RNA (dsRNA) induces degradation of the corresponding mRNA, leading to protein depletion and a loss-of-function phenotype. Double-stranded RNA is cut by a dsRNA-specific RNAse, Dicer, into small dsRNA molecules of 20–23 nucleotides (siRNAs or short-interfering RNAs). A multiprotein RNA-induced silencing complex (RISC) uses the siRNAs specifically to break down the corresponding target mRNA and prevent translation (Caplen, 2002). The RNAi knock-down of genes in *C. elegans* can be accomplished by simply injecting dsRNA into the gonad, by soaking animals in a bath containing dsRNA or by feeding *C. elegans* with bacteria that produce dsRNA. Gene-specific phenotypes can be observed either in the treated animals or in the next generation, and the induced phenotype can be maintained as long as animals are exposed to dsRNA. The time required to go from genotype to phenotype correlates with the time required to induce an RNAi effect for a given gene, enabling a novel, high-throughput and genome-wide reverse genetics approach.

Several research institutes and companies have built libraries containing either dsRNA or bacteria that produce dsRNA, representing all genes of the *C. elegans* genome (Tabara *et al.*, 1998; Timmons and Fire, 1998; Fire, 1999; Fraser *et al.*, 2000; Gonczy *et al.*, 2000; Kamath *et al.*, 2001; Devgen and Exelixis, personal communication). Such libraries are formatted in 96-well plates and are compatible with multiwell screening robotics. An important advantage of the RNAi approach as a drug discovery tool is that focused libraries of gene families, e.g. all G-protein-coupled receptors (GPCRs) or all putative druggable targets, can be assembled. The dsRNA material can be transferred to 96-well plates containing *C. elegans* animals and phenotypes can be scored over several days. A number of RNAi screens have been performed in *C. elegans* to identify genes linked to lethality, behavioral effects or developmental defects. Results of such RNAi experiments representing approximately one-third of all predicted *C. elegans* genes can be found in WormBase and more extensive and specific screens are underway at a number of institutions and companies. The RNAi does not work for all genes with the same efficiency; neurons in particular seem to be refractory to RNAi. However, attempts to improve efficiency are ongoing to include the creation of mutant strains that are more sensitive to RNAi (Simmer *et al.*, 2002).

Genome-wide knock-out

A third possibility to obtain gene knock-outs is via the generation of large deletion mutations that can be identified by polymerase chain reaction (PCR) technology (Jansen *et al.*, 1997; Edgley *et al.*, 2002). The entire procedure can be semi-automated by dividing EMS-mutagenized *C. elegans* populations into

96-well pools and then screening the pools in a systematic manner. Once a population of a single well has been found positive for a deletion, the progeny can be grown up and individuals can be tested to obtain a mutant strain. *Caenorhabditis elegans* hermaphrodites offer advantages when analyzed in this fashion because isolation of individual hermaphrodites is sufficient to establish a line. Although mutagenesis is a random event, a deletion knock-out can be generated via sequential mutagenesis campaigns. The *C. elegans* gene knock-out consortium produces several hundred knock-outs per year for the establishment of a genome-wide knock-out library.

Genome-wide expression profiling

The pathogenic status of a cell not only reflects changes in the activity of single genes or proteins but also changes in the activities of a range of genes or proteins that contribute to various pathways. The underlying rationale of target hunts is the assumption that several genes in a network contribute to a disease and, more importantly, that the modulation of several gene activities can reverse the disease state. This opens several entry points for therapeutic approaches and allows for the selection of the druggable genes. Instead of generating knock-outs gene by gene and then testing for disease relevance, an overall snap-shot of the activity of all genes may simultaneously identify all relevant genes associated with a particular pathology. This can be achieved by using DNA chips or DNA microarrays for expression profiling, allowing a comparison of changes throughout the genome in pathogenic versus normal cells. We will discuss here the use of DNA chips and microarrays for gene identification but other applications are possible, such as in diagnostic, pharmacogenomic and toxicogenomic studies.

Several DNA microarrays containing 17871 genes or >90% of the *C. elegans* genome are available (Jiang *et al.*, 2001). *Caenorhabditis elegans* chips are made up of PCR fragments of 1–2 kb genomic DNA. The RNA from one sample is used to prepare Cy3-labeled cDNA, and RNA from another sample is used to prepare Cy5-labeled cDNA. These two cDNA probes are simultaneously hybridized to a single DNA microarray and the hybridization intensities are measured. *Caenorhabditis elegans* DNA chips have been used to profile expression throughout development. Comparison of RNA samples from each developmental stage to a mixed population sample has revealed a twofold change in expression levels in about 12 486 of the 17 871 genes evaluated. *Caenorhabditis elegans* chips have been made available to the *C. elegans* community and data from more than 30 collaborations have been collected to develop a gene expression topomap (Kim *et al.*, 2001b). Data from 553 experiments have been used to create a correlation matrix of all genes to establish functional groups of genes having similar expression profiles. The *C. elegans* gene expression topomap contains 44 gene mountains or functional groups including collagen, metabolic enzymes or germ-line-specific genes. In one interesting case, members of the *C. elegans* Wnt family have been distributed to either the embryonic or larval Wnt signaling pathway. In this way, mountains or functional groups enrich for genes of a certain process of interest. For drug research, it would be very useful to create a 'pharmaceutically tractable genome chip' containing all *C. elegans* orthologs that are likely to be druggable (Milburn, 2001).

Genome-wide protein interaction mapping

The elucidation of protein interactions is a key component in understanding protein function. Protein–protein interactions are an important facet of biological processes and their characterization can be used to identify the key modulators of a given gene of interest. A *C. elegans* genome-wide protein interaction map project has been launched (Walhout *et al.*, 2000a). Complementary DNA from open reading frames is cloned into yeast two-hybrid (Y2H) vectors using the Gateway recombinational cloning system (Walhout *et al.*, 2000b). These vectors (containing DNA binding domains) can be used as baits to screen a *C. elegans* cDNA library. The identified proteins then can be used for new Y2H screens or tested in a matrix (a vector carrying an activating domain and a vector carrying the DNA binding domain) to build protein interaction networks. Such an analysis has been conducted on 27 genes required for *C. elegans* vulva development. The 27 genes have been used as baits in extensive Y2H screens leading to the isolation of 124 interacting partners (Walhout *et al.*, 2000a). It is commonly accepted that Y2H activity is not strongly predictive of physiologically relevant protein–protein interactions. Confirmation studies are normally required. The utility of a genome-wide Y2H campaign is realized by the combination of Y2H data with knock-out and gene expression data. This combination allows for the description of a skeleton of genes that form a pathway or network. It is important to examine the extent to which interactions identified in the *C. elegans* genome predict for human protein interactions. As an example, conserved interactions (inter-ologs) have been used to identify *C. elegans* DNA damage response genes (DDR genes) (Boulton *et al.*, 2002). The *C. elegans* genome has been compared with the human sequences of DDR genes and 75 putative orthologs were identified. These were tested in a Y2H matrix. Seventeen of the 33 interactions that are known in humans have been detected, or stated otherwise; the experimental data indicate that at least 17 protein–protein interactions are conserved between human and *C. elegans*. A further eight, potentially novel, protein–protein interactions of human genes were indicated by the experimental results.

Annotation of the *C. elegans* genome

The open scientific culture of the *C. elegans* community in combination with the availability of the complete *C. elegans* genome has stimulated several institutes to coordinate their efforts in the study of gene function on a genome-wide scale. These data are integrated in WormBase (http://www.wormbase. org) (Stein *et al.*, 2001). WormBase is a database of genetic and molecular data for *C. elegans*. It was developed by an international consortium of biologists and computer scientists and was founded in 2000. WormBase includes information on all genes, including accession numbers, a summary of protein function or predicted function, literature, available mutants, clones and expressed sequence tags (ESTs). Furthermore, it contains data from many high-throughput gene validation approaches, such as micro-array expression data, expression pattern and RNAi knock-down data. It also features an anatomic *C. elegans* atlas, genetic maps and other analysis tools.

One interesting question is: how many genes actually exist in the genome? The number of genes comprising the *C. elegans* genome is predicted to be ca. 19 000, whereas the number of experimentally confirmed genes and ESTs is only 9503 (The *C. elegans* Sequencing Consortium, 1998; Reboul *et al.*, 2001). Open reading frame sequence tags (OSTs) can be used to verify rapidly which of the predicted open reading frames are 'real' genes, because they are less dependent on the expression level than ESTs. The OST experiments have provided experimental evidence for the presence of at least 17 300 genes (91%) in the *C. elegans* genome. A more recent report suggests that about 20% of the predicted genes may be pseudogenes. This estimate was obtained by using promoter green fluorescent protein (GFP) fusion constructs derived from genes that were recently duplicated, in an evolutionary sense (Mounsey *et al.*, 2002). The discrepancy between 'real genes' and gene prediction for even a simple genome reflects the complexity of gene annotation for a full genome. Current figures indicate that 350 eukaryotic sequencing projects are underway (Bernal *et al.*, 2001). Industrial-scale genome annotation approaches are required for drug discovery teams to make efficient use of the massive amount of raw sequencing data now being generated. The combination of a high density of available genetic data, the high-throughput technologies for gene function analysis and the compact genome of *C. elegans* make *C. elegans* an ideal model for the development of automated gene annotation projects (Bingham *et al.*, 2000; Eisenhaber *et al.*, 2000).

Target identification and validation strategies

Caenorhabditis elegans technologies for the rapid identification of disease-relevant functional targets can be enhanced further by their combination with emerging technologies. Forward genetics tools such as Mos1 transposon insertion technology promise a reduction in the time required for positional cloning, and techniques are being developed for the use of proteomics in *C. elegans* (Bessereau *et al.*, 2001; van Rossum *et al.*, 2001; Hirabayashi and Kasai, 2002). Many novel genes have been identified using a variety of non-*C. elegans*-based technology, but their function in disease-related processes is often poorly understood. *Caenorhabditis elegans* technologies offer a range of solutions for the functional characterization of genes, but it would exceed the scope of this review to describe all of them. Importantly, a target can be shown provisionally to be a functionally relevant target for the treatment of disease by analysis in a mammalian assay, but the ultimate proof can only be established via a positive outcome from a clinical trial. One may therefore question the merits of gene identification or validation in *C. elegans*. There are two answers to this question. Firstly, *C. elegans* technology has proven utility in rapidly reducing large pools of potential targets, arising for example from an expression analysis assay, into a manageable pool of a dozen or so genes. These genes then can be examined in more time-intensive and laborious mammalian assays. Secondly, targets arising from *C. elegans* target hunts have been selected based on their ability to modulate disease-relevant biological pathways that mimic those present in humans. Moreover, the data collected during a *C. elegans* target hunt can be used to facilitate the interpretation of mammalian data or to help develop new validation strategies in mammalian assays.

3.3 Lead discovery

Drug discovery, as we know it today, started in the late 19th century when some of the essential fundamentals of chemistry and pharmacology were established. In 1815, F. W. Sertürner isolated morphine from opium extract and suggested a role for the active components in plants (Sertürner, 1817). Kekulé's benzene theory of 1865 stimulated dye research (Drews, 2000). By 1870, Avogadro's atomic hypothesis had been confirmed and a Periodic Table of the elements was established. In the following years, Paul Ehrlich and John Newport Langley proposed a link between drugs and their action on specific components in tissues (Maehle *et al.*, 2002). In 1907, Paul Ehrlich postulated the existence of 'chemoreceptors' for drugs, which led to our modern understanding of drug action.

Until the early 1960s, drug research involved the individual testing of drug candidates one by one in whole-animal assays. This was a tedious and labor-intensive approach, yet many successful drugs for CNS, cardiovascular diseases and cancer were identified in this manner. Later, the use of isolated organs and tissues was introduced, but by the 1980s the rate of new chemical entity discovery began to decline, indicating a need for novel concepts to accelerate the drug discovery process. Three major developments have revolutionized drug screening in the last 10–15 years: advances in structural biology have facilitated the development of techniques for the rapid testing of compounds on isolated proteins, such as enzymes, receptors, etc., the introduction of combinatorial synthetic methods has made it possible to generate large compound libraries; and innovations in engineering and automation technology have enabled high-throughput *in vitro* screening of hundreds of thousands of compounds in a short time. It is now possible to screen 100 000 compounds in an assay on a single day. Nevertheless, these advances have increased neither the number of chemical entities entering clinical trials nor the delivery of new drugs to the market (Horrobin, 2001). In other words, the translation of activity identified *in vitro* via these technologies to relevant *in vivo* activity has been inefficient. Several explanations have been proposed for this observation, including a lack of bioavailability and a suboptimal pharmacokinetic profile, which can cause hits to fail in animal tests or require prolonged cycles of lead optimization. Another explanation could be that the protein configurations used in many assay systems are sufficiently different from the *in vivo* state, leading to spurious or artificial results (Horrobin, 2001). Additionally, the underlying molecular mechanism of many diseases is poorly understood or may be multifactorial, which can lead to the introduction of inappropriate targets into screening campaigns.

These inefficiencies in target selection and hit identification can be mediated, in many cases, by the use of *C. elegans* compound screens. Hits identified from *C. elegans* compound screens are, *de facto*, bioavailable because the animal ingests the compound via feeding. Thus, the compound must be absorbed across the intestine and then must diffuse to the target to generate activity measured via a positive readout. The readouts used in *C. elegans* screens are commonly functional, such as changes in pharynx pumping rates or other categories of movement. Thus, positive hits must not only bind to the target but also be selective and potent enough to cause a measurable cellular response. An additional attraction of *C. elegans* screens is that the assays can be run in mutant strains carrying mutations in disease-related pathways, thus guaranteeing the disease relevance of hits.

A further unique advantage of *C. elegans*, particularly when compared with other model organisms, is that its size and robustness make it amenable to high-throughput and fully automated assay systems. On the following pages, we will describe a *C. elegans* high-throughput screen and discuss the relevance of *C. elegans*-derived hits for medical research.

The compound library

An essential requirement for lead discovery is access to a compound library of appropriate size, having a well-rationalized composition and compounds of high purity: 'one can only get out, what one has put in'. The majority of biotechnology companies depend on chemical libraries that are purchased 'off the shelf' from companies or universities, rather than libraries that have been tailored to suit the target class of interest. Such libraries often arise from the early days of combinatorial chemistry and, as such, are random libraries often containing mixtures of compounds with suboptimal solubility and bioavailability characteristics. Positive hits from such libraries are difficult to confirm and hits often disappear after deconvolution. Roger Lahana hit the nail on the head when he wrote: 'When trying to find a needle in a haystack, the best strategy might not be to increase the size of the haystack' (Lahana, 1999). Combinatorial chemistry is generally defined as the synthesis of libraries of compounds containing all possible combinations of reagents or building blocks (Rose, 2002). Although combinatorial chemistry has advanced and incorporates techniques such as automated parallel synthesis, solution-phase synthesis and solid-phase extraction, a trend reversal towards the use of more intelligently designed, high-quality libraries has emerged. In general, a highly diverse library, or one tailored to a specific target class, may increase the chance of finding hits. A fingerprint of descriptors is used to characterize the chemical diversity of libraries. The descriptors can be derived from a two-dimensional representation of the molecular structures of the library components, such as molecular weight, atom counts and hydrophobicity ($\log P$ value), or from their three-dimensional conformations, including dipole moment and shape (Livingstone, 2000). Chemical descriptors are used to define a multidimensional space around molecules. The distance to the next neighboring molecule and the distribution of all molecules within the space of the library provide a means to evaluate the diversity of a given library (Patterson *et al.*, 1996).

Given the enormous size of the organic chemical space ($> 10^{18}$ compounds), the aim of reaching high diversity should be tempered by insuring the 'drug-likeness' of the compounds. A popular approach is to filter compounds based on the 'rule-of-five' as defined by Lipinski and co-workers at Pfizer (Lipinski *et al.*, 2001). They analyzed 2245 drugs from the World Drug Inventory that were reported to have reached the phase II level of clinical evaluation. The compounds shared the following characteristics: a molecular weight of < 500, < 10 hydrogen-bond acceptors, < 5 hydrogen-bond donors and had $\log P$ values of < 5. It was shown that compounds that fulfill only two of the four criteria are likely to be poorly absorbed across the gut wall.

The use of chemical descriptors would be enhanced if they could be combined with biological descriptors. This has proved difficult because it is only through the interaction of hits with targets that the biological activity of compounds is revealed. Thus, the prediction of biological activity based on a compound's chemical structure remains largely an art. We will revisit this issue when discussing the value of a hit derived from a *C. elegans* screen.

Caenorhabditis elegans is amenable to high-throughput screening

High-throughput screening (HTS) is the complement to combinatorial chemistry and genome-wide target identification. It allows the screening of chemical libraries containing several hundreds of thousands of compounds against a wide range of novel targets in a robust and timely manner. If a non-mammalian model system is to have a significant impact on drug discovery in a time-effective way, it must be amenable to HTS. *Caenorhabditis elegans* can be grown in liquid and handled efficiently in 96-well or 384-well plates and is the only multicellular model organism whereby a population of hundreds of animals can fit into a single well. Inter-animal variations between members of a *C. elegans* population commonly used in laboratory settings are low, which is an advantage over mammalian animal populations. This low variability leads to the high reproducibility of assays and allows the application of statistical analyses such as the z-factor calculation of assay variability. Because most *C. elegans* biology has been established from the analysis of mutants that have been identified through genetic screening, appropriate assays to screen tens of thousands of animals are available. Thus, high-throughput rates and miniaturization can be readily achieved with *C. elegans* assays. Another important prerequisite for high-throughput library screening is a low compound concentration format. *Caenorhabditis elegans* takes up considerable amounts of compound through normal drinking processes, which allows compound screens to be performed at concentrations of 1–30 μM (Devgen, personal communication).

Assay design

The design of a *C. elegans* assay depends naturally on the selected target and the biological process of interest. The challenge for assay design is to ensure the relevance of an assay for a particular disease, therefore we will classify *C. elegans* assays by the type of genetically engineered animal used. The easiest assay type employs wild-type animals but, as with all *C. elegans in vivo* assays, a sufficient specific phenotype or pathway endpoint to track or measure the biology of interest is required. In our example of the *C. elegans* depression model, enhanced pharynx pumping is strongly correlated to increased serotonergic tonus at the synapse of the *C. elegans* pharynx. As shown in a previous section, this phenotype has been used successfully for target identification. In other cases, genetically mutated *C. elegans* disease models can be used for library screens. The principle is to knock down a disease-related gene and to screen for compounds that revert the disease-related phenotype to normal. An excellent example is the previously mentioned *C. elegans* model for type II diabetes. Type II diabetes or insulin resistance is characterized by reduced insulin signaling and thus potential therapeutics should enhance insulin signaling. The *C. elegans* model of insulin signaling carries a specific mutation in the *daf-2* gene, which is the *C. elegans* ortholog of the human insulin receptor (Gottlieb and Ruvkun, 1994; Kimura *et al.*, 1997). The *daf-2* mutants can be restored to wild type or 'cured' by genetically induced inhibition of phosphatase and tensin homolog, a negative regulator of the insulin signaling pathway (Ogg and Ruvkun, 1998; Gil *et al.*, 1999; Butler *et al.*, 2002). This *daf-2* mutant has been used by Devgen as a tool to screen for compounds that enhance insulin signaling.

The third way in which to genetically engineer a *C. elegans*-based compound screen is to express the desired human target in the animal. The *C. elegans* assay can be a preferred assay over a cell-based assay for the screening of 'tough' targets such as ion transporters, ligand-gated ion channels, voltage-gated ion channels and channels with accessory chains. These targets require a complex tissue environment. Voltage-gated ion channels open by a 'gating mechanism' upon change of the membrane potential and transport ions across the cell membrane. Although more than 300 human ion channels have been predicted, ion channels account for only 5% of the molecular targets of marketed drugs (Drews, 2000; Venter *et al.*, 2001). One of the reasons has been the lack of HTS technologies. The golden standard for studying ion channels is patch-clamp electrophysiology, which works at very low throughput. Over the last few years the trend has changed by the development of technologies such as the fluorimetric imaging plate reader (FLIPR), flux assays and HTS patch-clamp platforms, and ion channels have experienced a renaissance (reviewed by Owen and Silverthorne, 2002). One of the issues of the non-patch-clamp technologies is the lack of voltage control under physiologically relevant conditions. This can be overcome by expressing human voltage-gated ion channels in wild-type *C. elegans* or in animals defective for the *C. elegans* ortholog of the corresponding channel. This approach differs importantly from a standard overexpression-cell-based assay because expression of the human channel in *C. elegans* is required to achieve functionality. The advantage of this approach for compound screening is that the screen is performed on the actual and functionally active human target in an *in vivo* set-up. Running the assay on the transgene-negative strain or on wild-type *C. elegans* can easily filter out compounds that do not act directly on the transgene, so-called false positives. In addition, electrophysiology in *C. elegans* is a well-established technology

(Raizen and Avery, 1994; Davis *et al.*, 1995; Franks *et al.*, 2002). Whole-animal *C. elegans* electrophysiology and patch-clamping of target *C. elegans* tissue and cells allow functional characterization of the human or insecticidal ion channel in *C. elegans* and the pharmacological characterization of hit compounds. As an example, Devgen has performed high-throughput compound screens on ligand-gated ion channels. After *C. elegans* hit filtering, these compounds have been tested and confirmed in Xenopus (frog) oocyte voltage-clamp electrophysiology.

Assay development

Assay development is the art of establishing an experimental procedure to perform hundreds and thousands of tests in a highly reproducible and quantitative manner. For *C. elegans* assays, the same principles and goals applies as for any assay development program, such as the need for robustness, reproducibility, sensitivity, a procedure with only a few simple steps, ease of assay validation, reagent supply, up-scaling, assay automation and cost effectiveness (Bronson *et al.*, 2001). In the following, we shall highlight two *C. elegans*-specific challenges in assay development: the production of *C. elegans* animals and the automation of phenotypic readouts.

The first challenge for a *C. elegans* production unit is to deliver millions of *C. elegans* animals for each screening day, with all animals in the same condition. In the case of the 'drinking assay', this demands that every animal is in the same feeding state and has the same feeding activity, because we use feeding as an indirect measure of the serotonergic tonus. In other words, millions of animals have to behave in nearly the same way. Scaling up of a *C. elegans* population, from a few plates (corresponds to tens of thousands of animals) sufficient for one experiment to populations of several millions of animals that must be delivered day by day for several weeks of a screening campaign, requires sophisticated logistics and the utmost stringency in adhering to the culture protocol. This can be achieved by establishing and monitoring every parameter that influences drinking, such as the quantity and quality of food, the developmental stage of the animal, ambient temperature, medium components, etc. (Devgen, personal communication).

The second challenge is to enable the analysis of phenotypes in a high-throughput format. Even a phenotype that is seemingly easy to measure, such as live versus dead, limits the compound throughput to a few thousand per day because the examiner must analyze well by well. The assay would become even more work intensive if the readout had to be quantified and if the population in each well had to be counted. The drinking assay incorporates a fluorescent measurement as a readout for pharynx pumping. This measurement can be used in a plate reader and, as such, is amenable to

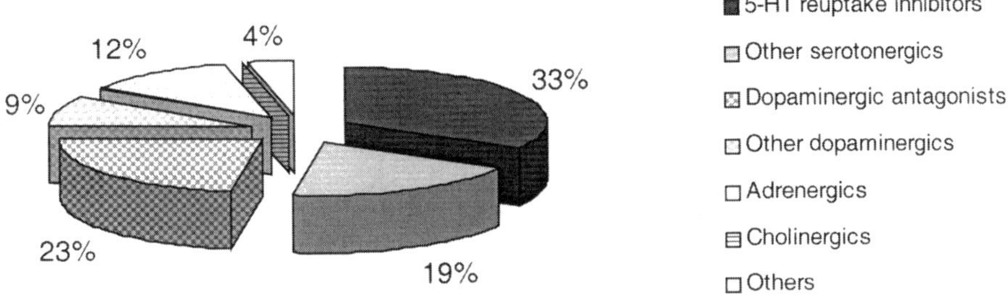

Figure 3.5 Distribution of 'drinking assay' hits of the learning set for CNS drugs. A learning set of ca. 250 CNS-related drugs has been tested in the 'drinking assay'. The distribution of hits to the various modes of action has been analyzed. The calculated percentage is the number of hits acting on a particular mode of action relative to the total number of hits

HTS. Typically, the ratio of the fluorescence signal of a population of normally pumping *C. elegans* animals versus that of a population with increased pharynx pumping is used to optimize the assay for robustness, sensitivity and reproducibility. The quality of the assay can be evaluated using the z'-factor as statistical parameter, which integrates the signal dynamic range and the data variation (Zhang *et al.*, 1999). An increase of the z' value means an increase of the assay quality. Typical z' values for *C. elegans* assays lie between 0.2 and 0.5, which is very high for a whole-animal assay. Another compelling example wherein a fluorescence marker for a behavioral phenotype is used is the 'chitinase assay' developed by Pharmacia & Upjohn (Gurney *et al.*, 2000). The 'chitinase assay' measures *C. elegans* egg-laying behavior indirectly through the chitinase activity produced by hatching larvae. Egg-laying activity is a useful endpoint to study CNS-related processes and is measured by counting the eggs laid within a defined time interval. Because *C. elegans* embryos secrete chitinase to permit hatching out of the chitin-containing eggs, the total chitinase activity of a well reflects the amount of hatching larvae, which is proportional to the amount of laid eggs. Pharmacia has developed this assay to screen some 10 000 compounds on a *C. elegans* model for Alzheimer's disease.

Automated image acquisition is an important way to screen phenotypes in a high-throughput manner. The hardware used is similar to that used for cell-based image acquisition systems and comprises an inverted microscope, a scanning stage, a charge-coupled device (CCD) camera and robotics for plate handling. In contrast to high-throughput fluorescence imaging systems such as the FLIPR, *C. elegans* image acquisition systems have a high requirement for resolution and for sophisticated image analysis software, because *C. elegans* animals are much richer in phenotypes than cells. High-content image acquisition is often sacrificed at the expense of throughput. Image-based

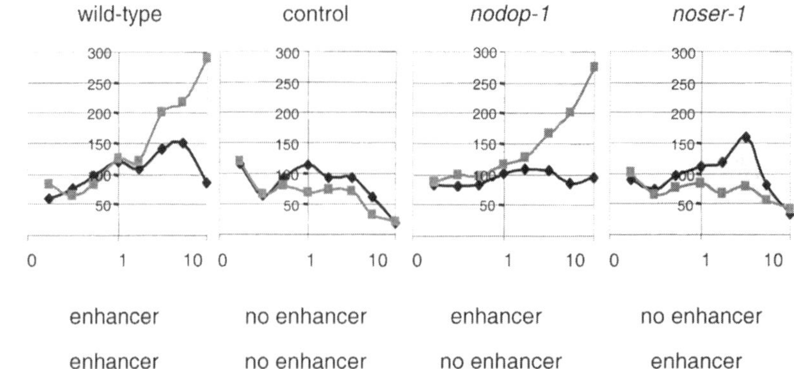

Figure 3.6 Hit filtering with *C. elegans* mutants. The two compounds SER1 (■), a putative serotonin reuptake inhibitor, and DOP1 (♦), a putative dopamine receptor antagonist, enhance drinking in the 'drinking assay'. The compound concentration (in μM) is given on the *x*-axis and the relative fluorescence (in %) is given on the *y*-axis; 100% corresponds to no effect on drinking for a given strain. The control strain has only minimal levels of the neurotransmitters serotonin and dopamine. Any compound that increases serotonergic signaling via the serotonin reuptake transporter cannot enhance drinking in this strain. Similarly, dopamine receptor antagonists cannot enhance drinking because the dopaminergic signaling is already reduced in this strain. The strain *nodop-1* lacks only dopaminergic signaling. A serotonergic compound can enhance drinking in this strain whereas a dopaminergic compound cannot. The results of the two compounds on the strain *nodop-1* suggest that only SER1 is a serotonergic compound. For the strain *noser-1* the situation is reversed. This strain lacks serotonergic signaling, hence SER1 cannot enhance drinking. Compound DOP1 enhances drinking and is therefore unlikely to be a serotonergic compound

screens with *C. elegans* only offer real advantages if data on complex phenotypes such as cell migration in a living animal, movement pattern and morphology can be generated.

A special type of *C. elegans* automation technology is COPAS™ from Union Biometrica, USA. This *C. elegans* animal sorter works like a conventional fluorescence-activated cell sorter (FACS). Different from conventional cell sorting, COPAS™ is able to sort tube-shaped living animals of lengths varying from 70 to 1300 µm. COPAS™ records four parameters per object passing the sorting chamber: the time of flight to measure length, the extinction to discriminate transparent larvae from darker adults and two fluorescent parameters (www.unionbio.com). Sophisticated software, the COPAS Profiler™, has been developed to sort fluorescence-labeled animals by the position of the fluorescence signal along the body axis. The COPAS™ platform is equipped with a plate handling system and operates quickly enough to be plugged into a high-throughput screening process.

Compound learning set for assay validation

In addition to confirming the quality of a *C. elegans* assay, the assay must also be pharmacologically validated. Devgen has used a learning set of about 250 CNS drugs to include drugs with mode of actions that modulate drinking and drugs that should not affect drinking. These classes of drugs have been used to validate the *C. elegans* 'drinking assay'. Serotonin reuptake inhibition should increase drinking and consequently >85% of the tested SSRIs and >75% of all 5-HT reuptake inhibitors have been shown to enhance drinking rates in the assay. Because dopamine negatively regulates drinking, inhibitors of dopaminergic signaling should also enhance drinking, but indirectly. Forty-five percent of the dopamine antagonists tested have been shown to influence drinking rates. The third important neurotransmitter that increases drinking rates is acetylcholine. Consequently, none of the tested antagonists were detected in the screen. A range of unrelated CNS compounds have been tested, including adrenergic antagonists, opioids and histaminergics. Fifteen percent of the adrenergic modulators were shown to be enhancers of drinking rates and must be considered as false positives. The conclusion is that the drinking screen for enhancers is highly sensitive because it identifies most of the SSRIs for which the assay has been configured. It should be noted that this *C. elegans* assay is able to identify reliable human drugs for a specific target and that the assay is another example of the high level of conservation between *C. elegans* and human pharmacology.

Another question is the number of false positives that can be expected. Thirty-three percent of all hits obtained in the 'drinking assay' are the desired SSRIs (Figure 3.5). In addition, 51% of the hits act specifically on the biology under investigation because a serotonin receptor agonist also increases drinking. Although these hits may be still of interest, we will show later how to filter out dopaminergics and serotonin receptor modulators. The remaining 16% of the hits are false positives, which need to be filtered out as well. The data validate the drinking assay as highly robust, sensitive and selective. We will describe the *C. elegans* screening platform and a screening campaign with the drinking assay.

The *C. elegans* screening platform

The HTS equipment used in a *C. elegans* screening unit includes the same robotics and technology found in standard HTS laboratories but with one important difference (Seethala, 2001): instead of targets or cells being present in the wells for screening, the wells are filled with living animals swimming through the medium. The growth and handling requirements associated with the use of living *C. elegans* animals limits the time available for screening to 3 days per week and per campaign, assuming no weekend shifts. The logistics of animal production dictates the timing of the screening campaigns, because animals must be grown up in advance of the screens and back-up cultures must be prepared to limit failure rates. A single day of screening with the drinking assay is composed of the following steps: dispensing the *C. elegans* animals into 96-well plates, adding compounds from the screening library, adding the dye-mix, incubation and fluorescence analysis with a plate reader. These steps are executed by robotics to achieve a throughput of about 30 000 data-points per day, where a data-point is defined as the fluorescent measurement on a per-well basis (Devgen, personal communication). Screening quality is monitored constantly because each plate contains 80 test samples plus eight positive and eight negative controls. Additional quality control plates are screened that contain only control samples. The z' factor is used as the acceptance criterion for a screening batch. Data processing is similar to standard HTS. The most interesting question is the hit rate. The biological meaning of a hit from a *C. elegans* screen is slightly different from that of an *in vitro* screen. Hits that meet the technical requirement of activity in the assay are defined as the 'positives'. The positives are first checked for autofluorescence and any such compounds are removed from the positive pool. Dose–response curves are then generated to confirm the positives. Positives having dose–response curves that meet certain criteria are defined as 'hits'. Hit rates of 0.2–2% are typical. As an example, Devgen has screened a 22 000-member chemical library in the 'drinking assay' and achieved a hit rate of 0.22% for drinking enhancers. Standard rates for hits and false positives have been determined for the assay by the use of Devgen's learning set (a collection of compounds used to validate the assay). We will now demonstrate how to select specific SSRIs from the hit list.

What is the meaning of a *C. elegans* hit?

A typical biochemical HTS assay employs only the desired target and will therefore identify 'on-target' hits. Knowledge of a hit's mode of action is required to support medicinal chemistry efforts and is needed later in the lead optimization phase to support the selection of relevant animal models. Mode-of-action information is also supportive of regulatory approval for marketing. Because *C. elegans* screens, by their *in vivo* nature, are not on-target screens, hits from the assays must be assigned to a mode of action prior to the hit to lead optimization. Before we explain the various routes to assign a mode of action, we would like to discuss how one estimates how many targets, and specifically which targets, have contributed to the measured phenotype. Theoretically, all druggable genes that are expressed during the course of an *in vivo* assay could serve as potential targets for a given compound. For humans,

10 000–15 000 human drug targets are predicted (Drews, 2000; Bailey *et al.*, 2001): the evaluation of discrete protein domains, such as the kinase domain, leads to the estimated presence of a few thousand druggable *C. elegans* targets. Under the conditions of a real-life chemical screen, the number of potential targets is significantly lower because *C. elegans* compound assays are optimized toward a particular biological process. For example, the 'drinking assay' for enhancers of drinking enriches for serotonin signaling agonists, as shown by analysis of the learning set. This is because serotonin is the key neurotransmitter that increases pharynx pumping frequency. Inhibition of glutaminergic and dopaminergic signaling, which downregulates pumping, could also increase pharynx pumping. Therefore, dopamine receptor antagonists have also been identified in the learning set, although at a lower yield rate because dopamine acts indirectly on the pharynx. Other pathways such as Ca^{2+} homeostasis or metabolism could be influenced in the presence of chemicals, leading to a change in pharynx pumping. However, blocking the serotonin reuptake transporter or activating the serotonin receptor are the most efficient ways to increase pharynx pumping.

Hit filtering

We will return to the outcome of the drinking screen to demonstrate a hit-calling assay or hit-filtering assay to identify SSRIs. *Caenorhabditis elegans* hits can be filtered through the use of strains carrying mutations in the desired pathway. This 'mutant filtering' approach works in the same way as the resistance genetics approach. A receptor knock-down mutant is resistant to a receptor agonist and, depending on the test, the hit acts either up- or downstream of this receptor. The use of sets of mutants allows the researcher to focus in on the position within a pathway where a compound acts. A selection of compounds from the drinking screen having potential anti-depressive activity was accomplished with a set of serotonin and dopamine pathway mutants. This allowed for the selection of hits acting on or downstream from the serotonin reuptake receptor. In this example, six mutant *C. elegans* strains were used to focus in on serotonin reuptake transporters. The use of a large range of mutant strains in this 'mutant filtering' approach increases the accuracy of predicting the compound's site of action within the pathway. The precision of this approach depends on the knowledge of the pathway and the corresponding biology. The upstream components of the serotonin pathway, including the receptors, are well understood and allow for the relatively simple selection of mutants useful for hit filtering. Knowledge of the downstream components of this pathway is limited, thus prediction of the mode of action for compounds acting downstream entails extensive experimentation.

An example is given in Figure 3.6. Two hits that enhance drinking were tested on several mutants in the serotonergic and dopaminergic pathway. The activity profile of the hit compound 'SER1' links it to the serotonergic pathway and excludes it from the dopaminergic pathway. For hit compound 'DOP1' the interpretation is reversed. Via the 'mutant filtering' approach we have assigned 40% of the hits from the drinking screen as SSRIs. This result is slightly better than expected from the learning set analysis. To confirm the results of the hit filtering, a selection of the hits were submitted to pharmacological profiling. Pharmacological profiling encompasses a battery of *in vitro* binding assays in which the compound of interest competes with a reference drug. This provides evidence on which target site a given compound acts. The selected hits from the drinking screen example have all been shown to test positively in relevant *in vitro* studies that incorporate the human serotonin uptake transporter.

Plugging *C. elegans* into drug discovery

We have demonstrated how to set up a *C. elegans* assay and how to perform a compound screening campaign with *C. elegans*. Many of the technologies and principles are identical to other HTS *in vitro* or cell-based systems. Actually, an observer would not notice any differences in the process as long as he did not look into a well. We have used an example of a screening project to show that assay quality, throughput and hit follow-up are comparable to cell-based screening. Finally, we demonstrated that it is possible to identify compounds that are active on the desired human target with a *C. elegans* HTS assay. We have chosen this example because the *C. elegans* and the human biology of serotonergic signaling are well understood and because pharmacological tools such as human drugs and follow-up *in vitro* pharmacological assays are available to validate the assay. However, as described earlier in the introduction to CNS disease assays using *C. elegans*, the R&D market seeks novel targets and novel chemistry that are distinguishable from the SSRIs. The question is: Can *C. elegans* support the needs of the market? We believe that the answer is in the affirmative because the modulation of pharynx pumping in *C. elegans* has been established as an *in vivo* model for serotonergic synapse function that is useful for target and for hit identification. Therefore, this model can be tailored and refined to discover novel mechanisms related to synapse function that are relevant for the treatment of depression and other CNS disorders. Moreover, *C. elegans* mutant strains or transgenic animals can be employed to refocus the assay to identify novel targets. For example, a screen using mutants having a defective serotonin reuptake transporter MOD-5 would only reveal compounds acting downstream of or in parallel to the transporter but not compounds acting on the transporter itself. The SSRIs that mediate the most efficient biological route toward an increase of serotonin at the synapse would not be detected. Such an assay would provide a means to screen for novel chemistry and novel modes of action in depression where such compounds, by definition, could not be SSRIs.

The next step is the process of hit to lead optimization. A model organism researcher may underestimate the efforts required for the development of a hit into a lead compound, yet it is recognized that hit optimization is a highly critical phase in drug development, and a clinical candidate has to pass a range of animal, toxicity and absorption, distribution, metabolism and excretion (ADME)/pharmacokinetics (PK) tests. Although the pharmacological conservation between *C. elegans* and humans is striking, toxicity and ADME/PK mechanisms are very species specific. *Caenorhabditis elegans* would not be an appropriate model for these preclinical studies because even vertebrate models are often not sufficiently predictive. However, there is not much practical value in using *C. elegans* assays to identify novel chemistry in an *in vivo* context and then to filter out the hits in subsequent biochemical assays. Any advantage that would have been achieved by screening in an animal model might be lost if follow-on *in vitro* assays were used to assess activity. Therefore, it is advisable to run *C. elegans* tests in parallel to animal model tests during hit optimization.

Caenorhabditis elegans is a fairly recent addition to the group of model systems used in the pharmaceutical industry. The future will show whether targets and drugs identified in *C. elegans* will gain a position in the pharmaceutical market place. We are entering a path toward a new way to conduct drug discovery.

3.4 Acknowledgment

We thank the Devgen team that has helped to move *C. elegans* into drug discovery.

3.5 References

Alaoui-Ismaili, M. H., Lomedico, P. T. and Jindal, S. (2002). Chemical genomics: discovery of disease genes and drugs. *Drug Discov. Today* 7, 292–294.

Bailey, D., Zanders, E. and Dean, P. (2001). The end of the beginning for genomic medicine. *Nat. Biotechnol.* **19**, 207–209.

Barnes, T. M., Kohara, Y., Coulson, A. and Hekimi, S. (1995). Meiotic recombination, noncoding DNA and genomic organization in *Caenorhabditis elegans. Genetics* **141**, 159–179.

Barr, M. M. and Sternberg, P. W. (1999). A polycystic kidney-disease gene homologue required for male mating behaviour in *C. elegans*. *Nature* **401**, 386–389.

Bernal, A., Ear, U. and Kyrpides, N. (2001). Genomes OnLine Database (GOLD): a monitor of genome projects world-wide. *Nucleic Acids Res.* **29**, 126–127.

Bessereau, J. L., Wright, A., Williams, D. C., Schuske, K., Davis, M. W. and Jorgensen, E. M. (2001). Mobilization of a *Drosophila* transposon in the *Caenorhabditis elegans* germ line. *Nature* **413**, 70–74.

Bessou, C., Giugia, J. B., Franks, C. J., Holden-Dye, L. and Segalat, L. (1998). Mutations in the *Caenorhabditis elegans* dystrophin-like gene *dys-1* lead to hyperactivity and suggest a link with cholinergic transmission. *Neurogenetics* **2**, 61–72.

Bingham, J., Plowman, G. D. and Sudarsanam, S. (2000). Informatics issues in large-scale sequence analysis: elucidating the protein kinases of *C. elegans*. *J. Cell Biochem.* **80**, 181–186.

Blier, P. and de Montigny, C. (1994). Current advances and trends in the treatment of depression. *Trends Pharmacol. Sci.* **15**, 220–226.

Boulton, S. J., Gartner, A., Reboul, J., Vaglio, P., Dyson, N., Hill, D. E. and Vidal, M. (2002). Combined functional genomic maps of the *C. elegans* DNA damage response. *Science* **295**, 127–131.

Brenner, S. (1974). The genetics of *Caenorhabditis elegans*. *Genetics* **77**, 71–94.

Bronson, D., Hentz, N., Janzen, W. P., Lister, M. D., Menke, K., Wegrzyn, J. and Sittampalam, G. S. (2001). Basic consideration in designing high-throughput screening assays. In *Handbook of Drug Screening*, R. Seethala and P. B. Fernandes (eds), pp. 5–30. New York: Marcel Dekker.

Brownlee, D. J. and Fairweather, I. (1999). Exploring the neurotransmitter labyrinth in nematodes. *Trends Neurosci.* **22**, 16–24.

Buller, R. and Legrand, V. (2001). Novel treatments for anxiety and depression: hurdles in bringing them to the market. *Drug Discov. Today* **6**, 1220–1230.

Butler, M., McKay, R. A., Popoff, I. J., Gaarde, W. A., Witchell, D., Murray, S. F., Dean, N. M., *et al.* (2002). Specific inhibition of PTEN expression reverses hyperglycemia in diabetic mice. *Diabetes* **51**, 1028–1034.

Caplen, N. J. (2002). A new approach to the inhibition of gene expression. *Trends Biotechnol.* **20**, 49–51.

Chang, C. and Sternberg, P. W. (1999). *C. elegans* vulval development as a model system to study the cancer biology of EGFR signaling. *Cancer Metastas. Rev.* **18**, 203–213.

Choy, R. K. and Thomas, J. H. (1999). Fluoxetine-resistant mutants in *C. elegans* define a novel family of transmembrane proteins. *Mol. Cell* **4**, 143–152.

Coppen, A. J. (1967). The biochemistry of affective disorders. *Br. J. Psychiatry* **113**, 1237–1264.

Croston, G. E. (2002). Functional cell-based uHTS in chemical genomic drug discovery. *Trends Biotechnol.* **20**, 110–115.

Culetto, E. and Sattelle, D. B. (2000). A role for *Caenorhabditis elegans* in understanding the function and interactions of human disease genes. *Hum. Mol. Genet.* **9**, 869–877.

Czachura, J. F. and Rasmussen, K. (2000). Effects of acute and chronic administration of fluoxetine on the activity of serotonergic neurons in the dorsal raphe nucleus of the rat. *Naunyn Schmiedebergs Arch. Pharmacol.* **362**, 266–275.

Davis, M.W., Somerville, D., Lee, R. Y., Lockery, S., Avery, L. and Fambrough, D. M. (1995). Mutations in the *Caenorhabditis elegans* Na,K-ATPase alpha-subunit gene, *eat-6*, disrupt excitable cell function. *J. Neurosci.* **15**, 8408–8418.

Davis, M. W., Fleischhauer, R., Dent, J. A., Joho, R. H. and Avery, L. (1999). A mutation in the *C. elegans* EXP-2 potassium channel that alters feeding behavior. *Science* **286**, 2501–2504.

de Montigny, C., Blier, P., Caille, G. and Kouassi, E. (1981). Pre- and postsynaptic effect of zimelidine and norzimelidine on the serotonergic system: single cell studies in the rat. *Acta Psychiatr. Scand.* **63**, 79–80.

Drews, J. (2000). Drug discovery: a historical perspective. *Science* **287**, 1960–1964.

Edgley, M., D'Souza, A., Moulder, G., McKay, S., Shen, B., Gilchrist, E., Moerman, D., *et al.* (2002). Improved detection of small deletions in complex pools of DNA. *Nucleic Acids Res.* **30**, e52.

Eisenhaber, B., Bork, P., Yuan, Y., Loffler, G. and Eisenhaber, F. (2000). Automated annotation of GPI anchor sites: case study *C. elegans. Trends Biochem. Sci.* **25**, 340–341.

Faber, P. W., Alter, J. R., MacDonald, M. E. and Hart, A. C. (1999). Polyglutamine-mediated dysfunction and apoptotic death of a *Caenorhabditis elegans* sensory neuron. *Proc. Natl. Acad. Sci. USA* **96**, 179–184.

Fire, A. (1999). RNA-triggered gene silencing. *Trends Genet.* **15**, 358–363.

Franks, C. J., Pemberton, D., Vinogradova, I., Cook, A., Walker, R. J. and Holden-Dye, L. (2002). Ionic basis of the resting membrane potential and action potential in the pharyngeal muscle of *Caenorhabditis elegans. J. Neurophysiol.* **87**, 954–961.

Fraser, A. G., Kamath, R. S., Zipperlen, P., Martinez-Campos, M., Sohrmann, M. and Ahringer, J. (2000). Functional genomic analysis of *C. elegans* chromosome I by systematic RNA interference. *Nature* **408**, 325–330.

Garcia, E. P., Gatti, E., Butler, M., Burton, J. and De Camilli, P. (1994). A rat brain Sec1 homologue related to Rop and UNC18 interacts with syntaxin. *Proc. Natl. Acad. Sci. USA* **91**, 2003–2007.

Gengyo-Ando, K., Kamiya, Y., Yamakawa, A., Kodaira, K., Nishiwaki, K., Miwa, J., Hori, I., *et al.* (1993). The *C. elegans unc-18* gene encodes a protein expressed in motor neurons. *Neuron* **11**, 703–711.

Gil, E. B., Malone, L. E., Liu, L. X., Johnson, C. D. and Lees, J. A. (1999). Regulation of the insulin-like developmental pathway of *Caenorhabditis elegans* by a homolog of the PTEN tumor suppressor gene. *Proc. Natl. Acad. Sci. USA* **96**, 2925–2930.

Gonczy, P., Echeverri, G., Oegema, K., Coulson, A., Jones, S. J., Copley, R. R., Duperon, J., *et al.* (2000). Functional genomic analysis of cell division in *C. elegans* using RNAi of genes on chromosome III. *Nature* **408**, 331–336.

Gottlieb, S. and Ruvkun, G. (1994). *daf-2, daf-16* and *daf-23*: genetically interacting genes controlling Dauer formation in *Caenorhabditis elegans. Genetics* **137**, 107–120.

Gurney, M. E., Geary, T. G., Ellebrock, B. R. and Thoams, E. M. (2000). A nematode drug screen for modulators of mammalian disorders. *Patent Application WO 00/73493 A2* (Pharmacia & Upjohn).

Habeos, I. and Papavassiliou, A. G. (2001). Type 2 diabetes mellitus and worm longevity: a transcriptional link to cure? *Trends Endocrinol. Metab.* **12**, 139–140.

Han, M. and Sternberg, P. W. (1990). *let-60*, a gene that specifies cell fates during *C. elegans* vulva induction, encodes a ras protein. *Cell* **63**, 921–931.

Hara, M. and Han, M. (1995). Ras farnesyltransferase inhibitors suppress the phenotype resulting from an activated ras mutation in *Caenorhabditis elegans. Proc. Natl. Acad. Sci. USA* **92**, 3333–3337.

Hengartner, M. O. and Horvitz, H. R. (1994). *C. elegans* cell survival gene *ced-9* encodes a functional homolog of the mammalian proto-oncogene bcl-2. *Cell* **76**, 665–676.

Hirabayashi, J. and Kasai, K. (2002). Separation technologies for glycomics. *J. Chromatogr. B* **771**, 67–87.

Hirschfeld, R. M., Keller, M. B., Panico, S., Arons, B. S., Barlow, D., Davidoff, F., Endicott, J., *et al.* (1997). The National Depressive and Manic-Depressive Association consensus statement on the undertreatment of depression. *JAMA* **277**, 333–340.

Holsboer, F. (1999). The rationale for corticotropin-releasing hormone receptor (CRH-R) antagonists to treat depression and anxiety. *J. Psychiatr. Res.* **33**, 181–214.

Horrobin, D. F. (2001). Realism in drug discovery – could Cassandra be right? *Nat. Biotechnol.* **19**, 1099–1100.

Hosono, R. and Kamiya, Y. (1991). Additional genes which result in an elevation of acetylcholine levels by mutations in *Caenorhabditis elegans*. *Neurosci. Lett.* **128**, 243–244.

Hosono, R., Hekimi, S., Kamiya, Y., Sassa, T., Murakami, S., Nishiwaki, K., Miwa, J., *et al.* (1992). The unc-18 gene encodes a novel protein affecting the kinetics of acetylcholine metabolism in the nematode *Caenorhabditis elegans*. *J. Neurochem.* **58**, 1517–1525.

Jakubowski, J. and Kornfeld, K. (1999). A local, high-density, single-nucleotide polymorphism map used to clone *Caenorhabditis elegans cdf-1*. *Genetics* **153**, 743–752.

Jansen, G., Hazendonk, E., Thijssen, K. L. and Plasterk, R. H. (1997). Reverse genetics by chemical mutagenesis in *Caenorhabditis elegans*. *Nat. Genet.* **17**, 119–121.

Jazwinska, F. C. (2001). Exploiting human genetic variation in drug discovery and development. *Drug Discov. Today* **6**, 198–205.

Jiang, M., Ryu, J., Kiraly, M., Duke, K., Reinke, V. and Kim, S. K. (2001). Genome-wide analysis of developmental and sex-regulated gene expression profiles in *Caenorhabditis elegans*. *Proc. Natl. Acad. Sci. USA* **98**, 218–223.

Kamath, R. S., Martinez-Campos, M., Zipperlen, P., Fraser, A. G. and Ahringer, J. (2001). Effectiveness of specific RNA-mediated interference through ingested double-stranded RNA in *Caenorhabditis elegans*. *Genome Biol.* **2**, RESEARCH0002.

Karp, J. E., Kaufmann, S. H., Adjei, A. A., Lancet, J. E., Wright, J. J. and End, D. W. (2001). Current status of clinical trials of farnesyltransferase inhibitors. *Curr. Opin. Oncol.* **13**, 470–476.

Kaufmann, S. H. and Hengartner, M. O. (2001). Programmed cell death: alive and well in the new millennium. *Trends Cell Biol.* **11**, 526–534.

Kim, J., Poole, D. S., Waggoner, L. E., Kempf, A., Ramirez, D. S., Treschow, P. A. and Schafer, W. R. (2001a). Genes affecting the activity of nicotinic receptors involved in *Caenorhabditis elegans* egg-laying behavior. *Genetics* **157**, 1599–1610.

Kim, S. K., Lund, J., Kiraly, M., Duke, K., Jiang, M., Stuart, J. M., Eizinger, A., *et al.* (2001b). A gene expression map for *Caenorhabditis elegans*. *Science* **293**, 2087–2092.

Kimura, K. D., Tissenbaum, H. A., Liu, Y. and Ruvkun, G. (1997). *daf-2*, an insulin receptor-like gene that regulates longevity and diapause in *Caenorhabditis elegans* [see comments]. *Science* **277**, 942–946.

Kuwabara, P. E. and O'Neil, N. (2001). The use of functional genomics in *C. elegans* for studying human development and disease. *J. Inherit. Metab. Dis.* **24**, 127–138.

Kwok, P. Y. (2001). Methods for genotyping single nucleotide polymorphisms. *Annu. Rev. Genomics Hum. Genet.* **2**, 235–258.

Lahana, R. (1999). How many leads from HTS? *Drug Discov. Today* **4**, 447–448.

Lander, E. S., Linton, L. M., Birren, B., Nusbaum, C., Zody, M. C., Baldwin, J., Devon, K., *et al.* (2001). Initial sequencing and analysis of the human genome. *Nature* **409**, 860–921.

Larsen, P. L. and Clarke, C. F. (2002). Extension of life-span in *Caenorhabditis elegans* by a diet lacking coenzyme Q. *Science* **295**, 120–123.

Lepine, J. P., Gastpar, M., Mendlewicz, J. and Tylee, A. (1997). Depression in the community: the first pan-European study DEPRES (Depression Research in European Society). *Int. Clin. Psychopharmacol.* **12**, 19–29.

Levitan, D., Doyle, T. G., Brousseau, D., Lee, M. K., Thinakaran, G., Slunt, H. H., Sisodia, S. S., *et al.* (1996). Assessment of normal and mutant human presenilin function in *Caenorhabditis elegans*. *Proc. Natl. Acad. Sci. USA* **93**, 14940–14944.

Lewis, J. A., Wu, C. H., Berg, H. and Levine, J. H. (1980a). The genetics of levamisole resistance in the nematode *Caenorhabditis elegans*. *Genetics* **95**, 905–928.

Lewis, J. A., Wu, C. H., Levine, J. H. and Berg, H. (1980b). Levamisole-resistant mutants of the nematode *Caenorhabditis elegans* appear to lack pharmacological acetylcholine receptors. *Neuroscience* **5**, 967–989.

Lindblad-Toh, K., Winchester, E., Daly, M. J., Wang, D. G., Hirschhorn, J. N., Laviolette, J. P., Ardlie, K., *et al.* (2000). Large-scale discovery and genotyping of single-nucleotide polymorphisms in the mouse. *Nat. Genet.* **24**, 381–386.

Link, C. D. (1995). Expression of human beta-amyloid peptide in transgenic *Caenorhabditis elegans*. *Proc. Natl. Acad. Sci. USA* **92**, 9368–9372.

Link, C. D. (2001). Transgenic invertebrate models of age-associated neurodegenerative diseases. *Mech. Ageing Dev.* **122**, 1639–1649.

Lipinski, C. A., Lombardo, F., Dominy, B. W. and Feeney, P. J. (2001). Experimental and computational approaches to estimate solubility and permeability in drug discovery and development settings. *Adv. Drug Deliv. Rev.* **46**, 3–26.

Livingstone, D. J. (2000). The characterization of chemical structures using molecular properties. A survey. *J. Chem. Inf. Comput. Sci.* **40**, 195–209.

Maehle, A. H., Pruell, C. R. and Halliwell, R. F. (2002). The emergence of the drug receptor theory. *Nature Rev. Drug Discov.* **1**, 637–641.

Middlemiss, D. N., Price, G. W. and Watson, J. M. (2002). Serotonergic targets in depression. *Curr. Opin. Pharmacol.* **2**, 18–22.

Milburn, J. (2001). Beyond the genome: turning data into knowledge. *Drug Discov. Today* **6**, 881–883.

Miura, M., Zhu, H., Rotello, R., Hartwieg, E. A. and Yuan, J. (1993). Induction of apoptosis in fibroblasts by IL-1 beta-converting enzyme, a mammalian homolog of the *C. elegans* cell death gene *ced-3*. *Cell* **75**, 653–660.

Montgomery, M. K. and Fire, A. (1998). Double-stranded RNA as a mediator in sequence-specific genetic silencing and co-suppression [see comments]. *Trends Genet.* **14**, 255–258.

Mounsey, A., Bauer, P. and Hope, I. A. (2002). Evidence suggesting that a fifth of annotated *Caenorhabditis elegans* genes may be pseudogenes. *Genome Res.* **12**, 770–775.

Murphy, M. A., Schnall, R. G., Venter, D. J., Barnett, L., Bertoncello, I., Thien, C. B., Langdon, W. Y., *et al.* (1998). Tissue hyperplasia and enhanced T-cell signalling via ZAP-70 in c-Cbl-deficient mice. *Mol. Cell Biol.* **18**, 4872–4882.

Nass, R., Hall, D. H., Miller, D. M., III and Blakely, R. D. (2002). Neurotoxin-induced degeneration of dopamine neurons in *Caenorhabditis elegans*. *Proc. Natl. Acad. Sci. USA* **99**, 3264–3269.

Ogg, S. and Ruvkun, G. (1998). The *C. elegans* PTEN homolog, DAF-18, acts in the insulin receptor-like metabolic signaling pathway. *Mol. Cell* **2**, 887–893.

Owen, D. and Silverthorne, A. (2002). Channelling drug discovery. *Drug Discov. World* **3**, 48–61.

Patterson, D. E., Cramer, R. D., Ferguson, A. M., Clark, R. D. and Weinberger, L. E. (1996). Neighborhood behavior: a useful concept for validation of 'molecular diversity' descriptors. *J. Med. Chem.* **39**, 3049–3059.

Perkins, L. A., Hedgecock, E. M., Thomson, J. N. and Culotti, J. G. (1986). Mutant sensory cilia in the nematode *Caenorhabditis elegans*. *Dev. Biol.* **117**, 456–487.

Petrie, R. X., Reid, I. C. and Stewart, C. A. (2000). The *N*-methyl-D-aspartate receptor, synaptic plasticity and depressive disorder. A critical review. *Pharmacol. Ther.* **87**, 11–25.

Raizen, D. M. and Avery, L. (1994). Electrical activity and behavior in the pharynx of *Caenorhabditis elegans*. *Neuron* **12**, 483–495.

Ranganathan, R., Cannon, S. C. and Horvitz, H. R. (2000). MOD-1 is a serotonin-gated chloride channel that modulates locomotory behaviour in *C. elegans*. *Nature* **408**, 470–475.

Ranganathan, R., Sawin, E. R., Trent, C. and Horvitz, H. R. (2001). Mutations in the *Caenorhabditis elegans* serotonin reuptake transporter MOD-5 reveal serotonin-dependent and -independent activities of fluoxetine. *J. Neurosci.* **21**, 5871–5884.

Reboul, J., Vaglio, P., Tzellas, N., Thierry-Mieg, N., Moore, T., Jackson, C., Kohara,Y., *et al.* (2001). Open-reading-frame sequence tags (OSTs) support the existence of at least 17 300 genes in *C. elegans*. *Nat. Genet.* **27**, 332–336.

Riddle, D., Blumenthal, T., Meyer, B. and Priess, J. (1997). *C. elegans II*. New York: Cold Spring Harbor Laboratory Press.

Rose, S. (2002). Statistical design and application to combinatorial chemistry. *Drug Discov. Today* **7**, 133–138.

Saria, A. (1999). The tachykinin NK1 receptor in the brain: pharmacology and putative functions. *Eur. J. Pharmacol.* **375**, 51–60.

Sawin, E. R., Ranganathan, R. and Horvitz, H. R. (2000). *C. elegans* locomotory rate is modulated by the environment through a dopaminergic pathway and by experience through a serotonergic pathway. *Neuron* **26**, 619–631.

Seethala, R. (2001). Screening platforms. In *Handbook of Drug Screening*, R. Seethala and P. B. Fernandes (eds), pp. 31–67. New York: Marcel Dekker.

Sertürner, F. W. (1817). *Gilbert's Ann. Phys.* **25**, 56.

Simmer, F., Tijsterman, M., Parrish, S., Koushika, S., Nonet, M., Fire, A., Ahringer, J., *et al.* (2002). Loss of the putative RNA-directed RNA polymerase RRF-3 makes *C. elegans* hypersensitive to RNAi. *Curr. Biol.* **12**, 1317.

Sonnhammer, E. L. and Durbin, R. (1997). Analysis of protein domain families in *Caenorhabditis elegans*. *Genomics* **46**, 200–216.

Stein, L., Sternberg, P., Durbin, R., Thierry-Mieg, J. and Spieth, J. (2001). WormBase: network access to the genome and biology of *Caenorhabditis elegans*. *Nucleic Acids Res.* **29**, 82–86.

Sternberg, P. W. and Han, M. (1998). Genetics of RAS signaling in *C. elegans*. *Trends Genet.* **14**, 466–472.

Sulston, J. (1988). Cell lineage. In *The Nematode Caenorhabditis elegans*, W.B. Wood (ed.), pp. 123–156. New York: Cold Spring Harbor Laboratory Press.

Swan, K. A., Curtis, D. E., McKusick, K. B., Voinov, A. V., Mapa, F. A. and Cancilla, M. R. (2002). High-throughput gene mapping in *Caenorhabditis elegans*. *Genome Res.* **12**, 1100–1105.

Sze, J. Y., Victor, M., Loer, C., Shi, Y. and Ruvkun, G. (2000). Food and metabolic signalling defects in a *Caenorhabditis elegans* serotonin-synthesis mutant. *Nature* **403**, 560–564.

Szymkowski, D. E. (2001). Too many targets, not enough target validation. *Drug Discov. Today* **6**, 397.

Tabara, H., Grishok, A. and Mello, C. C. (1998). RNAi in *C. elegans*: soaking in the genome sequence. *Science* **282**, 430–431.

The *C. elegans* Sequencing Consortium (1998). Genome sequence of the nematode *C. elegans*: a platform for investigating biology. *Science* **282**, 2012–2018.

Thase, M. E. (1992). Long-term treatments of recurrent depressive disorders. *J. Clin. Psychiatry* **53**, 32–44.

Timmons, L. and Fire, A. (1998). Specific interference by ingested dsRNA. *Nature* **395**, 854.

Tissenbaum, H. A. and Guarente, L. (2002). Model organisms as a guide to mammalian aging. *Dev. Cell* **2**, 9–19.

van Rossum, A. J., Brophy, P. M., Tait, A., Barrett, J. and Jefferies, J. R. (2001). Proteomic identification of glutathione S-transferases from the model nematode *Caenorhabditis elegans*. *Proteomics* **1**, 1463–1468.

Vaux, D. L., Weissman, I. L. and Kim, S. K. (1992). Prevention of programmed cell death in *Caenorhabditis elegans* by human bcl-2. *Science* **258**, 1955–1957.

Venter, J. C., Adams, M. D., Myers, E. W., Li, P. W., Mural, R. J., Sutton, G. G., Smith, H. O., *et al.* (2001). The sequence of the human genome. *Science* **291**, 1304–1351.

Walhout, A. J., Sordella, R., Lu, X., Hartley, J. L., Temple, G. F., Brasch, M. A., Thierry-Mieg, N., *et al.* (2000a). Protein interaction mapping in *C. elegans* using proteins involved in vulval development. *Science* **287**, 116–122.

Walhout, A. J., Temple, G. F., Brasch, M. A., Hartley, J. L., Lorson, M. A., van den, H. S. and Vidal, M. (2000b). GATEWAY recombinational cloning: application to the cloning of large numbers of open reading frames or ORFeomes. *Methods Enzymol.* **328**, 575–592.

Wang, D. G., Fan, J. B., Siao, C. J., Berno, A., Young, P., Sapolsky, R., Ghandour, G., *et al.* (1998). Large-scale identification, mapping and genotyping of single-nucleotide polymorphisms in the human genome. *Science* **280**, 1077–1082.

Ward, S. (1973). Chemotaxis by the nematode *Caenorhabditis elegans*: identification of attractants and analysis of the response by use of mutants. *Proc. Natl. Acad. Sci. USA* **70**, 817–821.

White, J. (1986). *The Structure of the Nervous System of the Nematode Caenorhabditis elegans*. Cambridge: Cambridge University Press.

White, J. (1988). The anatomy. In *The Nematode Caenorhabditis elegans*, W. B. Wood, (ed.), pp. 81–122. New York: Cold Spring Harbor Laboratory Press.

Wicks, S. R., Yeh, R. T., Gish, W. R., Waterston, R. H. and Plasterk, R. H. (2001). Rapid gene mapping in *Caenorhabditis elegans* using a high density polymorphism map. *Nat. Genet.* **28**, 160–164.

Wittenburg, N., Eimer, S., Lakowski, B., Rohrig, S., Rudolph, C. and Baumeister, R. (2000). Presenilin is required for proper morphology and function of neurons in *C. elegans*. *Nature* **406**, 306–309.

Yoon, C. H., Lee, J., Jongeward, G. D. and Sternberg, P. W. (1995). Similarity of *sli-1*, a regulator of vulval development in *C. elegans*, to the mammalian proto-oncogene c-cbl. *Science* **269**, 1102–1105.

Yu, M., Lin, J., Khadeer, M., Yeh, Y., Inesi, G. and Hussain, A. (1999). Effects of various amino acid 256 mutations on sarcoplasmic/endoplasmic reticulum Ca^{2+} ATPase function and their role in the cellular adaptive response to thapsigargin. *Arch. Biochem. Biophys.* **362**, 225–232.

Zhang, J. H., Chung, T. D. and Oldenburg, K. R. (1999). A simple statistical parameter for use in evaluation and validation of high throughput screening assays. *J. Biomol. Screen.* **4**, 67–73.

Zheng, X. F. and Chan, T. F. (2002). Chemical genomics: a systematic approach in biological research and drug discovery. *Curr. Issues Mol. Biol.* **4**, 33–43.

Zobel, A. W., Nickel, T., Kunzel, H. E., Ackl, N., Sonntag, A., Ising, M. and Holsboer, F. (2000). Effects of the high-affinity corticotropin-releasing hormone receptor 1 antagonist R121919 in major depression: the first 20 patients treated. *J. Psychiatr. Res.* **34**, 171–181.

Zwaal, R. R., Van Baelen, K., Groenen, J. T., van Geel, A., Rottiers, V., Kaletta, T., Dode, L., *et al.* (2001). The sarco-endoplasmic reticulum Ca^{2+} ATPase is required for development and muscle function in *Caenorhabditis elegans*. *J. Biol. Chem.* **276**, 43557–43563.

4

Drosophila as a Tool for Drug Discovery

Hao Li and Dan Garza

Comparative genomics of humans and *Drosophila* demonstrates a high degree of conservation both at the level of molecular building blocks (genes) and at the level of disease-relevant pathways (gene networks). This conservation provides the basis for at least four major areas of *Drosophila* application in drug discovery: discovery of drug targets, mechanism-of-action studies, compound screening and genotoxicity tests. This chapter discusses how fly models of human diseases can be established and utilized in the four areas of drug discovery. We also include a brief discussion of the available experimental tools and high throughput tools that need to be developed further.

4.1 *Drosophila* as a model organism for biomedical science

Introduction

One of the major paradigms in today's basic biomedical research is to use experimentally tractable model organisms to study human gene function. For nearly a century, the fruit fly *Drosophila melanogaster* has been utilized as a genetic system to study a variety of basic biological processes. Several features make *Drosophila* attractive as a model organism for genetic and biomedical research. First, *Drosophila* is easily cultured in the laboratory. Flies are small (ca. 1 mm), have a life cycle of less than 2 weeks and grow on simple cornmeal/

Model Organisms in Drug Discovery. Edited by Pamela M. Carroll and Kevin Fitzgerald.
© 2003 John Wiley & Sons, Ltd. ISBN 0 470 84893 6

yeast/molasses media. Flies are also prolific: a single male and single female (a pair mating) can produce more than 100 progeny. Second, *Drosophila* has a relatively simple karyotype, with only four pairs of chromosomes. Third, the extensive use of *Drosophila* as a model organism has produced an invaluable knowledge base concerning *Drosophila* development and anatomy, as well as an extensive set of genetic tools.

The experimental advantages of *Drosophila* as a model system would have little benefit for drug discovery were it not for the demonstration of conservation between *Drosophila* and humans. This conservation includes gene sequence conservation and, more importantly functional conservation of regulatory and biochemical pathways, so that knowledge gained using *Drosophila* can be applied to humans. The conservation of disease genes and disease-related pathways provides a key impetus for adopting *Drosophila* as a tool for drug discovery.

Four major areas of *Drosophila* application in drug discovery are discussed in this chapter. One is drug target discovery. The aim here is to utilize the advantages of *Drosophila* as a genetic model system to understand better the molecular mechanisms of human diseases and identify new and potentially novel disease-related genes. The second major area is for the determination of the mechanism of action (MOA) of selected compounds. The goal is the identification of unknown cognate target molecules and target molecular pathways for compounds already known to have desirable pharmacological effects on disease phenotypes. A third major area, still largely undeveloped, is compound screening, in which whole organism-based, disease-associated phenotypes and/or *Drosophila* cell-based assays are used as a primary or a secondary screen against a compound library. The fourth is using flies to test and study the genetic toxicity of pharmaceutical compounds. This chapter describes: why highly relevant fly models of human diseases can be established; how to use fly models in the four areas of drug discovery; and the experimental tools that are currently available or need to be developed.

Comparative genomics of humans and *Drosophila*

Based on extensive studies over the past century, it has been firmly established that many biological processes, including those that are directly relevant to human diseases, are highly conserved between humans and *Drosophila*. The conservation is at the gene level (similar molecular building blocks) and at the genetic circuitry level (similar architecture of gene networks or pathways). The homology at both levels is critically important in using *Drosophila* as a model system in drug discovery, because it enables relatively reliable extrapolation of information from one system to the other.

Conservation of gene sequences and molecular functions

The *Drosophila* genome is predicted to contain about 14 000 genes, as compared to about 35 000 genes in the human genome. The surprisingly small difference in the number of genes belies an even smaller difference in the number of protein families and protein domains, because only 7% of the 1300 InterPro families present in human genome are absent in fly genome (Lander *et al.*, 2001). Very often, several human genes are represented by a single *Drosophila* gene, therefore the function of single copy genes in flies can be dissected genetically without the masking effects of redundant copies as in mammals.

Comparative genomic analysis of flies and mammals showed that as many as 50% of fly genes have mammalian homologs (BLASTP $E < 10^{-10}$) (Rubin *et al.*, 2000). In a systematic analysis of about 1200 human disease genes encoding proteins in the Online Mendelian Inheritance in Man (OMIM) database, 670 are found to have homologs (BLASTP $E < 10^{-10}$) in *Drosophila* (Reiter *et al.*, 2001; Chien *et al.*, 2002). In an initial comparative analysis of fly and human genomes, about 2800 human–fly orthologs are found based on an unambiguous one-to-one relationship (Venter *et al.*, 2001). With improved bioinformatics tools, the number of ortholog pairs is likely to increase. The extensive similarity in gene sequences, protein families and protein domains between human and *Drosophila* genomes demonstrates that humans and *Drosophila* utilize a conserved repertoire of molecular building blocks.

The observed sequence conservation reflects the underlying conservation of molecular function. Well before the *Drosophila* and human genome sequences became available, *Drosophila* researchers had established that there was considerable conservation of gene function experimentally. *Drosophila* led the way for the cloning and functional analysis of genes that play critical roles in a number of developmental and cellular processes. Many of the identified genes were found to be conserved evolutionarily, and functional conservation was demonstrated directly through the rescue of *Drosophila* mutations by the corresponding mammalian homolog (or parts thereof). One of the first examples of this approach was analysis of the homeotic gene function in *Drosophila*. The homeotic genes were identified originally on the basis of mutations that lead to changes in cellular identity, such that one organ or tissue type is replaced by another. Cloning and subsequent characterization of these genes led to the identification of a protein domain called the homeobox that was highly conserved across all higher organisms (Scott *et al.*, 1989; Affolter *et al.*, 1990). Functional conservation was demonstrated for the *Deformed* (*Dfd*) gene by rescue of *Dfd* mutant phenotypes using the corresponding mammalian gene (Malicki *et al.*, 1990; McGinnis *et al.*, 1990). Similar cross-phylum rescue experiments since have been successfully carried out with genes involved in various developmental processes, including embryonic brain development blood cell development and eye development (Oliver and Gruss, 1997; Fossett and Schulz, 2001; Reichert, 2002).

Conservation of signaling pathways

Over the past 20 years, *Drosophila* researchers have demonstrated that the observed molecular functional conservation at the gene level often reflects the functional conservation of associated developmental, genetic, biochemical and signaling pathways between *Drosophila* and higher organisms. It is this pathway conservation that provides the impetus for the utilization of *Drosophila* in drug discovery.

There are numerous examples of pathway conservation and only three are described here briefly to illustrate the point. First, the insulin signaling pathway is highly conserved; almost all members of the pathway in humans have their counterparts in flies, such as insulin, insulin receptor, IRS, PI3K, PTEN, PDK1, PKB, TSC1, TSC2, TOR, S6K, all the way to the downstream forkhead transcription factors (Stocker and Hafen, 2000; Lasko, 2002). Second, the signal transduction pathways involving receptor tyrosine kinases, RAS, mitogen-activated protein kinases (MAPKs) and transcription factors such as Ets-type proteins are conserved (Matthews and Kopczynski, 2001; Rebay, 2002). Third, the WNT signaling pathways are conserved, homologs are found for Wnt, Fz, GSK/ZW3, Amadillo/beta-catenin, APC and Tcf genes and they occupy the same positions in the hierarchy (Nusse, 1999; Moon *et al.*, 2002).

Limitations

There are limitations to the use of *Drosophila* as a model system that must be borne in mind in order to make best use of the system. Sequence conservation does not always mean functional conservation of genes or pathways. For example, the sex determination pathways are largely not conserved between flies and humans (Marin and Baker, 1998). Yet the *sex-lethal* gene, which encodes a splicing factor and is at the top of the fly sex determination pathway, has several good homologs in humans such as HUD1, -3 and -4, with almost 50% identify over half of the protein sequences. The HUD1, -3 and -4 proteins do not have known roles in the sex determination pathway of humans, thus conservation at both the gene level and the pathway level are important if a *Drosophila* pathway is going to be used to model a human disease pathway for drug discovery. Additionally, approximately 50% of fly genes that do not have human homologs can give significant experimental 'noise', such as finding these genes in a genetic screen, which has little value for drug discovery.

Conservation is more limited at the cellular, tissue, organ and system levels. Many mammalian cell types are not found in flies, e.g. chondrocytes and erythocytes. It is also not clear whether fly fat body cells are more closely related to hepatocytes or adipocytes in mammals. In some cases differences in signaling outputs from pathways must be recognized and taken into account. For example, although the SREBP pathway is conserved between flies and humans, the pathway controls cholesterol homeostasis in humans and controls saturated fatty acid and phospholipid biosynthesis (such as palmitate to maintain membrane integrity) in the fly (Dobrosotskaya *et al.*, 2002; Seegmiller *et al.*, 2002). Thus, whether a disease pathway can be modeled in flies and to what extent it can be modeled must be decided carefully on a case-by-case basis.

Using *Drosophila* for drug target identification and validation

Drug targets can be broadly defined as molecules in a human body whose functions can be modulated by pharmacological agents to treat diseases. In practice, the majority of the 500 drugs in market are targeting proteins, which are encoded by their cognate genes. The most prevalent use of *Drosophila* in drug discovery is for drug target identification. This is a logical extension of the long history of academic research using *Drosophila* as a genetic system to identify genes controlling biological processes. Target identification implies that the newly discovered genes were not known previously to have roles in a particular disease pathway. Target validation implies that there is some experimental evidence for an association between a gene and a disease pathway, but, additional evidence is needed to substantiate the linkage. *Drosophila* provides a model system for the identification and validation of candidate genes for drug discovery through the use of relatively low-cost, high-efficiency forward and reverse genetic screens. Ultimately, these candidate genes must be used to identify the corresponding mammalian genes and follow-up assays must be performed in mammalian cellular assays or transgenic models. Thus, one could argue that *Drosophila* is used in this regard as an efficient genetic system for indirect functional annotation and prioritization of human genes as potential drug targets.

Forward genetics

Forward genetics (from mutant phenotypes to genes) involves identification of mutations that cause or modify specific phenotypes, followed by identification of the genes in which the mutations have occurred. This has been the major approach used in flies to dissect disease-related pathways and identify candidate genes for further drug discovery efforts. The scalability of laboratory culture makes flies amenable to large-scale forward genetic screens, and *Drosophila* is one of the few higher eukaryotic model organisms in which a forward genetic screen to identify phenotypic modifiers can be carried out to statistical saturation. Some of the best examples of forward genetic screens are their use for the discovery of most pattern formation genes (Nusslein-Volhard and Wieschaus, 1980; Lewis, 1985; Roush, 1995).

Forward genetic screens usually follow a defined series of steps:

1. A fly disease model is produced, typically using mutations in genes previously identified as core components of a disease pathway. A disease model is a collection of well-characterized mutant phenotypes (morphological, biochemical or physiological).

2. A modifier screen is carried out to identify mutations in other genes that modify (enhance or suppress) one or more of the defined mutant phenotypes. The mutations to be screened may be generated using either chemical mutagenesis or transposon mutagenesis (see Section 4.2; mutagenesis). Follow-up of a primary screen may involve re-testing the modifiers in the original screen to eliminate those producing only small or variable effects, and various secondary screens to eliminate genetic background and other non-specific effects.

3. The modifier mutations are mapped and the affected genes are identified. For mutations generated by chemical mutagenesis, this involves meiotic mapping using visible morphological markers (low resolution) and/or single-nucleotide polymorphism (SNP) markers (higher resolution), and confirmation of the mutation through transgene rescue and/or sequencing of the mutant allele. For mutations generated by transposon mutagenesis this may involve amplification and sequencing of DNA flanking the transposon insertion site or may simply require bioinformatic analysis if one of the predefined transposon insertion collections is utilized. The resulting modifier genes are then filtered to eliminate those for which mammalian homologs do not exist.

4. The relative positions of the modifiers within the disease pathway are established and their biochemical/molecular functions are studied. This step is the most difficult because determining the function of modifier genes within a pathway often requires a significant research effort and a variety of different experimental approaches. However, functional analysis provides the best validation for the modifiers and therefore provides the best filter to eliminate those genes whose participation in a pathway cannot be firmly established.

5. Although identification and validation steps can be carried out using *Drosophila*, the mammalian homologs of the identified fly genes also must

be tested and validated in a mammalian system. This raises the issue of how far to proceed with the functional analysis of modifiers before testing them in a mammalian assay system, and the answer must be considered on a case-by-case basis. When high- or medium-throughput mammalian cellular assays are available, filtering of identified modifiers through functional analysis in *Drosophila* is not critical. When mammalian assay systems are unavailable or of low throughput (such as the production of transgenic mouse models), then the modifiers must be prioritized carefully.

Forward genetic screens, especially those based on chemical mutagenesis, have been favored historically by academic researchers largely because of the relatively random distribution of mutations obtained. Because this approach does not require preconceived assumptions about the molecular structure and functions of the modifiers or the pathways that may affect the disease phenotypes, it is more likely than other approaches to identify novel drug targets. However, this approach does have several limitations. First, it is less effective for identifying genes that are functionally redundant. Second, a significant proportion of the identified modifier genes are not expected to have practical value in drug discovery because about 50% of *Drosophila* genes do not have mammalian homologs. Third, current mutation mapping methods require up-front investment of substantial human effort for the identification of the modifier genes, particularly if chemical mutagenesis is used.

Reverse genetics

Reverse genetics (from genes to phenotypes) involves the manipulation/mutation of specific genes to create phenotypes. This approach has gained increased popularity as a result of the availability of the complete genome sequence of *Drosophila*.

The *Drosophila* homolog of a disease-causing human gene now can be readily identified and cloned. The genomic map location, expression patterns of transcripts and proteins of the fly gene are obtained using standard molecular biology methods. Loss-of-function (LOF) mutations could be generated by several means. Classically, one makes predictions about the mutant phenotypes, such as lethality or changes in tissue-specific morphology. Random mutations are generated by chemical or transposon mutagenesis and are screened against chromosomal deficiencies encompassing the target gene. Molecular analysis is then carried out to identify the mutations that disrupt the structure and function of the gene of interest. Because the classical approach is laborious and dependent on assumptions regarding the mutant phenotype, alternative methods have been developed for the direct disruption of genes. These include gene knock-out by homologous recombination and synthetic sequence-specific zinc finger nucleases, and gene knock-down by a transgene expressing dsRNAi (see Section 4.2; mutagenesis). Loss-of-function mutations can be generated also by expression of a dominant-negative mutant form of the protein. In addition to LOF mutations, gain-of-function (GOF) mutations are also informative about a gene's function. Gain-of-function mutations can be made by overexpression of the wild-type gene or by expression of a constitutively active form of the protein. Once the LOF and/or GOF mutant flies are obtained, their phenotypes are investigated to gain insights into the cellular and molecular pathways underlying the human disease. The analysis forms the foundation for the fly model of the human disease. The disease model is usually a starting point for forward genetic screens. The disease model also may prove useful for investigating the functional importance of other candidate human genes that are related to the human disease by only limited evidence, such as differential expression. A pioneering example of reverse genetic approaches in *Drosophila* is its use in the dissection of signaling events triggered by receptor tyrosine kinases (Shilo, 1992).

Reverse genetics can be applied also to study some dominant human diseases caused by mutations that produce toxic proteins that have no fly homologs. For example, Huntington's disease, a dominantly inherited neurodegenerative disease, is caused by expansion of polyglutamine-encoding CAG trinucleotide repeats in the *huntingtin* gene. The expanded polyglutamine sequence in the *huntingtin* gene causes a dominant GOF that is neurotoxic and not present in the wild-type protein. Expression of the polyglutamine-containing peptide in *Drosophila* has been shown to cause progressive neurodegeneration and nuclear inclusions, both characteristic pathological features of the human disease (Kazemi-Esfarjani and Benzer, 2000; Marsh *et al.*, 2000). Based on the fly models, forward genetic screens have been done and modifiers have been discovered, such as the fly orthologs of the human heat shock chaperon protein Hsp40 and myeloid leukemia factor 1 (MLF1) (Kazemi-Esfarjani and Benzer, 2002). These modifiers previously were not known to be linked to neurodegenerative diseases, demonstrating the ability of such screens to identify novel genes and thereby increase our understanding of disease pathways.

Genome-scale coverage

At the genomic scale, one would like to have LOF and GOF mutations in each of the 14 000 fly genes. This would provide the ultimate resource for identification of modifiers, mapping mutations and a variety of other applications (see below). In fact, the *Drosophila* research community has been working towards this goal for nearly a century. Thousands of mutant stocks are currently available from public fly stock centers and many more can be obtained through various academic laboratories. Several thousand transposon-insertion mutations have been made and their genomic insertion sites determined by the *Drosophila* Gene Disruption Project. Currently, the publicly available mutant collection covers as much as 45% of the predicted genes in the *Drosophila* genome. Further increases in coverage may come next from large-scale analysis of specific gene families, such as kinases and phosphatases, using reverse genetic methods. For drug discovery efforts it would be sufficient to have mutations only in fly genes with human homologs, thus eliminating background 'noise' in genetic screens and focusing on the most useful genes.

Genome-scale genetic approaches

Pathway kit analysis

In addition to the initial discovery of modifiers of a particular disease pathway, it is important to understand the specificity of the modifier genes. The information will indicate potential uses of the modifier genes in treating other diseases and also indicate their potential side-effects. One means for addressing this is the use of a 'pathway kit' to obtain the activity spectrum of a putative target gene in most disease-relevant pathways. The 'pathway kit' approach may involve several steps. First, it is necessary to establish a collection of important *Drosophila* strains that can be used to test the involvement of a modifier mutant/gene in the pathways. They can be LOF and/or GOF mutants of previously characterized core components in each of the pathways. Second, a limited number of strains that are most diagnostic for each pathway will be used collectively as the 'first-pass filter kit'. Third, only if tests are positive through the strains of the first-pass filter are more test strains for a particular pathway used as a secondary test to obtain detailed knowledge. The value of this approach will increase exponentially as the *Drosophila* knowledge base expands over the next few years to include more genetic and protein–protein interaction data, whole-genome RNA interference analysis and large datasets from transcriptome, proteome and metabolome analysis.

Whole-genome double-stranded RNA interference method

The development of RNA interference (RNAi) technology has opened up a number of new possibilities for functional analysis of the *Drosophila* genome. Because *Drosophila* tissue culture cells tolerate long double-stranded (ds)RNA and take up these molecules without a requirement for transfection, it is possible to knock down gene expression levels in cultured *Drosophila* cells at relatively low cost and high efficiency (Clemens *et al.*, 2000; Worby *et al.*,

2001). In mammalian cells, equivalent short interfering (si)RNA experiments require the synthesis of short siRNA molecules for each gene or the production of specific plasmids capable of expressing siRNA molecules, as well as transfection of either the siRNA or DNA from each siRNA expression vector. Double-stranded RNAi in *Drosophila*-cultured cells can be used for genetic screens if a disease-relevant pathway is active in the cultured cell line and robust, specific assays can be developed. Such assays include reporter gene expression, cell morphology, and biochemical markers. An additional advantage of using *Drosophila* for whole-genome RNAi experiments is that a higher success rate is expected relative to mammalian cells due to the lower genetic redundancy of the fly genome. Furthermore, it has been shown that multiple genes can be knocked down simultaneously, facilitating the analysis of redundant genes and the understanding of genetic hierarchies and biochemical pathways (Dobrosotskaya *et al.*, 2002), as well as the cross-talk between different pathways. Perhaps the greatest advantage of carrying out such analyses in *Drosophila* is that the target genes identified by this approach can be verified rapidly in transgenic *Drosophila* using reverse genetics methods.

This approach does have limitations. First, dsRNAi-mediated knock-down of gene expression will not be effective for those proteins having a long half-life. Second, knock-down may not produce the desired assay output if the cell line utilized is not active for the pathways under analysis, although this limitation should be surmountable if appropriate proof-of-concept and/or pilot experiments are carried out on different cell lines, or by developing RNAi methods for primary cells or tissues. Lastly, again, mammalian homologs of the identified *Drosophila* genes must be tested in mammalian assay systems.

Fly genes that are valuable for drug discovery: where to make the cutoff?

One of the major challenges of using *Drosophila* for drug discovery today is not whether new genes can be identified for a disease pathway but how to prioritize these genes to feed into the drug discovery pipeline. Not all fly genes are created equal, especially from a drug discovery standpoint. Their degree of homology to human genes varies, as well as their biochemical functions. Based on the homology to human genes, *Drosophila* genes can be classified according to their practical values as follows:

1. *High-confidence orthologs.* Orthologs can be defined heuristically as bidirectional BLASTP best hits. There are about 4000 fly–human ortholog pairs (BLASTP $E < 10^{-30}$) (Gilbert, 2002). Good orthologs should have a high degree of sequence homology. However, it is difficult to quantify 'high degree'. As an example, using the BLASTP E value of

$< 10^{-75}$ as cutoff identifies about 2200 orthologs. This class of fly genes has high predictive power for the function of the corresponding human genes. However, many of these genes encode proteins for basic cellular machinery, such as basic transcription, splicing, translation and replication apparatus. Many of them are less likely to be drug targets.

2. *Low-confidence orthologs.* This class of genes still has a one-to-one relationship with their human counterparts but, because the level of sequence homology is lower, the confidence level about information transfer to human genes is lower.

3. *High-confidence homologs.* This class of genes has equivalent homology to multiple human genes, and homology is throughout most of the encoded protein sequences. It is important to note that two genes may not be considered good homologs if the homology between the two protein sequences is restricted only to a small region and the containing protein domain is prevalent in the genomes, such as the protein kinase domain or Ankyrin repeat. This one-to-many class of fly genes makes it difficult, if not impossible, to predict which human homolog is more relevant. It requires commitment of considerable resources for experimental determination in mammalian systems. On the other hand, the single corresponding fly gene may carry out all or some of the functions of one or more of the human homologs, so functional analysis of the fly gene can reveal insights into the function of the human genes while avoiding the masking effects of functional redundancy in mammals.

4. *Low-confidence homologs.* This class of fly genes maintains the one-to-many property but with lower sequence homology. The double-negatives make this gene class less attractive.

5. *Insect-specific genes.* The insect-specific genes have received less attention from the pharmaceutical industry than from agricultural industry. However, these genes can provide drug targets for insect-borne diseases, such as malaria or dengue fever, by aiming at the homologs in the insect vectors.

Drugability of target proteins: valuable filter or moving target?

The value of fly genes in drug discovery is not solely determined by their sequence homology to human genes. The biochemical/molecular functions of the encoded proteins have a strong influence on their values in the near future. Among the 500 targets of marketed drugs today, the majority belong to a limited number of protein families (Drews, 2000). For example, 45% of known targets are receptors, 28% are enzymes, 11% are hormones and factors, 5% are ion channels and 2% are nuclear receptors, with only 9% falling into other categories. Except for hormones and factors, which themselves are used as drugs, most of the known drugs for these targets are low-molecular-weight compounds. For historical, economical, biological, chemical and pharmacological reasons, small-molecule drugs continue to be the favorite of the drug industry. The feasibility of developing specific small-molecule agonists or antagonists to modulate the biochemical/molecular functions of a protein or a protein domain is known as the 'drugability' of the protein or protein domain. Besides the few known druggable protein families, there are also known protein domains with very low drugability, such as those involved in protein–protein interactions. A protein's drugability is also influenced by its subcellular location, which affects its accessibility to drugs. Thus, when prioritizing fly genes it is critical to consider their drugability property. Unfortunately, there is not enough information at present to establish reliable drugability scores for all of the known protein domains. It is also important to note that new approaches to drug design and drug screening are constantly being developed and many gene products currently considered to be less than ideal may become druggable targets in the near future.

Using *Drosophila* for the study of the mechanism of action of known drugs

Drugs in the market or in the late stages of clinical trials have demonstrated therapeutic activity. However, for some of these drugs the target molecules or target pathways are not clearly defined. The same is true for some natural product drugs that have a strong *in vivo* effect. Owing to difficult synthesis and some undesirable physicochemical properties they are not suited for further preclinical development. Because the target molecules of these drugs are validated by virtue of the effectiveness of these drugs, it is of tremendous value to identify them. Some good examples of this type of drug include the antidepression drugs tianeptine and bupropion (Vaugeois *et al.*, 1999; Meyer *et al.*, 2002), the antiepileptic drugs topiramate and zonisamide (Smith *et al.*, 2000; Leppik, 2002), antihyperlipidemia drugs (Zhu *et al.*, 2002) and the marine sponge-derived antitumor drugs bengamides and phorboxazoles (Thale *et al.*, 2001; Uckun, 2001). With the target molecules in hand, the time frame can be dramatically shortened for making the second-generation drugs with higher specificity, higher potency, reduced side-effects and lower cost. With knowledge of the target pathway, drugs can be developed for different target molecules in the same disease pathway, perhaps with better therapeutic value, either in general or in certain patient populations. In addition, marketed drugs can be used in new indications because of shared pathways, saving time and resources in development.

It has been recognized that fly genetics can be used for the identification of target molecules and target pathways of selected compounds, the so-called 'mechanism-of-action (MOA)' studies (Matthews and Kopczynski, 2001). The rationales are as follows. First of all, because of the general conservation between fly and human genomes, a compound's human target is likely to have a fly homolog. Second, a compound-induced specific phenotype is due to changed activity of its target molecule, thus it should be similar to the mutant phenotype of the target gene. Third, genetic screens can be performed to find mutants that suppress or enhance compound-induced phenotypes; some of them should have mutations in the target molecule that affect the compound–target interaction, or in other genes in the same pathway. Fourth, by finding the mutant genes and subsequent analysis, the target genes and pathways can be discovered.

The most important step in a MOA study of a compound is to find compound-induced phenotypes that are specific, reliable and easily detectable. Commonly used phenotypes include viability, fertility, behavior and morphology. Phenotypes based on cellular markers may give better specificity than gross phenotypes. Analysis of whole-genome expression profiles induced by the compound can provide compound-specific 'fingerprints' and transcriptional markers (Hughes *et al.*, 2000). A dose–response curve should be established. The best phenotype–dose pair for screening should be the one that gives the largest phenotypic difference at the lowest compound dosage and with minimal variation. If available, a structural derivative series of a compound with different levels of bioactivity should be used to verify the specificity of a phenotype.

One major theoretical concern of a fly MOA study is the potential high background noise. For example, the phenotypic effect of a compound is affected not only by its interaction with the target(s) but also strongly by its absorption, distribution, metabolism and excretion (ADME). Thus, mutations that suppress or enhance compound-induced phenotypes may be located in genes affecting ADME rather than in the target molecule or target pathway. Currently, there is no strong evidence to suggest that the knowledge of the fly ADME mechanism can be extrapolated to the human ADME mechanism. Thus, for the time being, mutations affecting a drug's fly ADME are best considered as noises in MOA studies. Because the molecular machinery involved in ADME is not specific to any particular drug, methods can be developed to separate mutations that affect ADME and other non-specific mutations from the mutations that specifically affect drug target(s) and target pathways. For example, several specificity tests can be established by using a few highly specific compounds with well-known targets and associated phenotypes. Mutations that can affect phenotypes induced by several of these compounds are unlikely to be in the target molecules or target pathways of the drug under study.

One excellent example of a *Drosophila* MOA study is analysis of the cocaine sensitization mechanism. Behavior sensitization, in which repeated exposure to cocaine leads to increased severity of response, has been linked to cocaine addiction and enhanced drug craving in humans. However, the biological basis of sensitization is not well understood. A cocaine-induced phenotype in flies was first established by showing that repeated cocaine exposure leads to stereotyped behavior and behavior sensitization, similar to those seen in mammalian animal models (McClung and Hirsh, 1998). Mutations in fly tyrosine decarboxylase gene (TDC) and circadian genes were found to suppress the behavior sensitization (Andretic *et al.*, 1999; McClung and Hirsh, 1999). Tyrosine decarboxylase converts tyrosine to tyramine, so the sensitization failure of TDC mutant flies could be rescued by feeding the flies with tyramine. Tyrosine decarboxylase is induced after cocaine exposure but not in circadian mutant flies, indicating that the circadian genes are regulators of TDC induction. Recently, it was shown that cocaine sensitization in the mouse is also dependent on circadian rhythm and on one of the mouse circadian genes, *period1* (Abarca *et al.*, 2002). These studies indicate that drugs modulating circadian gene products and TDC may be beneficial in treating cocaine addiction.

Using *Drosophila* for compound screening and chemical genetics

Because the case for using *Drosophila* in drug target discovery is compelling, one may argue that *Drosophila* could be used for compound screening. However, little has been done in this regard and the value of this approach remains to be investigated. In principle, flies can be used in the same way as they are used for genetic screening – starting with a defined phenotype that is relevant for a conserved disease-related pathway and screening for compounds that modify the phenotype. Before we discuss the potential value of flies in compound screening and related technical issues, it is necessary to have some basic understanding of the current methodology of compound screens.

Current approaches for compound screens

In today's drug discovery process, most of the compound screens and lead optimizations are initiated after target identification. Two basic approaches are used: *in vitro* purified target-based assays and cell-based assays (the majority being mammalian cells) (Moore and Rees, 2001; Johnston, 2002). In an *in vitro* purified target-based assay the hit/lead compound–target interaction is assayed based on direct binding affinity, effect on ligand displacement or effect on target molecular function such as enzymatic activity. In a cell-based assay, compound–target interaction is assayed indirectly based on engineered readout that selectively represents the target activity. The purified target-based assay is generally favored because it offers higher throughput, greater exposure to chemical diversity, direct and detailed knowledge of the kinetic or chemical MOA and a simple structure–activity relationship (SAR) against the purified target. Cell-based assays are often used as secondary or tertiary assays to examine the effect of compounds on the target in a more relevant cellular environment and to select compounds with better cellular penetration, activity and stability. In addition, cell-based screens can discriminate between agonist, allosteric modulator and antagonist activity that binding assays cannot, as well as provide information on the acute cytotoxicity of compounds. The two-step serial method applies to many intracellular targets such as enzymes.

In general, a cell-based assay is not favored for primary screens because the cell membrane limits the screen range of pharmacophores, and hit compounds may be found due to effects on other unknown molecules in the cells that give the same readout, which makes subsequent SAR study difficult. However, when a purified target-based assay is not feasible, a cell-based assay is used instead, such as for voltage-gated ion channels, orphan receptors, other targets expressed in the cell membrane, targets requiring assembly of a complex that is difficult to reconstitute *in vitro* and for assaying changes in the subcellular localization of a target. There is a general requirement for an assay in a high-throughput screen to have an adequate dynamic range to separate strongly active and weakly active compounds from the background noise (Zhang *et al.*, 1999). Optimization for cell-based assays can sometimes be very challenging. Because of these limitations, non-mammalian-cell-based assays sometimes provide unique opportunities.

In the absence of identified targets, cellular assays based on functional readout can still be used for compound screens. In fact, the cell-based assay is one of the oldest methods to generate lead compounds, and many drugs in the market today were identified by this approach many decades ago (Moore and Rees, 2001).

Potential benefits of compound screens in Drosophila

With some understanding of the current compound screen methodology, we can now ask what value *Drosophila* cell/organism-based screens might offer and when it is appropriate to use this approach. *Drosophila*, as well as *Drosophila* cell lines, are made up of sophisticated machinery with a highly interconnected network of dynamic molecular processes that are regulated by internal and external signals. These evolutionary conserved machinery and processes give stereotypic structural and physiological outputs. Thus, *Drosophila* could be thought of as an alternative assay platform containing most of the assay components and by which compound activity is assessed. This argument follows the same underlining logic used for developing successful robust non-mammalian assays, such as the amphibian melano-phore-based assay for human G protein-coupled receptors (Nuttall *et al.*, 1999).

When an *in vitro* purified target-based assay is feasible and a good relevant mammalian cell-based secondary assay is available, there is generally no strong rationale for a *Drosophila* approach. However, if there is difficulty in making a usable mammalian cell-based assay and there is a well-established fly-based functional readout for the target or a robust readout can be engineered quickly, then a fly-based assay may be useful.

A *Drosophila*-based assay has the advantage of having a property against 'assay drift'. In mammalian cell-based assays, due to multiple passages of cell culture, there could be substantial loss of cellular response and changes of assay statistics. This is most likely due to the combination effect of genetic instability, manifested as an accumulation of aberrations in genetic material that cannot be got rid of by mitosis, and genetic selection in cell cultures. In this regard, a *Drosophila* whole-organism-based assay is far more stable. This is because the fly culture is maintained by sexual reproduction – a process that requires the stability of chromosome number and structure and involves meiotic recombination. Most *Drosophila* mutant strains maintain their original phenotypes even after many years, and any second-site genetic modifiers that do accumulate can be removed by outcrossing to a wild-type strain for a few generations. In cases where there is selection against a phenotype produced using the multiple engineered components (transgenes) necessary for an assay, these components can be maintained separately in two different fly strains and brought together by mating in just one generation. For example, by using the binary Gal4/UAS system (see Section 4.2; analytical tools) the driver (Gal4 transgene) and the responder (UAS transgene) can be maintained separately and then crossed to overexpress a target protein in the progeny that produces a phenotype such as lethality. Compounds that inhibit the activity of the target protein and/or pathway can be identified by virtue of their ability to reverse the lethal phenotype.

The idea of using a disease pathway phenotype of *Drosophila* to look for chemical modifiers, in very much the same logic as using a genetic screen for genetic modifiers, is also worth exploring. In this case, the target in the compound screen is the disease pathway and not a specific gene product. If the assay phenotype is sufficiently validated, the chemical modifiers discovered should have significant relevance. There are several unique features in this approach. One obvious advantage of a chemical modifier screen is that it can overcome the problem of genetic redundancy in a genetic screen. Second, in an organism-based screen the delivery of compounds will not be specific to a particular tissue or cell type, nor would it be uniform for all cells. Thus, the effect of a specific chemical inhibitor is expected to mimic that of a LOF mutation but with varied effectiveness in different tissue and cell types due to local concentrations of the compound. By controlling compound dosage, one can adjust the degree of functional loss in the target – a situation analogous to the creation of an allelic series of mutations in a gene. Another unique feature is that hit compounds are selected not only by their potency against the targets but also by pharmacological properties such as absorption, membrane permeability and cellular/organism stability. Thus, the hit rate might be low but the value of hits is high.

Technical hurdles for carrying out compound screens in **Drosophila**

There are a number of technical challenges in using organism-based assays for screening large chemical diversity. First, even though we have achieved in-house success in sorting and dispensing embryos and larvae into 96-well plates using instruments available in the market (Li *et al.*, 2001), growing flies in a 96-well format has not yet been optimized. In particular, the culture medium needs to be modified to be compatible for automation, compound addition, larval growth and adult viability. Second, machine-readable assay phenotypes need to be developed. Traditional morphological phenotypes such as bristle number, roughness of eyes or behavior are difficult to adapt to a high-throughput compound screen. For example, the classical readout for circadian rhythm is the rhythmic behavior of local motor activity, which is not very easy to scale up. By using the rhythmic activity of the promoter of the *period* gene to drive the expression of the firefly luciferase, the throughput of the circadian rhythm readout is dramatically increased (Plautz *et al.*, 1997). Third, most compound libraries formated for *in vitro* screens and cell-based screens may not be useful for compound screens in *Drosophila*. This results from the fly's tolerance limit for dimethylsulfoxide (DMSO), combined with a requirement for higher compound concentrations in fly growth medium relative to cell culture media. For example, 40 µM rapamycin and 2 mM cycloheximide have been used to delay *Drosophila* larval development (Britton and Edgar, 1998; Oldham *et al.*, 2000) and 40 mM G418 is typically used for selecting flies expressing a G418 resistance gene (Xu and Rubin, 1993). If a screen is to be done at 100 µM in 1% DMSO, a 10 mM library in 100% DMSO would be needed to keep the DMSO concentration at or below 1%, above which there is toxicity. As a comparison, the compound library provided by the National Cancer Institute contains compounds at a concentration of 1 mM in 100% DMSO.

Using sensitized assays may help to reduce the demand on compound concentration. For example, the effective rapamycin concentration for delaying larval development may be halved if using flies heterozygous for a null mutation in the rapamycin target gene dTOR, or in the S6K gene, which is the downstream target of dTOR (Britton and Edgar, 1998). Also, the $2\,\mu M$ histone deacetylase inhibitor SAHA was able to suppress, by 40%, the adult lethality caused by neuronal overexpression of the polyglutamine repeats in the Huntington's disease gene (Steffan *et al.*, 2001).

Chemical genetics

The use of *Drosophila* for compound screens is not limited to the goals of hit-to-lead programs. Compounds can be used as chemical tools for understanding the biology of disease pathways. This approach has been used for quite a long time in biology, especially in the field of neurobiology. For example, the scorpion charybdotoxin has been used extensively for the characterization of potassium channels (MacKinnon *et al.*, 1998). Recently, this strategy has been formalized as 'chemical genetics' (Mitchison, 1994; Crews and Splittgerber, 1999; Stockwell, 2000). The chemical structures of hit compounds from a large diversity screen may tell us what kinds of target molecules they hit in the pathways, based on compound–target knowledge databases. Alternatively, based on the concept that 'similar folds bind to similar ligands' (Breinbauer *et al.*, 2002), screens may be done using special compound libraries targeted to conserved protein domains/fold structures, such as the metallopeptidase and tyrosine kinase domains. This approach may provide information for an educated guess about the candidate genes of a disease pathway and provide impetus to initiate a genetic screen or test to look for those druggable target genes.

Pathway-targeted compound screens also should be done using *Drosophila*-cultured cells. This is because the same assay readout of a pathway in a cell line can be used for both compound screens and dsRNAi-based genetic screens. The integration of information from both types of screens will generate a compound–target pair hypothesis that can be tested rapidly using the same cellular assay – transgenic *Drosophila* – and equivalent assays in mammalian cells.

In summary, through careful evaluation, good experimental designs, technology improvement and integration of information from genetic screens, a *Drosophila*-based compound screen has the potential of adding significant value to the drug discovery process.

Using *Drosophila* for evaluating the genetic toxicity of drugs

Before a new drug goes to clinical trials its toxicity profiles must be determined to ensure the safety of patients. One potential toxic property of a drug is its genetic toxicity (genotoxicity), which represents the interaction of the drug with DNA and other cellular molecules that maintain the integrity of genetic material. In drug development, genotoxicity studies are used to prioritize lead compounds in early development, to get the necessary information to meet global regulatory requirements and to understand mechanistically potential adverse effects of marketed products during post-marketing surveillance.

There are several advantages in using *Drosophila* for developing fast and reliable assays to detect the genotoxic activities of compounds. *Drosophila* is a complex multicellular organism with a short generation time. It has well-documented stereotypic morphological features controlled by conserved genetic pathways that can be used as a sensitive readout for the effects of genotoxins. Many genetic and modern molecular biology tools are available in *Drosophila* to engineer assays. *Drosophila* can enzymatically activate promutagens and procarcinogens *in vivo*. *Drosophila* has long been used for identifying carcinogens and germline mutagens and for studying the mechanisms of chemical mutagenesis (Vogel *et al.*, 1999). The demonstration of highly conserved genomes between humans and *Drosophila* opens a new page for broader application of *Drosophila* in genotoxicology studies.

The classical *Drosophila* assay for mutagen testing is the sex-linked recessive lethal (SLRL) assay (OECD, 1984; EPA, 1998). This assay detects both point mutations and small deletions in the fly germline. It is a forward mutation assay capable of detecting mutations at about 800 loci on the X chromosome. Compounds are fed to adult male flies (P1 males) that are then individually mated to virgin females of a test strain with marked and multiple inverted X chromosomes to prevent meiotic recombination in the gametes of their progeny. The F1 female progenies are individually mated with their brothers. In the F2 generation, each culture is scored for the absence of wild-type males, which indicates the presence of an X-linked recessive lethal mutation in the F1 female, derived from a germ cell of the P1 male. The frequency of F1 females giving no F2 wild-type sons is correlated to the mutagenicity of the test compound. The SLRL test has been used for more than 700 chemicals, making it one of the well-established tests (Vogel *et al.*, 1999). It was found to have high specificity, meaning that a positive SLRL response of a chemical would predict it to be a mammalian mutagen and carcinogen (Foureman *et al.*, 1994a,b). However, the SLRL test has a relatively low sensitivity for mammalian genotoxins and thus would have a high false-negative rate.

The better *Drosophila* genotoxicity assays today are the somatic mutation and recombination tests (SMART) (Vogel *et al.*, 1999); SMART makes use of recessive markers for eye or wing imaginal disc cells to detect the mutagenic and recombinagenic activity of compounds by the loss of heterozygosity of the markers. The wing spot test uses two recessive wing bristle markers *mwh* (multiple wing hairs) and *flr* (flare) on the left arm of chromosome 3. The *mwh* mutation is located at the tip of the chromosome arm and the *flr* mutation is closer to the base of the arm. They are separated by 38 recombination map units (*mwh* 3-0.3 and *flr* 3-38.8), making it easy to detect many recombination events. Larvae that are *trans*-heterozygous for the two markers are treated chronically or acutely by oral administration of the test compounds. Mutant clones can be induced in wing disc cells of larvae that are under rapid proliferation. Such mutant clones will eventually differentiate into somatic spots on the wingblades of the adult flies. Single spots made of *mwh* and/or *flr* bristles are due to different genotoxic mechanisms: point mutation, deletion, chromosome breakage and mitotic recombination. Twin spots consisting of an *mwh* bristle area and an adjacent *flr* bristle area are produced only by mitotic recombination. The eye spot test uses the recessive marker *w* (white) mutation. Special strains with high cytochrome P-450-dependent enzyme activities have been developed to enhance the detection of mutagenic compounds that require metabolic activation. The mutational events detected by SMART are induced by compounds that directly or indirectly cause DNA damage (alkylation, deamination, cycloalkylation, crosslinking, nucleotide misincorporation, intercalation or strand scission) and directly or indirectly affect repair systems (i.e. nucleotide excision repair, post-replication repair or cross-link specific repair).

The SMART assays have been validated with all known classes of genotoxic chemicals, including more than 400 chemical compounds, and are shown to have good sensitivity (75–78%) and accuracy (83–86%). It is a simple one-generation test that has very low cost, thus multiple doses and protocols can be used. New and improved SMART assays based on classical genetics are continuously being developed. For example, modified eye spot tests have become available recently for the simultaneous detection of structural chromosome aberrations, homologous mitotic recombination, intrachromosomal (deletion/amplification) recombination and chromosome gain (non-disjunction) (Vogel and Nivard, 2000). New-generation SMART assays based on engineered fluorescent or luminescent markers may be developed to increase assay throughput by automation.

Drosophila genotoxicity assays, especially the SLRL test due to its long history (since 1927) and large knowledge database, have been used widely by academia, government agencies and in some cases by industries. For example, the SLRL test has been used to evaluate the genotoxicity of several drugs such as hydrochlorothiazide, busulfan, nitrofurantoin, indomethacin and budesonide (www.fda.gov). Some of the studies were done under the auspices of the National Toxicology Program (NTP) in the USA.

Drosophila assays are not yet used routinely in lead compound prioritization, in regulatory tests and in post-market product monitoring in the pharmaceutical industry. For example, in a recent survey of 352 marketed drugs from the 1999 Physician's Desk Reference, only eight were shown to be tested by the SLRL assay (Snyder and Green, 2001). Thus, there is a paradoxical phenomenon that, on the one hand, *Drosophila* is widely recognized for its predictive power for human gene function, including genes involved in safeguarding DNA and chromosomes, yet, on the other hand, it lacks broad use in testing the genotoxicity of pharmaceutical compounds. The likely explanations are: the classical SLRL assay is complex, requires special training and facilities not available in most toxicology laboratories in industry and has some weaknesses; a genotoxicity test requires extensive evaluation before it becomes a standard test and only recently has such a level of validation been achieved for the SMART assays; and, most importantly, *Drosophila* assays are not yet included in recommendations or guidelines issued by international regulatory agencies, such as the International Conference on Harmonization of Technical Requirements for Registration of Pharmaceuticals for Human Use (ICH). For example, in the ICH topic S2B, *Genotoxicity: a Standard Battery for Genotoxicity Testing of Pharmaceuticals*, three tests are recommended: a test for gene mutation in bacteria (such as the Salmonella Ames test); an *in vitro* test with cytogenetic evaluation of chromosomal damage with mammalian cells or an *in vitro* mouse lymphoma tk assay; and an *in vivo* test for chromosomal damage using rodent hematopoietic cells.

Even though it will take some time before *Drosophila* assays are included in the regulatory tests, they can be applied immediately in some toxicology studies in drug discovery. For example, for lead compound prioritization, rapid and inexpensive assays are preferred after *in silico* evaluation of the mutagenic potential of the compounds based on their chemical structures and prior experience. The SMART assays are perfectly fitted for this application, especially if higher throughput assay formats and fly handling methods are developed. *Drosophila* assays also may be used as complementary or confirmatory tests when ambiguous results are obtained by the standard guideline tests. In addition, as an excellent genetic, biochemical and molecular biological experiment system, *Drosophila* is well suited for studying the mechanisms of genotoxicity of an important compound.

4.2 Research tools in *Drosophila* studies

Information resources

A century of innovative research and community effort has given *Drosophila* biologists a wide array of research tools. One of the most important research tools is the extensive information resource. Currently, the compiled information is primarily made available through the Internet and books. We list some of the major ones here.

Internet resources

- Flybase (www.flybase.org): the most comprehensive database about *Drosophila*

- Berkeley *Drosophila* Genome Project (www.fruitfly.org): genomic sequences, cDNA collections, insertion sites of P element mutations, bioinformatic tools

- NCBI *Drosophila* genome (www.ncbi.nlm.nih.gov/cgi-bin/Entrez/map00?taxid = 7227)

- EuGene: Genomic Information for Eukaryotic Organisms (iubio.bio.indiana.edu:8089/): abridged comparison of genes between different eukaryotic genomes

- *Drosophila* homologs of human disease genes (homophila.sdsc.edu)

- Interactive fly (sdb.bio.purdue.edu/fly/aimain/1aahome.htm)

Book titles

- Biology of *Drosophila* (Demerec, 1994)

- The Genome of *Drosophila melanogaster* (Lindsley and Zimm, 1992)

- *Drosophila*: a Laboratory Handbook and Manual (Ashburner, 1989)

- The Development of *Drosophila melanogaster* (Bate and Martinez Arias, 1993)

- The Genetics and Biology of *Drosophila* (Ashburner and Novitski, 1976)

- *Drosophila*: a Practical Approach (Roberts, 1998)

- *Drosophila* Cells in Culture (Echalier, 1997)

- Fly Pushing: the Theory and Practice of *Drosophila* Genetics (Greenspan, 1997)

- *Drosophila* Protocols (Sullivan *et al.*, 2000)

Balancer chromosomes

Balancer chromosomes in *Drosophila* belong to a special set of utility chromosomes that have multiple inversions to suppress recombination with the homologous chromosome, they are homozygous lethal or sterile and they bear both dominant markers and recessive mutations that allow them to be followed easily during stock maintenance and genetic crosses. Balancers are of tremendous utility for the isolation of mutations and for the maintenance of mutant stocks, as well as for genetic experiments. Few other multicellular experimental organisms have balancer chromosomes.

Mutagenesis

Random mutagenesis

Random mutagenesis has been the cornerstone of *Drosophila* forward genetics (see Section 4.1; using *Drosophila* for drug target identification and validation). There are three major methods to induce mutations in flies: chemical mutagens, radiation and transposons (Ashburner, 1989; Greenspan, 1997; Roberts, 1998).

The most widely used chemical mutagen in *Drosophila* today is ethylmethanesulfonate (EMS), which is normally used to make point mutations although not all mutations derived from an EMS mutagenesis experiment are point mutations and chromosomal aberrations may occur. Because EMS only affects one strand of the DNA helix, an induced mutation may be reverted or fixed after an additional round of DNA replication. Very often EMS induces missense mutations and there is a 5–10% chance that the isolated mutations are conditional mutations. By combining EMS and the FLP/FRT mitotic recombination system (see later section on analytical tools), high-throughput F1 genetic screens can be performed easily.

X-ray radiation from X-ray tube and γ-ray radiation from ^{60}Co or ^{137}Cs are normally used to induce chromosome/DNA aberrations such as deletions, inversions, duplications and translocations.

The advantage of chemical- and radiation-based mutagenesis is their relative non-selectivity on DNA sequence context, therefore the induced mutations are more randomly distributed in the genome than other methods. If the aim of a genetic screen is to discover most of the genes that affect the assay phenotype, these types of mutagens, especially chemical mutagens, should be used. However, their major disadvantage is that the induced mutations need to be mapped before knowing which genes are mutated. Mapping mutations is a very time-consuming process, even with genome-wide SNP information (Hoskins *et al.*, 2001).

Engineered transposons, such as the *P* element, *hobo* element and *PiggyBac* element (Handler and Harrell, 1999; Horn and Wimmer, 2000; Horn *et al.*, 2002; Thibault, 2002), are used as mutagens for insertional mutagenesis. The great advantage of using transposons as mutagens is that they also serve as DNA 'tags' in the mutated gene. By using the inverted polymerase chain reaction (PCR) method (Takagi *et al.*, 1992), the location of mutations in the genome and linked genes can be rapidly determined. However, transposon insertion is sensitive to DNA sequence context, thus is less 'random' compared with chemical mutagens. The combination of the PiggyBac transposon and the FLP/FRT mitotic recombination system is likely to make a significant contribution to genetic screens in the near future (Nystedt *et al.*, 2002).

Targeted mutagenesis

The complete sequencing of the *Drosophila* genome heightens the need for targeted mutagenesis methods based on gene sequences. This is especially true for systematic functional analysis of gene families, such as kinases, phosphatases and proteases, which are highly relevant gene families for drug discovery. Only recently has targeted mutagenesis been achieved in *Drosophila*. These methods include homologous recombination, synthetic sequence-specific zinc finger nuclease (ZFN) and *in vivo* dsRNAi.

The lack of embryonic stem cell lines of *Drosophila* has hampered, for many years, the use of homologous recombination to knock-out genes. However, a method has been developed recently to induce homologous recombination *in vivo* (Gloor, 2001; Rong *et al.*, 2002). A transgene is made to carry a donor element with a DNA fragment from the target gene and a marker gene. The DNA fragment has engineered mutations and an I-SceI endonuclease recognition site in the middle; the entire donor element is flanked by flipase (FLP) recognition sites. A circular extrachromosomal donor DNA element is induced by expression of the FLP-site-specific recombinase *in vivo*. The I-SceI endonuclease, also expressed from a transgene, converts the circular DNA into a linear recombinogenic molecule. A successful homologous recombination event will insert the marker and mutations in the target gene. The frequency of targeting events in the germline depends on the target genes. In a study of five targeted genes, the homologous recombination frequency varies from 1 in 1500 gametes for one gene, to 1 in 34 000 gametes for another gene (Rong *et al.*, 2002). Owing to the need for a donor transgene with engineered mutation for transfer and a low homologous recombination frequency, the gene targeting is still too inefficient to meet the demands of drug discovery. Incorporating positive and negative selection schemes should relieve the screening burden and thus increase throughput (Gloor, 2001).

Another method for gene targeting is based on ZFN. A C_2H_2-type zinc finger can specifically bind to a DNA site with three nucleotides. Thus a collection of 64 zinc fingers are needed to bind any one of the 64 triplets. Because C_2H_2 zinc fingers act in a modular manner, by stringing several zinc fingers together it is possible to create a protein with multiple zinc fingers that can bind any sequence of interest (Beerli and Barbas, 2002). A ZFN is a chimeric protein with a non-specific DNA cleavage domain and zinc fingers for sequence-specific DNA binding. A pair of ZFNs, each with three zinc fingers fused to the non-specific endonuclease domain of the restriction enzyme FokI, forms a dimer through the endonuclease domain and leads to cleavage at a DNA target site specified by the six zinc fingers (Bibikova *et al.*, 2001). This method has been used successfully in *Drosophila* to generate target gene mutations at a designed site in the *yellow* gene through dsDNA cleavage and non-homologous end joining, which introduces deletions and insertions at the gap (Bibikova *et al.*, 2002). The germline mutation frequency was found to be 0.4% of male gametes at the target site. It would be expected that the mutation frequency would vary, depending on target sites, due to local chromatin structure, DNA modification and binding of other proteins. Compared with the homologous recombination method, ZFN-based mutagenesis has an even lower throughput due to the requirement of making and verifying two ZFNs for a given target gene. In addition, no selectable marker is inserted into the target gene; thus screening for mutations remains a major effort.

Targeted mutagenesis can be achieved also by the dsRNAi method (Hannon, 2002). Using a DNA fragment from the target gene, an expression construct can be made with an inverted repeat that will produce double-stranded hairpin RNAs. *In vivo*-produced hairpin RNAs effectively knock down the mRNA levels of the corresponding target genes (Fortier and Belote, 2000; Kennerdell and Carthew, 2000; Piccin *et al.*, 2001). The efficiency of dsRNAi-based gene knock-down depends on the levels and half-lives of target mRNAs and proteins, as well as the structure and expression level of the dsRNA hairpin and its sequences. The major advantages of *in vivo* dsRNAi-based gene knock-down, compared with homologous recombination and ZFN, are its simplicity and potential for industrial scale-up. In addition, currently available tissue-specific and inducible expression technologies can be easily incorporated. This is extremely important because it avoids the lethality often associated with LOF mutations, allowing analysis of gene function and functional dissection of signaling pathways in dispensable tissues such as the *Drosophila* eye. The major drawback is that it is a gene knock-down method rather than a gene knock-out method, thus null mutations (100% knock-down) are rarely, if ever, obtained. Consequently, when no phenotypic effect is seen for a targeted gene knock-down, one does not know if it is due to insufficient dsRNAi or to no effect of the gene on the phenotype.

Analytical tools

The Gal4/UAS binary system for gene expression

One of the most widely used techniques for controlled gene expression in *Drosophila* is the binary Gal4/UAS system (Fischer *et al.*, 1988; Brand and

Perrimon, 1993; Duffy, 2002). One module in this system is the gene encoding the yeast transcriptional activator GAL4. The *Gal4* gene is put under the control of characterized promoters and enhancers, such as the *hsp70* promoter, or is inserted randomly in the *Drosophila* genome to be driven by nearby genomic enhancers. Thus, GAL4 expression may be controlled spatially or temporally. Fly strains carrying the *Gal4* gene are often referred to as Gal4 drivers. The other module in the system is a gene of interest under the control of the GAL4 binding site UAS (upstream activation sequences). When the two modules are brought together in the progenies of a cross, the expression of the gene of interest will reflect the expression of the GAL4 driver. In the *Drosophila* research community, there are a large number of tissue-specific Gal4 drivers that display diverse spatial and temporal patterns of Gal4 expression (Calleja *et al.*, 1996; Gustafson and Boulianne, 1996; Manseau *et al.*, 1997; Mata *et al.*, 2000; Lukacsovich *et al.*, 2001) (http:// flystocks.bio.indiana.edu/gal4.htm). In addition, drug-inducible Gal4 drivers are established (Smith *et al.*, 1996; Osterwalder *et al.*, 2001; Roman *et al.*, 2001; Stebbins and Yin, 2001; Stebbins *et al.*, 2001; Klueg *et al.*, 2002). To generate a large set of UAS-controllable genes, a specialized P-element transposon bearing a UAS element fused to a minimal *Drosophila* promoter and transcription start site, known as an EP (enhancer/promoter) element, has been used. This EP element is inserted randomly throughout the *Drosophila* genome, and transcriptional activation of the EP element by GAL4 often leads to transcriptional activation and overexpression of adjacent genes (Rorth, 1996; Rorth *et al.*, 1998). There are currently more than 2000 publicly available EP strains.

The FLP/FRT and Cre/lox systems for site-specific recombination

The introduction of the yeast FLP/FRT system into *Drosophila* has greatly facilitated tissue-specific analysis of gene functions (Golic and Lindquist, 1989; Golic, 1991). This system utilizes the yeast site-specific recombinase, flipase (FLP), and its recognition target sequence FRT. There are many applications for this system. For example, by placing the FRT site at the base of the two homologous chromosome arms and expressing FLP using a tissue-specific enhancer, mitotic recombination can be induced in the proliferating tissue to generate mosaic animals with twin mitotic clones of cells (Xu and Rubin, 1993). One of the mitotic clones is homozygous for one of the FRT chromosome arms and the other mitotic clone is homozygous for the other FRT chromosome arm, whereas the non-FLP-expressing tissues are hetero- zygous for the two FRT chromosome arms. If one of the FRT chromosome arms has a recessive mutation in a gene of interest and the other arm has a cellular marker, the mutation's effect can be studied in the mitotic clones that are marked by the lack of the cellular marker. This negative labeling in a positive background is sometimes inadequate and a positive labeling method has been developed (see below). The FLP/FRT-based mosaic analysis method can be used to study even homozygous lethal mutations (Xu and Rubin, 1993).

A refinement of the FLP/FRT mosaic technique is to introduce a tissue-specific dominant cell lethal gene and a recessive cell lethal mutation on one of the two homologous FRT chromosome arms (the one that does not bear the mutation in study) (Stowers and Schwarz, 1999). For example, using this method and eye-specific FLP expression, the entire *Drosophila* eyes can be made with the same cells that are homozygous for the mutation, greatly improving the efficiency of studying specific mutations, as well as for genetic screens.

Because of the public availability of all five major chromosome arms with the FRT site at their bases, the F1 genetic screen based on mitotic recombination of FRT chromosomes has become a mainstay in today's fly genetic study (Xu and Harrison, 1994; Xu *et al.*, 1995; Theodosiou and Xu, 1998).

Another application of the FLP/FRT system is its use to create ON/OFF gene switches. In this case a gene encoding a phenotypic marker is inserted between the promoter and the coding region of a target gene, thus inactivating the target gene. At each end of the inserted marker gene are FRT recombination sites. Following FLP expression, the inserted marker gene is excised via FLP-mediated recombination and the target gene is activated by juxtaposition of the promoter with the coding region of the gene (Struhl *et al.*, 1993). Expression of the target gene therefore can be controlled spatially and temporally through regulation of FLP expression. Unlike the GAL4/UAS system, however, the changes in gene expression are permanent because they are mediated through DNA rearrangement. One important usage of the gene switch is for genetic ablation of specific cells by FLP/FRT-mediated expression of a cellular toxin gene. Owing to leakiness of basal promoters, many strong toxin genes cannot be introduced into the fly genome even by using inducible promoters.

An alternative site-specific recombination technique is the Cre/lox system in which Cre is a site-specific DNA recombinase from bacteriophage P1 and lox is the Cre recognition sequence (Sauer, 1998; Perkins, 2002). This system was initially used in mouse and later introduced into *Drosophila* (Siegal and Hartl, 1996).

The combination of GAL4/UAS and FLP/FRT systems, such as driving FLP expression with UAS (UAS-FLP) (Duffy *et al.*, 1998) gives *Drosophila* geneticists a versatile toolbox with which to design sophisticated experiments.

A further development of the two systems is the usage of GAL80, which is a GAL4 inhibitor (Lee and Luo, 2001). By putting a transgene ubiquitously expressing GAL80 in an FRT chromosome arm, and UAS-GFP and a mutation of interest on the other homologous FRT chromosome arm, the mitotic clones that are homozygous for the mutation of interest can be positively labeled in a negative background. This so-called 'MARCM' (mosaic analysis with a repressible cell marker) technique is especially important for studying groups of cells with complex morphologies, such as neurons (Lee and Luo, 2001). The GAL80 also may be used to stabilize a fly stock with a GAL4 driver/UAS-responder transgene pair that would otherwise be deleterious or lethal to the organism.

Transgenics

Drosophila germline transformation is typically carried out by direct injection of early-stage embryos with DNA vectors based on the P transposable element-derived vectors, although other transformation vectors based on other transposable elements, such as the Piggybac element, are now also available (Ashburner, 1989; Handler, 2002). Using the P element vectors, up to 20% of the fertile G0 flies produce one or more germline transformants that can be identified by the expression of genetic markers present in the vector. It takes about a month to get F1 transgenic flies and another generation (ca. 10 days) to have a stable transgenic line. A skilled technician can inject one or two constructs each day and follow up on the resulting embryos, flies and requisite genetic crosses. This level of throughput is significantly higher than the transgenic mice process. However, the throughput could be increased if embryo injections and genetic crosses are automated.

Automated fly sorting and dispensing

One of the important high-throughput technologies in the *Drosophila* field is the development of instrumentation and methods for automated sorting and dispensing of live embryos and larvae (Furlong *et al.*, 2001a; Li *et al.*, 2001). The principle of sorting is based on flow cytometry using multiple fluorescent markers and organism size as sorting parameters. This technology allows rapid isolation and enrichment of flies with unique properties defined by the multiparameters. This technology has been used to isolate a homogenous population of mutant embryos from mixed populations for an RNA profiling experiment (Furlong *et al.*, 2001b), for an enhancer trapping screen (Gisselbrecht *et al.*, 2002) and for an exon trapping screen (Morin *et al.*, 2001; Buszczak *et al.*, 2002; Quinones *et al.*, 2002). In addition, the ability to dispense accurately single or multiple embryos or larvae into 96-well plates and maintain good viability (Li *et al.*, 2001) bridges the gap between *Drosophila* biology and an industry standard high-throughput process. It should soon be possible to carry out experiments in *Drosophila* using a larger number of experimental variables than previously feasible, such as large-scale compound screen and large-scale study of interactions between gene function, metabolism, physiology and environment variables (e.g. diet and stress).

Cell culture

Drosophila-cultured cells have been studied extensively for the last 50 years (Echalier, 1997; Cherbas and Cherbas, 1998). Many different permanent cell lines and extensive primary cell culture methods have been developed. The most widely used permanent cell lines are the Kc cells and S2 cells, both derived from embryos. The Kc cells are considered to be like larval lymph gland cells and hemocytes (Cherbas and Cherbas, 1998). From the drug discovery standpoint, one of the most important usages of *Drosophila* cell culture is for high-throughput LOF genetic screens. This approach is feasible because of the demonstrated specificity, dose-dependency and perdurance of dsRNAi effects by low-cost long dsRNA (Worby *et al.*, 2001). Double-stranded RNA molecules against most of the fly genes can be synthesized through *in vitro* transcription using the full-length cDNA clones in the *Drosophila* Gene Collection (DGC) (Stapleton *et al.*, 2002). There is great potential in combining *in vivo* cell-specific markers, mass isolation of the marked cells, short-term primary cell culture and genome-wide dsRNAi to identify genes involved in disease-relevant pathways. As mentioned above, *Drosophila* cells may be useful also for MOA studies and compound screens. A major advantage of using *Drosophila* cell lines is that in each case the identified candidate genes can be introduced rapidly back into transgenic flies and studied further using the advanced genetic approaches available in *Drosophila*.

High-throughput tools to be developed

To meet the high demands of the functional genomics era, as well as the demand for shortening the drug development cycle, it is important to develop high-throughput technology in many areas of *Drosophila* research:

1. Fast and reliable methods are needed to maintain mutant stocks. A simple calculation of 14 000 genes with one LOF mutant allele and one GOF allele per gene would result in a stock of 28 000 mutants. (13 000 if genes without human homologs are excluded.)

2. Controlled and highly parallel genetic cross-technology is necessary (such as in 96-well format) if many modifier screens are to be performed against a

mutant library for all fly genes or a human homolog subset. A prerequisite is an automated method to separate and dispense adult males and females from the mutant library.

3. If chemical mutagens are used for genetic screens, high-throughput mutation mapping technology is essential. This could be developed based on SNPs (Hoskins *et al.*, 2001) and existing technologies such as single base extension plus matrix-assisted laser desorption–ionization time-of-flight (MALDI/TOF) mass spectrometry or invasive cleavage-based assays (Shi, 2002), as well as sample pooling for assessing allele frequency (Shi, 2002). Alternatively, if saturation of all of the genes of interest is achieved through a combination of insertional mutagenesis and the production of transgenic animals, most mutations could be mapped using complementation tests.

4. For high-throughput handling of live flies it is necessary to have a growth medium of special formulation that would be compatible with robotics, normal growth of flies through all stages of the life cycle and addition of testing materials such as compounds.

5. It is desirable to have high-throughput methods for delivering/injecting into flies DNA constructs, dsRNAs, compounds and other reagents.

6. High-throughput imaging systems equipped with real-time expert pattern recognition capabilities need to be developed to take advantage of many established fly assays that are based on morphological and behavioral phenotypes, or the assays need to be reformatted to be readable by off-the-shelf machines.

The basic concept underlying this wish list is the design and execution of methods necessary to transform *Drosophila* from a basic academic research tool into an industrial application tool.

4.3 Prospects

Drosophila research has contributed tremendously in the last 100 years to our conceptual understanding of biology, from the chromosome theory of heredity, determination of the structure of eukaryotic genes and the elucidation of genetic pathways through to genomics. Along the way, many experimental tools and assays have been established. The accumulated physical, methodological and informational resources provided by *Drosophila* scientists have ensured a strong foundation for the field. Current *Drosophila* research is moving forward at an increasing pace and now includes not only activities at academic institutions but also in the private sector. Many challenges are still ahead for applying *Drosophila* to drug discovery. These challenges include more direct modeling of human diseases, demonstrating more extensively the utility of *Drosophila* for MOA studies, genetic screens against phenotypes generated by drug treatment and compound screens. Further technological development is also necessary, especially in the areas of high-throughput biology. However, the utility of *Drosophila* in drug discovery is now firmly established and what remains to be determined is how much more this remarkably robust model system can provide for the drug discovery process.

4.4 Acknowledgments

We would like to thank R. Fernandez and M. Konsolaki for critical review of the manuscript and Q. Wang and X. Shi for technical help in establishing automated fly sorting and dispensing methods.

4.5 References

Abarca, C., Albrecht, U. and Spanagel, R. (2002). Cocaine sensitization and reward are under the influence of circadian genes and rhythm. *Proc. Natl. Acad. Sci. USA* **99**, 9026–9030.

Affolter, M., Schier, A. and Gehring, W. J. (1990). Homeodomain proteins and the regulation of gene expression. *Curr. Opin. Cell. Biol.* **2**, 485–495.

Andretic, R., Chaney, S. and Hirsh, J. (1999). Requirement of circadian genes for cocaine sensitization in *Drosophila*. *Science* **285**, 1066–1068.

Ashburner, M. (1989). *Drosophila: A Laboratory Handbook and Manual* (2 Vols). Cold Spring Harbor, New York: Cold Spring Harbor Laboratory Press.

Ashburner, M. and Novitski, E. (1976). *The Genetics and Biology of Drosophila*. London: Academic Press.

Bate, M. and Martinez Arias, A. (1993). *The Development of Drosophila melanogaster*. Plainview, New York: Cold Spring Harbor Laboratory Press.

Beerli, R. R. and Barbas, C. F., III (2002). Engineering polydactyl zinc-finger transcription factors. *Nat. Biotechnol.* **20**, 135–141.

Bibikova, M., Carroll, D., Segal, D. J., Trautman, J. K., Smith, J., Kim, Y. G. and Chandrasegaran, S. (2001). Stimulation of homologous recombination through targeted cleavage by chimeric nucleases. *Mol. Cell. Biol.* **21**, 289–297.

Bibikova, M., Golic, M., Golic, K. G. and Carroll, D. (2002). Targeted chromosomal cleavage and mutagenesis in *Drosophila* using zinc-finger nucleases. *Genetics* **161**, 1169–1175.

Brand, A. H. and Perrimon, N. (1993). Targeted gene expression as a means of altering cell fates and generating dominant phenotypes. *Development* **118**, 401–415.

Breinbauer, R., Vetter, I. R. and Waldmann, H. (2002). From protein domains to drug candidates – natural products as guiding principles in the design and synthesis of compound libraries. *Angew Chem. Int. Ed. Engl.* **41**, 2879–2890.

Brltton, J. S. and Edgar, B. A. (1998). Environmental control of the cell cycle in *Drosophila*: nutrition activates mitotic and endoreplicative cells by distinct mechanisms. *Development* **125**, 2149–2158.

Buszczak, M. H., Morin, X., Quinones, A. T., Chia, U. and Cooley, L. (2002). High throughput protein trapping in *Drosophila*. *Dros. Res. Conf. Proc.* **43**, 982A.

Calleja, M., Moreno, E., Pelaz, S. and Morata, G. (1996). Visualization of gene expression in living adult *Drosophila*. *Science* **274**, 252–255.

Cherbas, L. and Cherbas, P. (1998). Cell culture. In *Drosophila. A Practical Approach*, D.D.Roberts (ed.), pp. xx, 398. Oxford: IRL Press.

Chien, S., Reiter, L. T., Bier, E. and Gribskov, M. (2002). Homophila: human disease gene cognates in *Drosophila*. *Nucleic Acids Res.* **30**, 149–151.

Clemens, J. C., Worby, C. A., Simonson-Leff, N., Muda, M., Maehama, T., Hemmings, B. A. and Dixon, J. E. (2000). Use of double-stranded RNA interference in *Drosophila* cell lines to dissect signal transduction pathways. *Proc. Natl. Acad. Sci. USA* **97**, 6499–6503.

Crews, C. M. and Splittgerber, U. (1999). Chemical genetics: exploring and controlling cellular processes with chemical probes. *Trends Biochem. Sci.* **24**, 317–320.

Demerec, M. (1994). *Biology of Drosophila* (Facsimile edition). Cold Spring Harbor, New York: Cold Spring Harbor Press.

Dobrosotskaya, I. Y., Seegmiller, A. C., Brown, M. S., Goldstein, J. L. and Rawson, R. B. (2002). Regulation of SREBP processing and membrane lipid production by phospholipids in *Drosophila*. *Science* **296**, 879–883.

Drews, J. B. (2000). Drug discovery: a historical perspective. *Science* **287**, 1960–1964.

Duffy, J. B. (2002). GAL4 system in *Drosophila*: a fly geneticist's Swiss army knife. *Genesis* **34**, 1–15.

Duffy, J. B., Harrison, D. A. and Perrimon, N. (1998). Identifying loci required for follicular patterning using directed mosaics. *Development* **125**, 2263–2271.

Echalier, G. (1997). *Drosophila Cells in Culture*. San Diego, CA: Academic Press.

EPA. (1998). *OPPTS Harmonized Test Guidelines, Series 870 Health Effects Test Guidelines, 870.5275 Sex-linked Recessive Lethal Test in Drosophila melanogaster (August, 1998)*. Washington, DC: US Environmental Protection Agency, Office of Prevention, Pesticides and Toxic Substances.

Fischer, J. A., Giniger, E., Maniatis, T. and Ptashne, M. (1988). GAL4 activates transcription in *Drosophila*. *Nature* **332**, 853–856.

Fortier, E. and Belote, J. M. (2000). Temperature-dependent gene silencing by an expressed inverted repeat in *Drosophila*. *Genesis* **26**, 240–244.

Fossett, N. and Schulz, R. A. (2001). Functional conservation of hematopoietic factors in *Drosophila* and vertebrates. *Differentiation* **69**, 83–90.

Foureman, P., Mason, J. M., Valencia, R. and Zimmering, S. (1994a). Chemical mutagenesis testing in *Drosophila*. IX. Results of 50 coded compounds tested for the National Toxicology Program. *Environ. Mol. Mutagen.* **23**, 51–63.

Foureman, P., Mason, J. M., Valencia, R. and Zimmering, S. (1994b). Chemical mutagenesis testing in *Drosophila*. X. Results of 70 coded chemicals tested for the National Toxicology Program. *Environ. Mol. Mutagen.* **23**, 208–227.

Furlong, E. E., Andersen, E. C., Null, B., White, M. P. and Scott, M. P. (2001a). Patterns of gene expression during *Drosophila* mesoderm development. *Science* **293**, 1629–1633.

Furlong, E. E., Profitt, D. and Scott, H. P. (2001b). Automated sorting of live transgenic embryos. *Nat. Biotechnol.* **19**, 153–156.

Gilbert, D. G. (2002). euGenes: a eukaryote genome information system. *Nucleic Acids Res.* **30**, 145–148.

Gisselbrecht, S. S., Bayes, J., Etchin, J., Dell'Orfano, B., Ferrante, A. and Michelson, A. M. (2002). A rapid and efficient approach to vital enhancer trap screening in *Drosophila* embryos. *Dros. Res. Conf. Proc.* **43**, 143.

Gloor, G. B. (2001). Gene-targeting in *Drosophila* validated. *Trends Genet.* **17**, 549–551.

Golic, K. G. (1991). Site-specific recombination between homologous chromosomes in *Drosophila*. *Science* **252**, 958–961.

Golic, K. G. and Lindquist, S. (1989). The FLP recombinase of yeast catalyzes site-specific recombination in the *Drosophila* genome. *Cell* **59**, 499–509.

Greenspan, R. J. (1997). *Fly Pushing: The Theory and Practice of* Drosophila *Genetics*. Cold Spring Harbor, New York: Cold Spring Harbor Laboratory Press.

Gustafson, K. and Boulianne, G. L. (1996). Distinct expression patterns detected within individual tissues by the GAL4 enhancer trap technique. *Genome* **39**, 174–182.

Handler, A. M. (2002). Use of the piggyBac transposon for germ-line transformation of insects. *Insect Biochem. Mol. Biol.* **32**, 1211–1220.

Handler, A. M. and Harrell, R. A., II (1999). Germline transformation of *Drosophila melanogaster* with the piggyBac transposon vector. *Insect Mol. Biol.* **8**, 449–457.

Hannon, G. J. (2002). RNA interference. *Nature* **418**, 244–251.

Horn, C. and Wimmer, E. A. (2000). A versatile vector set for animal transgenesis. *Dev. Genes Evol.* **210**, 630–637.

Horn, C., Offen, N., Nystedt, S., Hacker, U. and Wimmer, E. A. (2002). piggyBac transposon mutagenesis and enhancer trapping to target novel gene loci. *Dros. Res. Conf. Proc.* **43**, 984C.

Hoskins, R. A., Phan, A. C., Naeemuddin, M., Mapa, F. A., Ruddy, D. A., Ryan, J. J., Young, L. H., *et al.* (2001). Single nucleotide polymorphism markers for genetic mapping in *Drosophila melanogaster*. *Genome Res.* **11**, 1100–1113.

Hughes, T. R., Marton, M. J., Jones, A. R., Roberts, C. J., Stoughton, R., Armour, C. D., Bennett, H. A., *et al.* (2000). Functional discovery via a compendium of expression profiles. *Cell* **102**, 109–126.

Johnston, P. (2002). Cellular assays in HTS. *Methods Mol. Biol.* **190**, 107–116.

Kazemi-Esfarjani, P. and Benzer, S. (2000). Genetic suppression of polyglutamine toxicity in *Drosophila*. *Science* **287**, 1837–1840.

Kazemi-Esfarjani, P. and Benzer, S. (2002). Suppression of polyglutamine toxicity by a *Drosophila* homolog of myeloid leukemia factor 1. *Hum. Mol. Genet.* **11**, 2657–2672.

Kennerdell, J. R. and Carthew, R. W. (2000). Heritable gene silencing in *Drosophila* using double-stranded RNA. *Nat. Biotechnol.* **18**, 896–898.

Klueg, K. M., Alvarado, D., Muskavitch, M. A. and Duffy, J. B. (2002). Creation of a GAL4/UAS-coupled inducible gene expression system for use in *Drosophila* cultured cell lines. *Genesis* **34**, 119–122.

Lander, E. S., Linton, L. H., Birren, B., Nusbaum, C., Zody, M. C., Baldwin, J., Devon, K., *et al.* (2001). Initial sequencing and analysis of the human genome. *Nature* **409**, 860–921.

Lasko, P. (2002). Diabetic flies? Using *Drosophila melanogaster* to understand the causes of monogenic and genetically complex diseases. *Clin. Genet.* **62**, 358–367.

Lee, T. and Luo, L. (2001). Mosaic analysis with a repressible cell marker (MARCM) for *Drosophila* neural development. *Trends Neurosci.* **24**, 251–254.

Leppik, I. E. (2002). Three new drugs for epilepsy: levetiracetam, oxcarbazepine and zonisamide. *J. Child Neurol.* **17**, 53–57.

Lewis, E. B. (1985). Regulation of the genes of the bithorax complex in *Drosophila*. *Cold Spring Harbor Symp. Quant. Biol.* **50**, 155–164.

Li, H. H., Wang, Q., Shi, X. and Zusman, S. (2001). Developing a *Drosophila* high throughput screen technology. *Dros. Res. Conf. Proc.* **43**, 949A.

Lindsley, D. L. and Zimm, G. G. (1992). *The Genome of Drosophila melanogaster*. San Diego, CA: Academic Press.

Lukacsovich, T., Asztalos, Z., Awano, U., Baba, K., Kondo, S., Niwa, S. and Yamamoto, D. (2001). Dual-tagging gene trap of novel genes in *Drosophila melanogaster*. *Genetics* **157**, 727–742.

MacKinnon, R., Cohen, S. L., Kuo, A., Lee, A. and Chait, B. T. (1998). Structural conservation in prokaryotic and eukaryotic potassium channels. *Science* **280**, 106–109.

Malicki, J., Schughart, K. and McGinnis, W. (1990). Mouse Hox-2.2 specifies thoracic segmental identity in *Drosophila* embryos and larvae. *Cell* **63**, 961–967.

Manseau, L., Baradaran, A., Brower, D., Budhu, A., Elefant, F., Phan, H., Phillip, A. V., *et al.* (1997). GAL4 enhancer traps expressed in the embryo, larval brain, imaginal discs and ovary of *Drosophila*. *Dev. Dynam.* **209**, 310–322.

Marin, I. and Baker, B. S. (1998). The evolutionary dynamics of sex determination. *Science* **281**, 1990–1994.

Marsh, J. L., Walker, H., Theisen, H., Zhu, Y. Z., Fielder, T., Purcell, J. and Thompson, L. M. (2000). Expanded polyglutamine peptides alone are intrinsically cytotoxic and cause neurodegeneration in *Drosophila*. *Hum. Mol. Genet.* **9**, 13–25.

Mata, J., Curado, S., Michon, A. M., Yoshida, S. and Ephrussi, A. (2000). Gain of function screen in the female germ line. *Dros. Res. Conf. Proc.* **41**, 372D.

Matthews, D. J. and Kopczynski, J. (2001). Using model-system genetics for drug-based target discovery. *Drug Discov. Today* **6**, 141–149.

McClung, C. and Hirsh, J. (1998). Stereotypic behavioral responses to free-base cocaine and the development of behavioral sensitization in *Drosophila*. *Curr. Biol.* **8**, 109–112.

McClung, C. and Hirsh, J. (1999). The trace amine tyramine is essential for sensitization to cocaine in *Drosophila*. *Curr. Biol.* **9**, 853–860.

McGinnis, N., Kuziora, M. A. and McGinnis, W. (1990). Human Hox-4.2 and *Drosophila* deformed encode similar regulatory specificities in *Drosophila* embryos and larvae. *Cell* **63**, 969–976.

Meyer, J. H., Goulding, V. S., Wilson, A. A., Hussey, D., Christensen, B. K. and Houle, S. (2002). Bupropion occupancy of the dopamine transporter is low during clinical treatment. *Psychopharmacology (Berlin)* **163**, 102–105.

Mitchison, T. J. (1994). Towards a pharmacological genetics. *Chem. Biol.* **1**, 3–6.

Moon, R. T., Bowerman, B., Boutros, M. and Perrimon, N. (2002). The promise and perils of Wnt signaling through beta-catenin. *Science* **296**, 1644–1646.

Moore, K. and Rees, S. (2001). Cell-based versus isolated target screening: how lucky do you feel? *J. Biomol. Screen* **6**, 69–74.

Morin, X., Daneman, R., Zavortink, M. and Chia, W. (2001). A protein trap strategy to detect GFP-tagged proteins expressed from their endogenous loci in *Drosophila*. *Proc. Natl. Acad. Sci. USA* **98**, 15050–15055.

Nusse, R. (1999). WNT targets. Repression and activation. *Trends Genet.* **15**, 1–3.

Nusslein-Volhard, C. and Wieschaus, E. (1980). Mutations affecting segment number and polarity in *Drosophila*. *Nature* **287**, 795–801.

Nuttall, M. E., Lee, J. C., Murdock, P. R., Badger, A. M., Wang, F. L., Laydon, J. T., Hofmann, G. A., *et al.* (1999). Amphibian melanophore technology as a functional screen for antagonists of G-protein-coupled 7-transmembrane receptors. *J. Biomol. Screen.* **4**, 269–278.

Nystedt, S., Horn, C., Barmchi, M. P., Wimmer, E. A. and Hacker, U. (2002). Insertional mutagenesis on FRT chromosomes using piggyBac based vectors. *Dros. Res. Conf. Proc.* **43**, 985A.

OECD. (1984). *Genetic Toxicology: Sex-linked Recessive Lethal Test in Drosophila melanogaster. OECD Guideline For Testing of Chemicals.* Paris: Organization for Economic Cooperation and Development.

Oldham, S., Montagne, J., Radimerski, T., Thomas, G. and Hafen, E. (2000). Genetic and biochemical characterization of dTOR, the *Drosophila* homolog of the target of rapamycin. *Genes Dev.* **14**, 2689–2694.

Oliver, G. and Gruss, P. (1997). Current views on eye development. *Trends Neurosci.* **20**, 415–421.

Osterwalder, T., Yoon, K. S., White, B. H. and Keshishian, H. (2001). A conditional tissue-specific transgene expression system using inducible GAL4. *Proc. Natl. Acad. Sci. USA* **98**, 12596–12601.

Perkins, A. S. (2002). Functional genomics in the mouse. *Funct. Integr. Genom.* **2**, 81–91.

Piccin, A., Salameh, A., Benna, C., Sandrelli, F., Mazzotta, G., Zordan, M., Rosato, E., *et al.* (2001). Efficient and heritable functional knock-out of an adult phenotype in *Drosophila* using a GAL4-driven hairpin RNA incorporating a heterologous spacer. *Nucleic Acids Res.* **29**, E55-5.

Plautz, J. D., Straume, M., Stanewsky, R., Jamison, C. F., Brandes, C., Dowse, H. B., Hall, J. C., *et al.* (1997). Quantitative analysis of *Drosophila* period gene transcription in living animals. *J. Biol. Rhythms* **12**, 204–217.

Quinones, A. T., Buszczak, M., Morin, X., Chia, W. and Cooley, L. (2002). Using protein traps to study genes expressed during oogenesis. *Dros. Res. Conf. Proc.* **43**, 608B.

Rebay, I. (2002). Keeping the receptor tyrosine kinase signaling pathway in check: lessons from *Drosophila*. *Dev. Biol.* **251**, 1–17.

Reichert, H. (2002). Conserved genetic mechanisms for embryonic brain patterning. *Int. J. Dev. Biol.* **46**, 81–87.

Reiter, L. T., Potocki, L., Chien, S., Gribskov, M. and Bier, E. (2001). A systematic analysis of human disease-associated gene sequences in *Drosophila melanogaster*. *Genome Res.* **11**, 1114–1125.

Roberts, D. B. (1998). *Drosophila. A Practical Approach*. Oxford: IRL Press.

Roman, G., Endo, K., Zong, L. and Davis, R. L. (2001). P[Switch], a system for spatial and temporal control of gene expression in *Drosophila melanogaster*. *Proc. Natl. Acad. Sci. USA* **98**, 12602–12607.

Rong, Y. S., Titen, S. W., Xie, H. B., Golic, M. M., Bastiani, M., Bandyopadhyay, P., Olivera, B. M., *et al.* (2002). Targeted mutagenesis by homologous recombinatlon in *D. melanogaster*. *Genes Dev.* **16**, 1568–1581.

Rorth, P. (1996). A modular misexpresslon screen in *Drosophila* detecting tissue-specific phenotypes. *Proc. Natl. Acad. Sci. USA* **93**, 12418–12422.

Rorth, P., Szabo, K., Bailey, A., Laverty, T., Rehm, J., Rubin, G., Weigmann, K., *et al.* (1998). Systematic gain-of-function genetics in *Drosophila*. *Development* **125**, 1049–1057.

Roush, U. (1995). Nobel prizes: fly development work bears prize-winning fruit. *Science* **270**, 380–381.

Rubin, G. M., Yandell, M. D., Wortman, J. R., Gabor Miklos, G. L., Nelson, C. R., Hariharan, I. K., Fortini, M. E., *et al.* (2000). Comparative genomics of the eukaryotes. *Science* **287**, 2204–2215.

Sauer, B. (1998). Inducible gene targeting in mice using the Cre/lox system. *Methods* **14**, 381–392.

Scott, M. P., Tamkun, J. W. and Hartzell, G. W., III (1989). The structure and function of the homeodomain. *Biochim. Biophys. Acta* **989**, 25–48.

Seegmiller, A. C., Dobrosotskaya, I., Goldstein, J. L., Ho, Y. K., Brown, M. S. and Rawson, R. B. (2002). The SREBP pathway in *Drosophila*: regulation by palmitate, not sterols. *Dev. Cell* **2**, 229–238.

Shi, M. M. (2002). Technologies for individual genotyping: detection of genetic polymorphisms in drug targets and disease genes. *Am. J. Pharmacogenom.* **2**, 197–205.

Shilo, B. Z. (1992). Roles of receptor tyrosine kinases in *Drosophila* development. *FASEB J.* **6**, 2915–2922.

Siegal, M. L. and Hartl, D. L. (1996). Transgene coplacement and high efficiency site-specific recombination with the Cre/loxP system in *Drosophila*. *Genetics* **144**, 715–726.

Smith, H. K., Roberts, I. J. H., Allen, M. J., Connolly, J. B., Moffat, K. G. and O'Kane, C. J. (1996). Inducible ternary control of transgene expression and cell ablation in *Drosophila*. *Dev. Genes Evol.* **206**, 14–24.

Smith, L., Price-Jones, M., Hughes, K., Egebjerg, J., Poulsen, F., Wiberg, F. C. and Shank, R. P. (2000). Effects of topiramate on kainate- and domoate-activated [^{14}C]guanidinium ion flux through GluR6 channels in transfected BHK cells using Cytostar-T scintillating microplates. *Epilepsia* **41**, 48–51.

Snyder, R. D. and Green, J. W. (2001). A review of the genotoxicity of marketed pharmaceuticals. *Mutat. Res.* **488**, 151–169.

Stapleton, M., Liao, G., Brokstein, P., Hong, L., Carninci, P., Shiraki, T., Hayashizaki, Y., *et al.* (2002). The *Drosophila* gene collection: identification of putative full-length cDNAs for 70% of *D. melanogaster* genes. *Genome Res.* **12**, 1294–1300.

Stebbins, M. J. and Yin, J. C. (2001). Adaptable doxycycline-regulated gene expression systems for *Drosophila*. *Gene* **270**, 103–111.

Stebbins, M. J., Urlinger, S., Byrne, G., Bello, B., Hillen, W. and Yin, J. C. (2001). Tetracycline-inducible systems for *Drosophila*. *Proc. Natl. Acad. Sci. USA* **98**, 10775–10788.

Steffan, J. S., Bodai, L., Pallos, J., Poelman, M., McCampbell, A., Apostol, B. L., Kazantsev, A., *et al.* (2001). Histone deacetylase inhibitors arrest polyglutamine-dependent neurodegeneration in *Drosophila*. *Nature* **413**, 739–743.

Stocker, H. and Hafen, E. (2000). Genetic control of cell size. *Curr. Opin. Genet. Dev.* **10**, 529–535.

Stockwell, B. R. (2000). Chemical genetics: ligand-based discovery of gene function. *Nat. Rev Genet.* **1**, 116–125.

Stowers, R. S. and Schwarz, T. L. (1999). A genetic method for generating *Drosophila* eyes composed exclusively of mitotic clones of a single genotype. *Genetics* **152**, 1631–1639.

Struhl, G., Fitzgerald, K. and Greenwald, I. (1993). Intrinsic activity of the Lin-12 and Notch intracellular domains *in vivo*. *Cell* **74**, 331–345.

Sullivan, W., Ashburner, M. and Hawley, R. S. (2000). *Drosophila Protocols*. Cold Spring Harbor, New York: Cold Spring Harbor Laboratory Press.

Takagi, S., Kimura, M. and Katsuki, M. (1992). A rapid and efficient protocol of the inverted PCR using two primer pairs. *Biotechniques* **13**, 176–178.

Thale, Z., Kinder, F. R., Bair, K. W., Bontempo, J., Czuchta, A. M., Versace, R. W., Phillips, P. E., *et al.* (2001). Bengamides revisited: new structures and antitumor studies. *J. Org. Chem.* **66**, 1733–1741.

Theodosiou, N. A. and Xu, T. (1998). Use of FLP/FRT system to study *Drosophila* development. *Methods* **14**, 355–365.

Thibault, S. T. (2002). The piggyBac transposon complements P as a tool for large scale forward mutagenesis. *Dros. Res. Conf. Proc.* **43**, 969C.

Uckun, F. M. (2001). Rationally designed anti-mitotic agents with pro-apoptotic activity. *Curr. Pharm. Res.* **7**, 1627–1639.

Vaugeois, J. M., Corera, A. T., Deslandes, A. and Costentin, J. (1999). Although chemically related to amineptine, the antidepressant tianeptine is not a dopamine uptake inhibitor. *Pharmacol. Biochem. Behav.* **63**, 285–290.

Venter, J. C., Adams, M. D., Myers, E. W., Li, P. W., Mural, R. J., Sutton, G. G., Smith, H. O., *et al.* (2001). The sequence of the human genome. *Science* **291**, 1304–1351.

Vogel, E. W. and Nivard, M. J. (2000). Parallel monitoring of mitotic recombination, clastogenicity and teratogenic effects in eye tissue of *Drosophila*. *Mutat. Res.* **455**, 141–153.

Vogel, E. W., Graf, U., Frei, H. J. and Nivard, M. M. (1999). The results of assays in *Drosophila* as indicators of exposure to carcinogens. *IARC Sci. Publ.* **146**, 427–478.

Worby, C. A., Simonson-Leff, N. and Dixon, J. E. (2001). RNA interference of gene expression (RNAi) in cultured *Drosophila* cells. *Sci. STKE* **95**, L1.

Xu, T. and Harrison, S. D. (1994). Mosaic analysis using FLP recombinase. *Methods Cell. Biol.* **44**, 655–681.

Xu, T. and Rubin, G. M. (1993). Analysis of genetic mosaics in developing and adult *Drosophila* tissues. *Development* **117**, 1223–1237.

Xu, T., Wang, W., Zhang, S., Stewart, R. A. and Yu, W. (1995). Identifying tumor suppressors in genetic mosaics: the *Drosophila lats* gene encodes a putative protein kinase. *Development* **121**, 1053–1063.

Zhang, J. H., Chung, T. D. and Oldenburg, K. R. (1999). A simple statistical parameter for use in evaluation and validation of high throughput screening assays. *J. Biomol. Screen.* **4**, 67–73.

Zhu, D., Ganji, S. H., Kamanna, V. S. and Kashyap, M. L. (2002). Effect of gemfibrozil on apolipoprotein B secretion and diacylglycerol acyltransferase activity in human hepatoblastoma (HepG2) cells. *Atherosclerosis* **164**, 221–228.

5

Drosophila – a Model System for Targets and Lead Identification in Cancer and Metabolic Disorders

Corina Schütt, Barbara Froesch and **Ernst Hafen**

Genetic and genomic research in model organisms has tremendously accelerated our understanding of the basic biological processes that lie at the heart of human disease. The genetic dissection of disease-relevant signaling pathways in *Drosophila* offers an ideal tool to identify novel drug targets for human diseases. In this chapter we outline the conservation of signaling pathways and the different tools available in *Drosophila* that are suitable for target identification and target validation, as well as for direct screening for biologically active low-molecular-weight compounds.

5.1 Evolutionary conservation of disease-related pathways in *Drosophila*

Analysis of the genetic basis of the development of invertebrate model organisms such as *Drosophila melanogaster* and *Caenorhabditis elegans* has confronted us with a stunning degree of conservation of basic developmental processes in invertebrates and disease-related processes in humans. Of more than 1000 genes associated with human diseases, 77% are conserved in

Model Organisms in Drug Discovery. Edited by Pamela M. Carroll and Kevin Fitzgerald.
© 2003 John Wiley & Sons, Ltd. ISBN 0 470 84893 6

Drosophila (Reiter *et al.*, 2001; see also http://homophila.sdsc.edu/). The evolutionary conservation is not limited to individual genes but often reflects the functional conservation of entire gene networks. The most striking conservation is observed in two different classes of gene networks. The first encodes interacting transcription factors that regulate fundamental biological processes. The *Pax-6/eyeless – sine oculis – eyes absent* network involved in the specification of eye tissue in vertebrates and invertebrates or the *Hox* genes involved in the specification of the body axis are two prominent examples (Maconochie *et al.*, 1996; Hanson, 2001; Gehring, 2002). The second class of gene networks encodes components of intracellular signaling pathways. In Figure 5.1A and 5.1B, two sets of signaling pathways are depicted. Signaling components shown in rectangular boxes represent components that have been genetically identified in *Drosophila*. Filled black boxes represent components that have been identified as oncogenes or tumor suppressor genes in humans.

The misregulation of signaling pathways is at the center of numerous human diseases. The directed treatment of these diseases requires precise knowledge of the components as well as their hierarchical interactions in a given pathway. Furthermore, effective treatment requires the identification of a key component (drug target) in this signaling pathway, whose inhibition by a low-molecular-weight drug will block the signaling pathway and therefore attenuate the disease condition. The evolutionary conservation of entire signaling pathways makes the genetic dissection of such pathways in *Drosophila* and other model organisms a valuble tool to identify such key components independent of their molecular nature. If a mutation in a given gene attenuates or blocks signaling through a given pathway, then blocking the function of the corresponding gene product with a low-molecular-weight compound should also block this signaling pathway. In this way, novel drug targets are identified genetically. Given the ease with which the *Drosophila* genome can be saturated for mutations with a specific phenotype, it is possible to identify most, if not all, of the functionally relevant components of a signaling pathway. This is further aided by the smaller genome size of model organisms, resulting in a lower degree of functional redundancy. This approach will be exemplified by three signaling pathways that are at the center of many cancers in humans: the WNT pathway, the Ras pathway and the insulin signaling pathway.

The WNT pathway

Members of the Wnt/Wingless (Wg) (WNT) protein family play key roles as signaling proteins in many organisms. Close to 100 *Wnt* genes have been isolated so far and they all seem to encode secreted molecules that determine the fate or growth of adjacent cells. Although biological processes such as

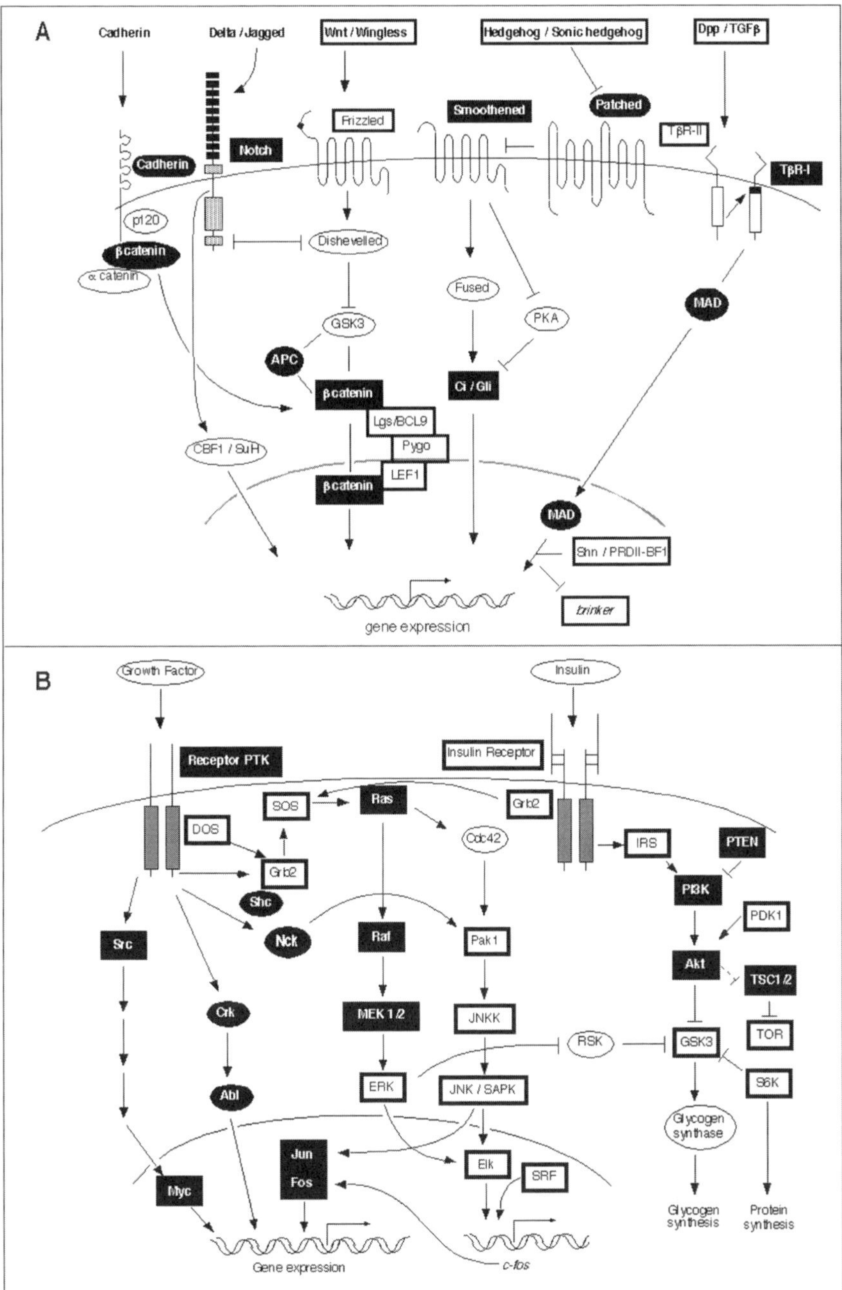

Figure 5.1 Schematic representations of conserved signaling pathways involved in human diseases. Components in rectangular boxes have been identified genetically in *Drosophila* first or the assignment to a specific pathway has been aided by genetic analysis in *Drosophila*. Filled black shapes represent components that have been identified as oncogenes or tumor suppressor genes. (See text for detail and references)

carcinogenesis in humans and embryogenesis in *Drosophila* seem to be very distinct processes, they both rely on cell communication via the WNT pathway (Peifer and Polakis, 2000). The first *Wnt* gene, mouse *Wnt-1*, was discovered nearly 20 years ago as a proto-oncogene. Furthermore, identification of the tumor suppressor APC (adenomatous polyposis coli) has linked colon cancer to WNT signaling. Tumor suppressor APC is a negative regulator of the WNT pathway and is mutated in most colorectal tumors. It is thought that inactivation of both APC alleles is one of the first steps occurring in tumorigenesis (Polakis, 2000).

Around 5% of the Western population develop colorectal malignancies during their lifetime. This not only leads to high medical costs but also to premature death. In over 85% of all human colon cancers, but also in some other cancers, the WNT pathway is aberrantly active and, as a result, the cells receive a continuous signal to proliferate (Morin, 1999; Bienz and Clevers, 2000; Polakis, 2000). Given the high frequency and severity of colon cancer and the fact that no good drug targets with enzymatic function have been identified so far, research in this field has high priority. Because the WNT pathway is highly conserved in flies and mammals, *Drosophila* can serve as an excellent model system.

Our current view of the Wg signal transduction pathway is largely based on genetic dissection of the pathway in *Drosophila* (Moon *et al.*, 2002). The secreted Wg protein acts via its receptor Fz (frizzled) and a short cascade of downstream components to stabilize armadillo (Arm), the *Drosophila* homolog of β-catenin, which together with pangolin/TCF (Pan/TCF) activates transcription of Wg-responsive genes (Figure 5.1). In the absence of a Wg signal, free cytoplasmic Arm/β-catenin is destabilized by the negatively acting multiprotein complex containing APC, axin and glycogen synthase kinase 3β (GSK3) (Cohen and Frame, 2001).

In the past, several genetic approaches to identify components of the Wg pathway have been taken in *Drosophila* by a number of groups, including that of K. Basler, a co-founder of The Genetics Company, Inc. We will describe two of these. First, screens for recessive lethal mutations identified essential components (i.e. *arm*) that caused a segment polarity phenotype similar to the loss of Wg function (Nüsslein and Wieschaus, 1980). Second, screens for suppressors of ectopic Wg signaling identified rate-limiting components in this pathway (Brunner *et al.*, 1997). In cancer cells the Wnt pathway is constitutively active, due to either the loss of the tumor suppressor gene *APC* or to activating mutations in *β-catenin* (Polakis, 2000). Drugs that interfere with positively acting components of the WNT pathway at the end of the cascade are thus attractive because such a block would disrupt the WNT transduction pathway irrespective of the nature of the original defects leading to WNT activation in various types of cancer cells. Ectopic WNT activation, as it occurs in cancer cells, was mimicked in *Drosophila* by overactivating the

Wg pathway in the developing eye (see Figure 5.2C). Expression of *wg* in a subpopulation of eye precursor cells (sev-wg) disrupts the regular arrangement of eye facets (ommatidia) resulting in a *rough* eye phenotype (Brunner *et al.*, 1997). In a screen for dominant suppressors of the rough eye phenotype, mutations in *armadillo* encoding the homolog of *β*-catenin and *pangolin/TCF* were identified (Brunner *et al.*, 1997). This provided the first functional evidence that pan/TCF is an essential transcription factor at the end of the WNT pathway. In addition, this screen produced mutations in two novel genes, *legless* (*lgs*), the *Drosophila* homolog of BCL9 (*B cell lymphoma 9* gene), and *pygopus* (*pygo*) (Kramps *et al.*, 2002; Parker *et al.*, 2002; Thompson *et al.*, 2002). The *lgs* gene functions as an adaptor protein for physically linking *pygo* to the *β*-catenin/TCF complex (Kramps *et al.*, 2002). Both genes fulfill the first criterion for a drug target in that they function at the same level or downstream of *β*-catenin. In addition, although overall sequence homology between human BCL9 and *Drosophila lgs* is low, a human BCL9 cDNA is able to rescue *lgs* mutant flies (Kramps *et al.*, 2002). The homology is concentrated to a few short amino acid stretches that are, however, arranged in a colinear fashion in BCL9 and *lgs*. Intriguingly, most mutations isolated in *lgs* map to these domains, suggesting that these are the functional domains (Kramps *et al.*, 2002).

The Ras pathway

H-*Ras* is one of the first oncogenes discovered. Since then it has been shown that the proto-oncogenes H-*Ras*, N-*Ras* and K-*Ras* are mutated in 30% of all human cancers (Bos, 1989). The Ras proteins are part of the large family of small GTPases that perform various signaling functions within the cell. Ras is inactive when it is bound to GDP but active when GDP is exchanged for GTP. Almost all the oncogenic mutations in *Ras* lock Ras in the GTP-bound active form (McCormick, 1997). Biochemical experiments in mammalian cell culture systems have shown that active Ras associates with the Raf serine/threonine kinase, which in turn associates with MAP or Erk kinase (MEK), which activates MAP kinase (also called ERK for extracellular signal-regulated kinase). Constitutively active Ras also associates with Phosphoinositol 3-kinase (PI3K) and an exchange factor for the small GTPase Ral (White *et al.*, 1995; Rodriguez-Viciana *et al.*, 1997). The simultaneous activation of these signaling pathways may contribute to its transforming potential (for a recent review, see Boettner and Van Aelst, 2002).

Two factors have complicated the understanding of the normal function of Ras in the cell. First, oncogenic mutations in *Ras* causing the constitutive activation of Ras are gain-of-function (GOF) mutations. It is impossible to deduce the normal function of a protein solely from a GOF phenotype. Second,

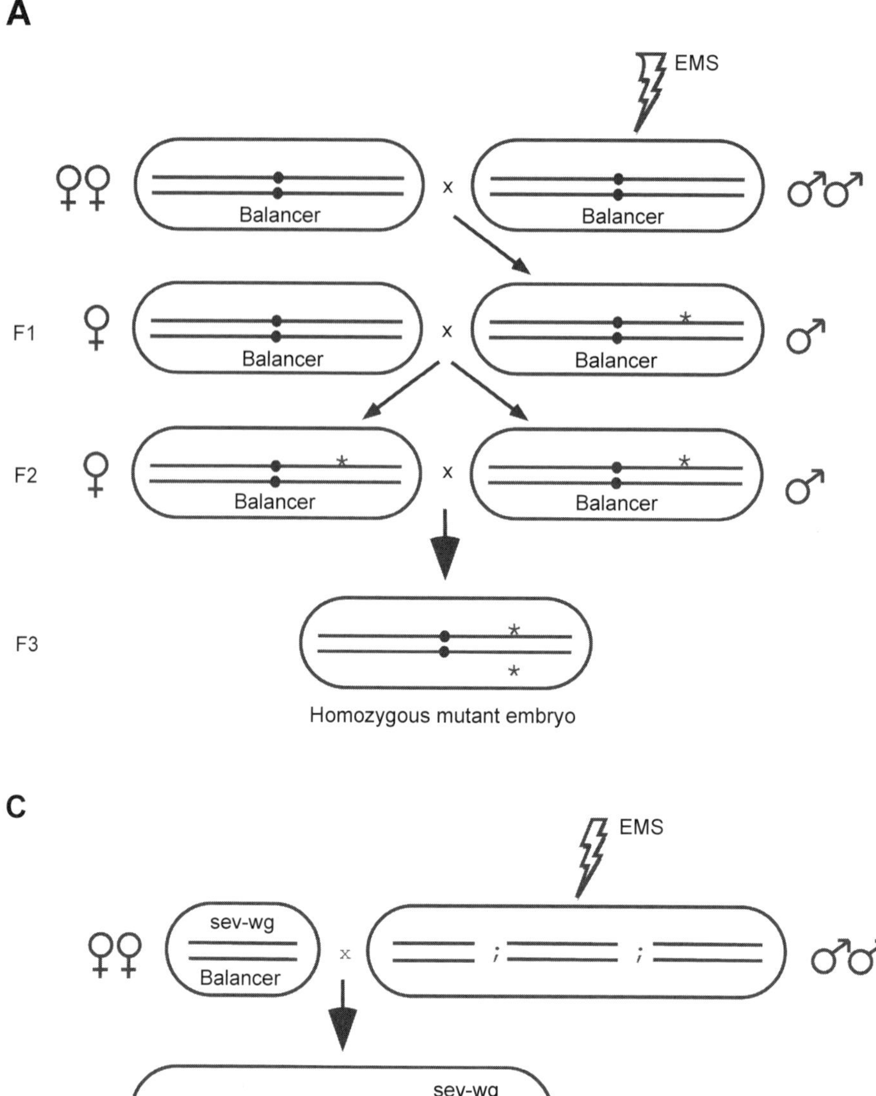

Figure 5.2 Four types of genetic screens. (A) Classical screen for recessive mutations as performed by C. Nüsslein-Volhard and E. Wieschaus. (B) Screen for recessive mutations in tissue-specific mosaic animals. Homologous recombination at the FRT sites produces two types of cell clones: cells homozygous mutant for one chromosome arm and cells homozygous for a cell lethal (cl) that kills the twin clone. (C) Dominant suppressors (or enhancers) induced anywhere in the genome modify a sensitized genetic background, in this

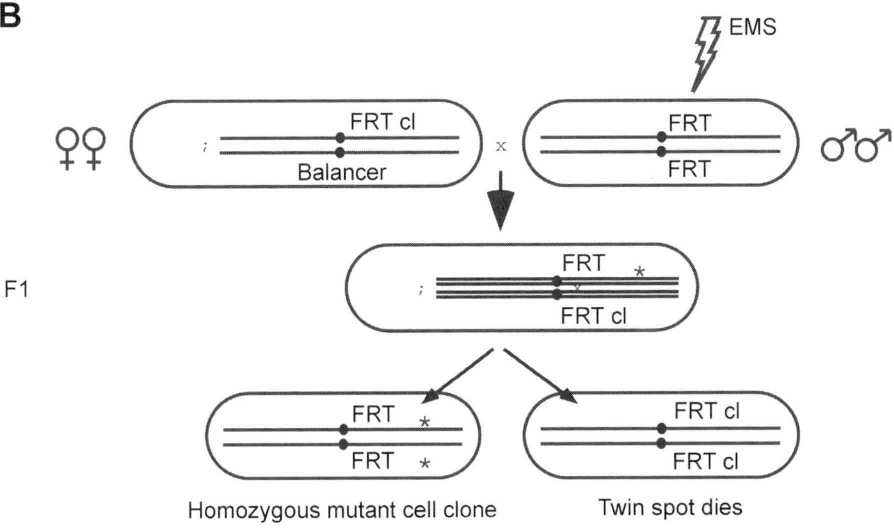

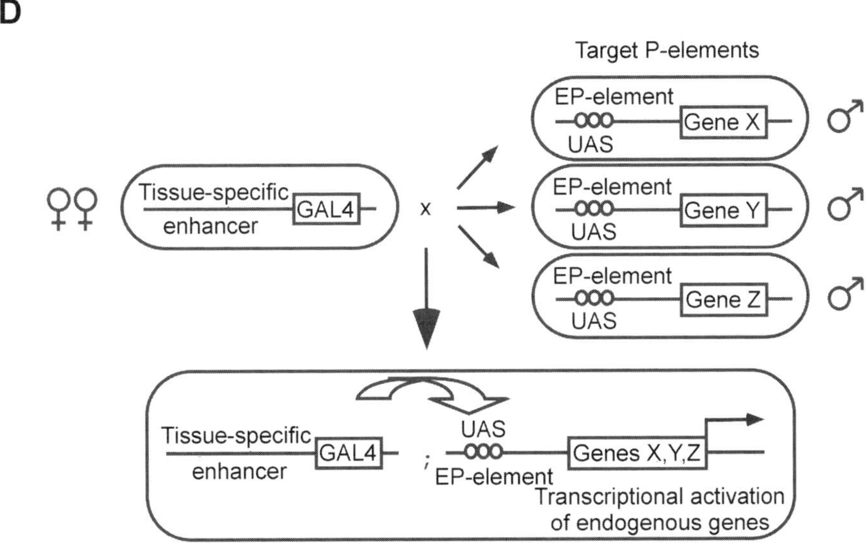

case the rough eye phenotype caused by ectopic activation of the Wnt pathway. (D) An EP overexpression screen. The EP element carries GAL4 binding sites (UAS sites) and a basal promoter that directs expression of any gene that happens to lie next to its insertion site. In combination with the tissue-specific expression of the GAL4 transcription factor that binds to the UAS element, the system is activated and allows the study of over- or misexpression phenotypes of tagged genes in the appropriate tissue. (See text for detail and references) it was unclear how Ras is activated by the ligand-induced stimulation of growth factor receptors. Genetic dissection of signaling pathways acting downstream of receptor tyrosine kinases in *Drosophila* and *C. elegans* have provided the necessary links and shed light on the normal function of Ras in the cells (Sternberg and Han, 1998; Wassarman *et al.*, 1995). In *Drosophila*, the link between receptors and Ras has been found by studying the differentiation of the R7 photoreceptor cell in the developing eye. The R7 photoreceptor cell depends on local activation of the Sevenless receptor tyrosine kinase in the precursor cells (Hafen *et al.*, 1987). To identify downstream factors, Simon *et al.* (1991) used a temperature-sensitive variant of Sevenless that, at an intermediate temperature, was sufficiently active to specify R7 cells. In this background, the reduction of any one of several rate-limiting components by mutating one copy of the corresponding genes (heterozygosity) would drop pathway activity below the threshold required for R7 specification and thus produce no R7 cells. Mutations in the *Drosophila* homolog of *Ras* (*Ras1*) and in the *Son-of-Sevenless* (*Sos*) gene were identified in this way (Simon *et al.*, 1991). The *Sos* gene encodes a guanine–nucleotide exchange protein that promotes the exchange of GDP for GTP on Ras and thus activates Ras. The SH2/SH3 domain adaptor protein Grb2/Sem-5/Drk, which links Sos to the activated receptor via binding of the Grb2 SH2 domain to the phosphorylated tyrosine motives in the receptor has been identified biochemically in mammalian cells and genetically in *C. elegans* and *Drosophila* (Clark *et al.*, 1992; Lowenstein *et al.*, 1992; Olivier *et al.*, 1993; Simon *et al.*, 1993). The discovery of the link between receptor tyrosine kinases and Ras exemplifies the complementary nature of the experimental approaches used. The biochemical analysis identified components physically interacting with the activated receptor. Genetic analysis in model organisms identified components whose functions are the key in coupling receptor activation with Ras activation.

The identification of functionally relevant components in the Ras signaling pathway by genetic screens has been repeated successfully by expressing activated forms of the Sevenless receptor, Ras1 or Raf in the developing eye (Olivier *et al.*, 1993; Therrien *et al.*, 1995, 1998; Dickson *et al.*, 1996; Karim *et al.*, 1996). In each case, pathway activity was placed at a threshold where too many R7 cells are recruited and thus the normally smooth, regular eye appears rough. Eye roughness becomes an easy measure of pathway activity. As in the Sevenless temperature-sensitive screen, rate-limiting effectors are identified as dominant suppressors (one functional gene copy is not enough) of the rough eye phenotype.

Like most signaling pathways, the Ras pathway controls a variety of different cellular responses. The question of whether the different cellular responses are triggered by different downstream effectors, as suggested by the studies with oncogenic Ras, has been addressed in *Drosophila*. Analysis of complete or partial loss of Ras1 function in the developing eye indicated that cell growth, cell survival and cell differentiation depend on different activity levels of Ras1 and are all mediated by the Raf MAP kinase pathway (Halfar *et al.*, 2001). Conversely, constitutively active forms of Ras1 activate MAP kinase and PI3K, as shown for oncogenic Ras. Activated Ras1 variants that activate either the MAP kinase or PI3K pathway are sufficient to promote growth but only activation of the MAP kinase pathway leads to stabilization of dMyc protein (Prober and Edgar, 2002). Therefore, it appears that the normal growth promoting function of Ras1 acts via dMyc. Constitutively active forms of Ras, however, can also promote growth via the PI3K growth pathway (see below) (Prober and Edgar, 2002).

What novel drug targets have come out of genetic dissection of the Ras pathway in *Drosophila*? In this case, the suitable targets should perform an essential function downstream of oncogenically activated Ras. Apart from confirming the essential role of the Raf-MEK-MAP kinase cascade, two novel components, KSR (kinase suppressor of *ras*) and CNK (connector enhancer of *ksr*), have come out of these screens (Therrien *et al.*, 1995, 1998). Each of these proteins performs an important function in mediating the Ras signal. Protein KSR contains an S/T kinase domain, although it is still unclear what the function of the kinase domain is in Ras signaling (Morrison, 2001). Multidomain protein CNK is involved in the subcellular localization of signaling components (Therrien *et al.*, 1998). Given the unclear role of the catalytic activity of KSR in Ras signaling and the absence of a similar catalytic function of CNK, it is not obvious how to develop low-molecular-weight inhibitors to block the function of these proteins. Like many other components in the pathway, they function as adaptors, scaffold or linkers in connection proteins. Devising a means to disrupt these essential protein–protein interactions is not straightforward but may provide access to more specific inhibitors than those blocking kinase activity.

The insulin signaling pathway

Body size and growth of animal cells depend on extracellular growth factors and hormones that activate intracellular signaling pathways and finally stimulate protein synthesis and other biosynthetic processes. Growth hormone (GH), for instance, plays a major role in stimulating postnatal mammalian growth. Children deficient in GH become dwarfs, whereas excessive GH production leads to overgrowth. Growth hormone stimulates growth largely by inducing the production of insulin-like growth factor 1 (IGF-1) (Butler and Roith, 2001). The insulin/IGF signaling pathways play a key role in the control of growth in both vertebrates and invertebrates. Insulin and IGF signal via their receptors, IR and IGFR, respectively, and the insulin receptor substrates (IRS). Proteins IRS1–4 are multifunctional adaptor proteins that link insulin and IGF signals to the Ras/MAP kinase pathway (proliferation) as well as to the PI3K pathway (metabolism, growth and survival) (Yenush and White, 1997).

In mammals, the primary role of insulin and its receptor IR is energy homeostasis via regulation of the level of glucose in the blood. However, mutations in the human IR gene also cause embryonic growth retardation (Saltiel and Kahn, 2001). The primary growth regulatory function is mediated by the IGF pathway (Nakae *et al.*, 2001). In mice, loss of *IRS1* function causes severe reduction in embryonic and postembryonic growth (Liu *et al.*, 1993). Loss of *IRS2* leads to hyperglycemia, increased body fat and female sterility (Burks *et al.*, 2000).

In *Drosophila* there is a single insulin-like receptor (*InR*) and a single IRS (*chico*) that control size, lipid metabolism and female fertility during development (Chen *et al.*, 1996; Böhni *et al.*, 1999; Brogiolo *et al.*, 2001). Similar to loss of *IGF1* or *IRS1* in mice, mutations in positive regulators of the *Drosophila* insulin pathway cause dramatic reduction in size. Surprisingly, flies mutant for *chico* show an alteration in energy stores. Although there is no significant difference in the levels of proteins and glycogen, lipid levels are increased nearly twofold (Böhni *et al.*, 1999). This is reminiscent of hypertriglyceridemia in *Irs*-deficient mice and increased levels of lipids in the blood of humans with diabetes (Burks *et al.*, 2000). Thus, the more ancestral insulin signaling pathway in *Drosophila* controls the physiological and growth processes of the mammalian insulin and IGF systems (Oldham and Hafen, 2003).

Several lines of evidence suggest a link between the activity of the insulin/IGF pathway and nutrient availability. The parallel between *chico*-mutant flies and flies reared under poor nutritional conditions is striking. Both the genetic defect and the environmental situation lead to developmental delay, smaller body size due to a reduction in cell size and cell number and female sterility. Similar phenotypes on growth and female fertility are observed in *Irs1*- and *Irs2*-mutant and starved mice (Thissen *et al.*, 1999). Finally, women who are underweight or suffer from diabetes show reduced fertility (Poretsky *et al.*, 1999). One obvious hypothesis for this connection is that nutrients control the expression of insulin/IGF (Ikeya *et al.*, 2002). It may well be that, initially, the insulin/IGF pathway evolved as a system to coordinate growth and reproduction with nutrient availability (Oldham and Hafen, 2003).

In addition to these functions in growth and metabolism, hyperactivation of the insulin/IGF signaling pathways is associated with a wide variety of cancers. Both *PI3K* and *Akt*, for instance, have been isolated as retroviral oncogenes (Brazil and Hemmings, 2001). Pentaerythritol tetranitrate (*PTEN*) is found to be lost in many tumors. Apart from p53, no other tumor suppressor has received as much attention as *PTEN* (Cantley and Neel, 1999).

The striking structural and functional conservation of the insulin/IGF signaling pathways during evolution establishes *Drosophila* as a valid model organism for the study of metabolic diseases such as diabetes and obesity, as well as for the study of growth disorders such as cancer (see next section).

5.2   Target identification/target validation strategies

The identification of targets is a bottleneck in drug development. The development of a drug until it is marketable costs $800 million on average (Tufts Center for the Study of Drug Development, 2001), and 80% of these costs are caused by the failure of most chemicals in any of the preclinical or clinical studies. Therefore, it is of tremendous interest to identify the promising targets and to get rid of the others at an early stage.

Taking *Drosophila* as a tool, disease-related pathways can be manipulated. Mutations that specifically interfere with a signal transduction pathway associated with a human disease point to the gene whose product performs an essential function and whose functional inactivation by a low-molecular-weight drug will attenuate the signaling pathway in much the same way as the mutation in the corresponding gene does.

The *Drosophila* system combines the advantages of cell cultures and *in vivo* experiments with higher animals. Short generation time, low breeding costs and a relatively small genome allow high- or at least medium-throughput screens. Such screens are usually performed to saturation levels (e.g. several alleles of the same genes are identified). In this way, all or most of the genes in the genome can be identified. Furthermore, because such screens are performed in living animals, any deleterious or lethal side-effects that may be tolerated by single cells in culture are not recovered. This is the first important step for target validation.

Forward genetic approaches

One of the most important tools that *Drosophila* provides is the ability to carry out large-scale genetic screens for mutations that affect a given process (St Johnston, 2002). It is possible to analyze the whole genome by saturation screens within a few months. This is unique within multicellular organisms. The more specific the screening strategy set-up and the more closely it reflects the misregulation of the signaling pathway in the disease condition, the more valuable the targets will be.

Choice of mutagen

Ethylmethanesulfonate (EMS), X-ray and P or EP elements are most frequently used for mutagenesis in *Drosophila*. Depending on the desired nature of the

Table 5.1 Comparison of different mutagens used in *Drosophila*

	EMS	X-ray	P-element	EP element
Nature of mutations	Point mutations Small deletions	Deletions Chromosome breaks Inversions Translocations	Loss-of-function mutations	Gain-of-function mutations
Advantages	Saturation screens	Fast Big deletions cytologically visible	Molecular anchor	Molecular anchor
Disadvantages	No molecular anchor	Several genes deleted	Hot spots for insertion	Hot spots for insertion
Special features	Most frequently used	Mainly null alleles	Reporter assays	Only method of efficiently inducing ectopic activation

mutations, the appropriate mutagen is chosen (see Table 5.1). Transposable elements such as P and EP insert into the genome almost randomly (Spradling *et al.*, 1999). At the insertion site, they have any of the following effects. When they integrate into coding regions, they destroy the corresponding gene, leading to a loss-of-function (LOF) mutation. This case, however, is rare because transposable elements have a preference to insert into non-coding sequences in the 5′ region of the gene. When P elements with a reporter gene driven by a minimal promoter come under the influence of an endogenous enhancer, the reporter will be expressed in the same way as the corresponding gene (enhancer trap) (O'Kane and Gehring, 1987). In this way, genes are identified based on their expression pattern rather than their LOF phenotype. Finally, EP elements containing an enhancer/promoter that is activated by the yeast transcription factor Gal4 are used to activate ectopically the nearby genes to produce a GOF mutation (Rorth *et al.*, 1998) (see later section on EP overexpression screens). Obviously, transposable elements have the advantage that they serve as a molecular tag to isolate flanking sequences. This makes the gene identification process very rapid. However, they are not suitable as a mutagen for saturation screens because the frequency with which the genes inactivate is low due to their preferential jumping into non-coding 5′ regions. Furthermore, not all the genes are targeted with P elements at the same frequency. There are hotspots and coldspots in the genome (Berg and Spradling, 1991).

The second widely used mutagen is EMS. Chemical mutagens induce mutations randomly. The frequency of mutations depends on the concentration of the mutagen and can reach one lethal hit per chromosome arm at high concentration (Lewis and Bacher, 1968). Using EMS, genome-wide saturation is reached readily. The degree of genome saturation is measured by the number of mutations (alleles) identified at different loci. For example, in a screen for suppressors of the rough eye phenotype caused by expression of an activated form of the Raf kinase, we identified, in a total of 300 000 flies, 45 individual mutations in *rolled*, encoding the *Drosophila* homolog of MAP kinase (Dickson *et al.*, 1996). For functional studies of a particular gene it is important to have several alleles with different effects on the final protein. In addition, when the identified mutations cluster in certain parts of the coding regions, this points to functionally important protein domains (see earlier section on WNT pathway, *lgs* alleles; Kramps *et al.*, 2002). The fact that it is not trivial to detect the mutations induced by EMS molecularly has been a disadvantage for a long time. However, with the availability of the entire genome sequence, new precise mapping strategies have become available (see later section on gene mapping strategies). For target identification, chemical mutagenesis using EMS is therefore the method of choice.

In the following sections we will discuss the art of designing screens: dominant or recessive screens, screens for LOF or GOF mutations, screens for null-alleles or hypomorphs, tissue-specific screens and modifier screens.

Screens for recessive mutations

Christiane Nüsslein-Volhard and Eric Wieschaus pioneered the recessive screens. Their work was revolutionary because it was the first mutagenesis in any multicellular organism that attempted to find most or all the genes that affect a given process (saturation screen). They identified most of the essential patterning genes that are required throughout embryonic development (Nüsslein and Wieschaus, 1980). Their groundbreaking work was honored with the Nobel Prize in 1995. For several reasons, however, screens for recessive mutations are limited to certain aspects of development and to special classes of genes:

1. Mutations in essential genes are homozygous lethal. The phenotypic classification is restricted to phenotypes that are visible during embryogenesis or larval development.

2. Only the first essential function of a particular gene can be identified. However, many genes are used several times during development. The *wg*

gene, for example, is required for early patterning of the embryo and was found in the famous Nüsslein–Volhard screen. Its later function in imaginal disc growth and patterning could not be detected by examining homozygous mutant animals.

3. Genes whose proteins possess overlapping (redundant) functions cannot be detected.

4. The screens for recessive mutations are laborious because it requires the establishment of independent lines in which the homozygous phenotype is detected only in the third generation (see Figure 5.2A).

Screens for recessive mutations in tissue-specific mosaic animals

To circumvent some of these problems, a method to identify genes in tissue-specific genetic mosaics was developed (Newsome *et al.*, 2000): the aim of the method is to generate flies that are homozygous for mutations on a chromosome arm in a tissue such as the adult eye. Because each fly receives a different set of mutagenized chromosomes from the mutagen-treated father, each individual fly will manifest the phenotype of a recessive mutation in this particular tissue. In the remaining tissues and in the germ-line, the fly is heterozygous for the same mutation. In contrast to the classical recessive screens over three generations, tissue-specific screens reveal the results in the first generation (see Figure 5.2B). Individual flies exhibiting the desired phenotype can be selected and established in a line. Furthermore, in many cases the mutant tissue survives to adulthood while homozygosity of the same mutation in the entire animal is lethal.

The principle of the method is depicted in Figure 5.2B. Homozygous mutant clones are produced by the Flp/FRT system (Xu and Rubin, 1993): Flp recombinase from yeast mediates site-specific recombination between its target sites, called FRT (Flp recombinase target). When these FRT sequences are integrated at identical positions on the homologous chromosomes, the Flp recombinase will mediate sister chromatid exchange and thereby induce site-specific mitotic recombination. When the Flp recombinase is induced in a cell heterozygous for a newly induced mutation on an FRT-bearing chromosome, mitotic recombination will generate two unequal daughter cells, one homozygous for the mutation and the other homozygous for the non-mutagenized chromosome. In subsequent divisions, each of these cells will develop into a clone of cells. To eliminate the clone of cells carrying the non-mutagenized chromosome, this chromosome is made to carry a recessive cell lethal mutation. Homozygosity for this mutation after mitotic recombination will eliminate this cell. Expression of Flp recombinase under the control of a tissue-specific promoter expressed in the early progenitor cells permits mitotic recombination to be restricted to a single tissue (see Figure 5.2B).

We will demonstrate the principle and efficiency of this type of screen with the example of the so-called 'pinhead screen': a screen to search for genes involved in cell growth and cell proliferation (Oldham *et al.*, 2000). In the pinhead screen, Flp recombinase was induced by an eye-specific enhancer (ey-Flp). This limits clone induction to the head capsule and prevents deleterious effects of the mutations in other tissues. Such flies were analyzed for mutations that affect cell growth and cell size. So-called pin- or bigheads were recovered when growth-promoting or growth-inhibiting genes were hit, respectively (see Figure 5.3D and 5.3E). In this screen, already known and novel components of the insulin pathway (*chico*), oncogenes (*PI3K, Akt, Tor*) and the tumor suppressors PTEN, TSC1 and TSC2 were identified (Oldham *et al.*, 2000; H. Stocker, S. Breuer and E. Hafen, unpublished results). In addition, we identified some 20 novel loci that either promote or inhibit growth. The corresponding genes are in the process of being characterized. Given their central role in the control of cellular growth, the novel growth-promoting genes are promising targets for anticancer therapy. The growth-inhibiting genes are potential tumor suppressor genes in humans and may serve as diagnostic markers.

The close link between cell growth and basic metabolism manifested by components of the insulin signaling pathways in mammals and *Drosophila* may also offer the opportunity to validate the products of the genes identified in this screen as target of metabolic disorders such as type 2 diabetes.

Dominant modifier screens

Most LOF mutations are recessive, which means that 50% of the wild-type protein is sufficient for normal function. When a particular process is already partially disrupted by another mutation, however, the amount of components in the same pathway may become rate-limiting. A sensitized genetic background therefore can be used to screen for dominant enhancers or suppressors of a particular mutation. The advantage of such a screen is its simplicity, because only one of the two alleles has to be mutant (F1 screen). Even more importantly, such dosage-sensitive genetic interactions are usually indicative of a specific association of the newly identified components with the sensitized signaling pathway. Examples of such screens have been discussed above in the context of the WNT and Ras pathways. For illustration of a dominant modifier screen, see Figures 5.2C and 5.3A–C.

EP overexpression screens

Traditionally, genes are characterized based on LOF phenotypes. However, it is estimated that two-thirds of all *Drosophila* genes have no obvious LOF phenotype (Miklos and Rubin, 1996). This is at least in part due to functional

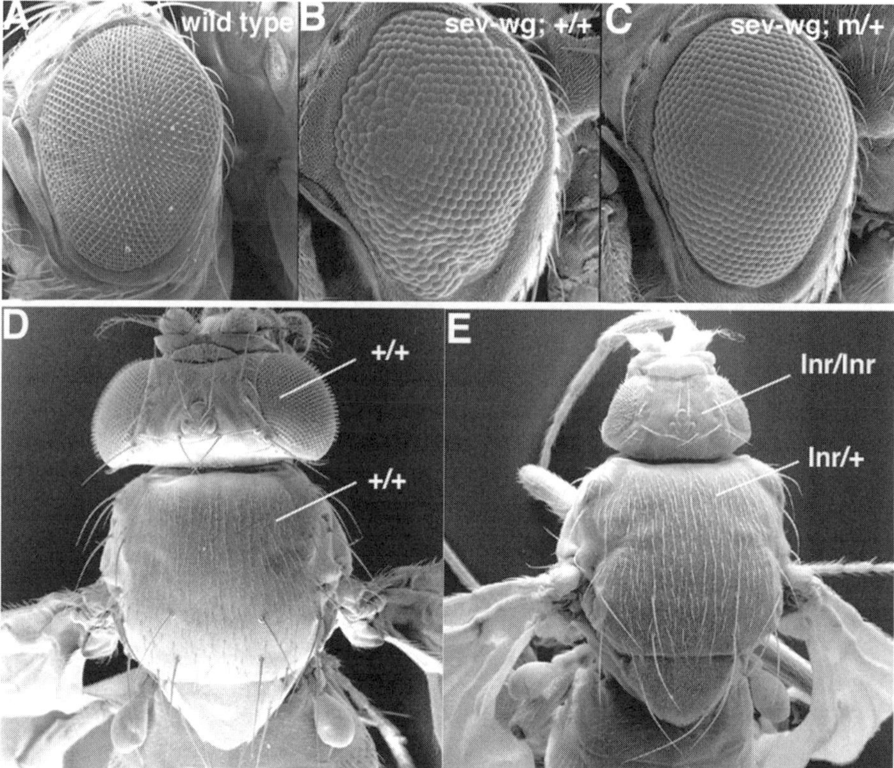

Figure 5.3 Genetic identification of components of disease-relevant signaling pathways in *Drosophila*. (A–C) Dominant suppression of the rough eye phenotype caused by the constitutive activation of the WNT pathway during eye development. (A) A wild-type eye. (B) An eye of a *sev-wg* transgenic fly. Unnatural expression of Wg protein in a subpopulation of eye progenitor cells results in activation of the WNT signaling pathway and thereby disrupting the regular arrangement of ommatidial units. The degree of pattern disruption (eye roughness) is critically dependent on WNT pathway activity. (C) Flies heterozygous of a mutation in any one of several genes coding for rate-limiting WNT signaling components (i.e. *β*-catenin, BCL9/*lgs*, *pygo*) make only 50% of the normal amount of this protein (only one functional gene copy). In this sensitized background, a 50% reduction is sufficient to suppress the eye roughness. In this way, novel components in signaling pathways can be identified genetically. Both B and C are shown at a larger magnification than A to highlight the irregular arrangement of the facets. (D, E) Example of a pinhead fly that is homozygous mutant for a growth-promoting gene (insulin receptor) in the head tissue. (D) Dorsal view of a wild-type fly. (E) Dorsal view of a genetically mosaic fly. Using the *ey-Flp cell lethal* technique (Newsome *et al.*, 2000), flies are rendered homozygous for randomly induced mutations in the head tissue. In the body and the germline they are heterozygous for the same mutation. Mutations in growth-promoting genes produce flies with small heads (pinheads) and mutations in growth-inhibiting genes produce flies with big heads. Complete loss of insulin receptor function (shown in E) permits eye cells to differentiate normally but they grow at a greatly reduced rate

redundancy. For these genes, over- or misexpression studies can provide unique functional information.

A modular misexpression system has been developed to carry out systematic GOF screens in *Drosophila*, called EP screens (Rorth *et al.*, 1998). The system is designed to allow conditional expression of genes upon insertion of a modified transposable element (EP element) (see Figure 5.2D). The EP element carries GAL4 binding sites (UAS sites) and a basal promoter that directs expression of any gene that happens to lie next to its insertion site. In combination with the tissue-specific expression of the GAL4 transcription factor that binds to the UAS element, the system is activated and allows the study of over- or misexpression phenotypes of tagged genes in the appropriate tissue. The modular EP system can be used either for a simple GOF screen as described above or it can be combined with a modifier screen to search for GOF suppressor mutations in a sensitized genetic background. In the latter screen, the association of gene products with a disease-relevant signaling pathway can be detected by virtue of their overexpression phenotype, even if overexpression of the same gene in a wild-type background does not produce a detectable phenotype. For example, overexpression of phosphatidyl-inositol-dependent protein kinase 1 (PDK-1) in the developing eye has no detectable effect. However, EP insertions in the PDK1 locus were identified as suppressors of the rough eye phenotype caused by overactivation of the Ras pathway in the eye (Rintelen *et al.*, 2001). These results suggest that overexpression of PDK1 antagonizes the Ras pathway. Indeed, similar cross-talk between the Ras and the PI3K pathway have been identified in mammalian cells (Rommel *et al.*, 1999; Zimmermann and Moelling, 1999).

Genes identified as suppressors of the phenotype by overactivation of a signaling pathway (e.g. Ras, WNT) in an EP screen encode potential negative regulators of this pathway and may thus correspond to tumor suppressor genes in humans. Conversely, genes identified in conventional suppressor screens in which a mutation in heterozygous condition suppresses the phenotype encode positive regulators of the pathway. These types of screens are therefore complementary and lead to the identification of positive as well as negative regulators of a signaling pathway. Positive regulators may serve as drug targets, whereas negative regulators may provide diagnostic markers for the classification of particular disease conditions.

Recessive modifier screens

Dominant modifier screens or EP screens cannot identify all the essential components in a signaling pathway. For some proteins that perform an essential function in a given pathway a 50% reduction in its amount may not be sufficient, even in a sensitized background. By screening large numbers of mutagenized chromosomes to reach multiple saturation of the genome it is possible to identify rare antimorphic (dominant-negative) mutations, which results in a more than 50% reduction of functional gene product. In a screen for dominant modifiers of an activated Raf kinase, Dickson *et al.* (1996) found five alleles of *hsp 83*, the *Drosophila* homolog of *hsp 90*. All these mutations are antimorphs (dominant-negative function) and show that the HSP90 protein plays an important role in modulating Raf activity. However, the chance of identifying such antimorphic mutations is rare and unpredictable.

A more reliable method for identifying genes whose products perform essential but not rate-limiting functions in a disease pathway is to screen for recessive suppressors. This is achieved by combining the tissue-specific recombination system (ey-FLP) with a genetically sensitized system. The phenotype of homozygous mutant eye tissue is not analyzed in a wild-type background but in a background of a hyperactivated signaling pathway that causes a rough eye phenotype. We will take the WNT pathway as example.

In many human cancers, the WNT pathway is constitutively active and, as a result, cells receive a continuous signal to proliferate. In a *Drosophila* model, ectopic activation of *wg* (encoding the *Drosophila* homolog of mammalian Wnt proteins) in the compound eye leads to uncoordinated cell growth and cell death, resulting in readily detectable small, rough eyes, which resembles the behavior of cancer cells. We performed a screen for recessive mutations that modifies the rough eye (C. S. and K. Basler, unpublished results). The screen is based on the *ey-Flp/FRT* technique, which induces homozygous mutant clones in the head (see earlier section on screens for recessive mutations). Blocking a critical downstream component of ectopic Wg transmission will suppress the dominant eye phenotype caused by *sev-wg*. The power of this screen is its stringency for Wg interacting genes and the possibility of identifying partially redundant genes whose products only become limiting in cells in which the WNT pathway is hyperactivated. Importantly, gene products identified in this way must not be essential for normal WNT signaling during development. Because if they were, the cells lacking this component owing to a homozygous mutation in the corresponding gene will not develop and contribute to the eye structure. In other words, this type of screen will only identify genes whose products are essential for abnormal WNT signaling but not for normal WNT signaling. From a purely functional point of view, these are the ideal drug targets. Inhibiting their function with a drug may block overactive WNT signaling in the cancer cell but will not interfere with normal WNT activity in other cells.

Whether such genes exist and whether they encode *drugable* proteins will be apparent when the first candidates from this screen have been characterized molecularly.

Gene mapping strategies

Ethylmethanesulfonate is the most commonly used mutagen in *Drosophila*. Because it primarily induces point mutations, identification of the affected gene is a tedious process. This is the main disadvantage of EMS compared with other mutagens such as the P and EP elements, which serve as a direct molecular tag for adjacent genes (see earlier section on choice of mutagens). Classical strategies to localize point mutations involve mapping with non-complementing chromosomal deficiencies and meiotic mapping relative to visible markers. These methods generally allow mapping of a mutation to a region of a few hundred kilobase pairs (kb), still containing dozens or even hundreds of genes. With the availability of complete genome sequences, however, new rapid and reliable strategies for gene mapping became possible. Single-nucleotide polymorphisms (SNPs) permit the mapping of mutations at a resolution not amenable to classical genetics (Berger *et al.*, 2001). We successfully used high-resolution SNP mapping by denaturing high-performance liquid chromatography (DHPLC) to identify EMS-induced mutations in several unknown genes within a short time (Nairz *et al.*, 2002).

The underlying principle of this technique is shown in Figure 5.4. In a first step, meiotic recombination between the mutation and a nearby visible marker on a standardized tester chromosome is induced. Such recombinant flies, chosen for fine mapping, are rare but are efficiently recovered by an appropriate crossing scheme. They are homozygous (no SNP) for the mutated chromosome on one side of the point of recombination and heterozygous for the mutated and the tester chromosome on the other side. Depending on the origin of the two strains used for mutagenesis and as a marker strain, the frequency of SNPs can vary. To facilitate work, the two strains should not be closely related. In a second step, an SNP map has to be established for the region between the marker and the mutation. The fragments chosen for amplification are derived from intergenic or intronic regions and possess an appropriate size of 800 bp. They should be spaced at intervals of approximately 20 kb. A slightly altered melting behavior of DNA heteroduplexes versus homoduplexes leads to a difference in retention time on ion-pair reversed-phase HPLC columns. Homoduplexes generally elute in one peak, whereas heteroduplexes produce two or more peaks. Between distantly related tester and mutant strains, an SNP is detected in approximately 70% of the fragments tested. Determination of the exact nature of the SNP by DNA sequencing is not required because the altered elution profile is sufficient to distinguish the two chromosomes at this position. Single recombinant flies are finally tested for the break points of recombination. In other words, we test between which SNPs the chromatography profile changes from homo- to heterozygosity. Ideally, recombinants are generated with two markers on either side of the mutation. The two closest recombination events to the

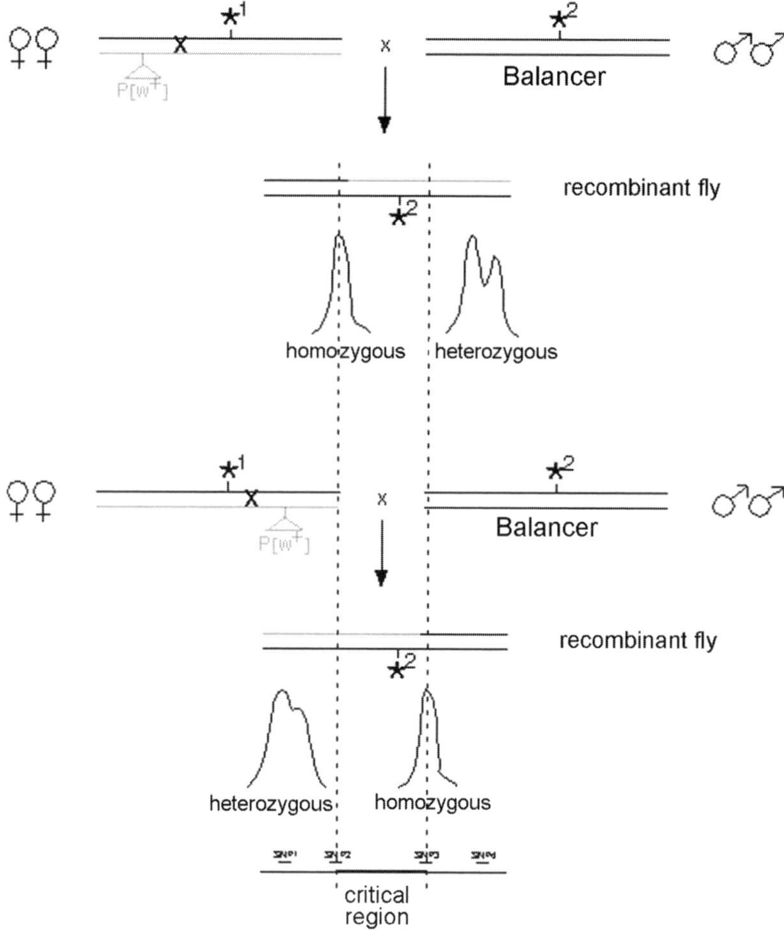

Figure 5.4   Single-nucleotide polymorphism (SNP) mapping of a gene ∗ with two alleles ∗[1] and ∗[2]. The ∗[1]/∗[2] combination is lethal. Two marker P elements on both sides of the mutation are used for recombination. Single recombinant flies are tested for their recombination profile. The two closest SNPs on either side of the mutation, SNP2 and SNP3, respectively, which are shown to be homozygous in at least one recombinant, define the critical region for the mutation

mutation from each side represent the maximal interval in which the mutation is located. In this way, the mutation in question can be mapped to a few tens of kilobase pairs. From the genes annotated in this region, the gene carrying the mutation is identified again by DHPLC. The DNA is extracted from flies heterozygous for different alleles and the original chromosomes used in the mutagenesis. Fragments uncovering the coding regions of the candidate genes are amplified and examined for an altered elution profile. Such a profile is indicative of a sequence difference between the mutant and the original chromosome. The mutations then have to be verified by DNA sequencing.

Reverse genetic approaches

Reverse genetic approaches are important to investigate the function of genes, of which only the sequence and maybe the expression profile is known. In organisms less amenable to forward genetic approaches, such as the mouse, reverse genetics is the predominant approach. The techniques developed in the mouse to inactivate genes by homologous recombination have revolutionized this model system for basic biological research and for drug discovery (Capecchi, 1989). Genes related to human diseases can be efficiently inactivated in the mouse and suitable disease models often ensue. Although gene targeting by homologous recombination is well established in mice and in yeast, this technique has been introduced only recently in *Drosophila* (Rong and Golic, 2000; Rong *et al.*, 2002). Are these techniques even necessary in model organisms such as *Drosophila*, given their powerful forward genetic tool kit discussed above? It is the rather negative answer to this question that may explain why it took so long for these techniques to be developed. We argue, however, that reverse genetic approaches are important also in *Drosophila*. With the completion of the human genome, there are increasing numbers of genes whose function needs to be studied and their products need to be assigned to particular biological pathways. Of more than 1000 genes associated with human diseases, more than 70% are conserved in *Drosophila* (Reiter *et al.*, 2001). For many of these, the function is not known in humans or in *Drosophila*. Having a LOF mutation in such a gene is a starting point (and not the final goal) of a functional analysis. The mutant phenotype cannot be analyzed in isolation. By testing for genetic interactions with other mutations, it may be possible to assign the corresponding gene product to a given signaling pathway and/or biological process. For this genetic characterization, which follows the initial discovery of the mutant phenotype, the *Drosophila* model offers strong advantages over more complex systems such as the mouse: short generation time, large collection of mutants and sensitized signaling pathways, to name just a few. Therefore, for *Drosophila* as a model in target discovery and target validation, reverse genetic approaches are important. In the following, we will discuss the advances in two of these techniques in *Drosophila*: gene targeting and post-transcriptional gene silencing using RNA interference (RNAi).

Targeted gene disruption

With the first successful gene targeting by homologous recombination 2 years ago, one of the important drawbacks in *Drosophila* research was eliminated. A general and efficient method to target basically every gene is now available (Rong and Golic, 2000; Rong *et al.*, 2002). Targeted gene disruption is extremely useful for genes that are linked to human diseases and for which no *Drosophila* mutants have been identified by classical means. This may be due to functional redundancy or to a phenotype that is too subtle to be identified in a forward genetic screen. Of the 13 000 genes in the *Drosophila* genome, only approximately one-third has been identified by forward genetic approaches based on an easily recognizable phenotype (Miklos and Rubin, 1996). In contrast to random mutagenesis and to RNAi (see next section), homologous recombination offers the unique possibility to introduce specific mutations in a particular gene and to study their effect.

The tumor suppressor gene *p53* regulates the cell cycle and apoptosis in response to a variety of cellular stress signals in mammals. Mutations resulting in the loss or inactivation of *p53* are the most common genetic lesions found in human cancers. Unraveling the complexity of p53 function in mammals may be aided by studying its function in a simpler system such as *Drosophila*. In *Drosophila*, a homolog of *p53* (*Dmp53*) was identified. It was among the first genes that were knocked out by homologous recombination (Rong *et al.*, 2002). Surprisingly, *Dmp53* knock-outs lack an obvious phenotype. Nevertheless, Dmp53 binds specifically to human p53 binding sites and overexpression of Dmp53 induces apoptosis in *Drosophila*. Inhibition of Dmp53 function by a dominant negative allele renders cells resistant to apoptosis induced by DNA damage (Ollmann *et al.*, 2000). Although not yet routine, the number of genes that have been inactivated in *Drosophila* is rising rapidly. This technique therefore fills an important gap in the genetic tool box of *Drosophila*.

Ribonucleic acid interference

It started from an accidental observation by Fire *et al.* (1998) in *C. elegans*. They observed that, upon injection, the sense RNA probe was more efficient in silencing gene function than the antisense probe. After realizing that the sense probe contained double-stranded RNA, it became rapidly apparent that double-stranded RNA was much more efficient in inactivating gene function than single-stranded RNA. Over the past few years, the mysteries of RNA interferences have been unraveled and the technique has been shown to work in most, if not all, organisms (Sharp, 1999; Hunter, 2000). The demonstration that it also works in human cells revolutionizes functional analysis in tissue culture (Elbashir *et al.*, 2001). For the first time, it is possible in this system to infer gene function not from overexpression (GOF) experiments but from experiments involving the reduction or loss of gene function (LOF). The RNAi technique has also made its mark in *Drosophila*. Since its first application in *Drosophila* (Kennerdell and Carthew, 1998; Misquitta and Paterson, 1999) RNAi has developed to a standard procedure to analyze gene functions in flies. Moreover, most *Drosophila* cell lines respond to RNAi

(Clemens *et al.*, 2000). With the availability of the whole genome sequence, it is possible in cell culture to carry out genome-wide screens by RNAi and to silence every single gene (K. Basler, personal communication). In *Drosophila*, double-stranded RNA is delivered to the embryo by injection, therefore RNAi screens have some of the same limitations as the screens for recessive mutations in that they will only be able to detect the earliest function of a gene during development. This problem can be overcome by generating an inducible transgenic construct coding for the double-stranded RNA (Kennerdell and Carthew, 2000).

The RNAi technique has been used successfully to study functionally redundant genes. It is possible to silence simultaneously several genes by injecting a mixture of double-stranded RNAs into a single animal. For example, RNAi helped to identify the Wg receptors *frizzled* and *frizzled 2* (Kennerdell and Carthew, 1998). In contrast to many other components of the Wg pathway that were found as mutants with a Wg phenotype, the situation for the Wg receptor was more complex because Wg has been shown to interact with Fz and Fz2 proteins in cell culture. However, various mutations in *fz* indicated that it plays no role in Wg signaling. For *fz2*, no mutation was available at that time. Thus RNAi with either *fz* or *fz2* alone had no effect, but silencing both genes together produced embryonic defects that mimic the loss of Wg function. This was the first demonstration that *fz* and *fz2* act in the Wg pathway and are functionally redundant. Here, the advantage of RNAi lays not only in the ease with which a double mutant situation is created but also in the fact that, by injecting double-stranded RNA into early embryos, both maternal and zygotic mRNAs are degraded. In fact, many mutations, particularly also in signaling pathways, do not show an embryonic phenotype because there is sufficient maternal mRNA in the egg to support gene function during the first 24 h of development. In the case of *fz* and *fz2*, RNAi was used for epistasis analysis to confirm the function of these genes in the WNT signaling pathway. Both Fz and Fz2 double-stranded RNA suppressed the phenotype caused by overexpression of Wg but did not when the WNT pathway was activated by the loss of GSK2/shaggy. This example demonstrates how versatile RNAi is, even for genetically well-characterized model organisms such as *Drosophila*.

After praising the method of RNAi interference, it is worth pointing out some of its limitations. First, like all the non-genetic methods of gene silencing, the degree to which gene function is inactivated by double-stranded RNA (either by embryo injection or by transgene expression) is variable both from animal to animal and within organisms. This variable penetrance and expressivity makes it difficult to identify a consistent phenotype, particularly if there is no clear indication of what to look for. For many genes that are studied by RNAi, this is precisely the problem. After all, mutations in these genes have not been identified in conventional genetic screens. Furthermore, in many cases RNAi does not completely inactivate gene function, thus creating partial LOF phenotypes. Although these may be helpful by revealing in which process the gene is most critical, interpretation of these phenotypes is difficult without knowing the complete null phenotype. In summary, RNAi is a useful and versatile method for the characterization of molecularly characterized genes. As a gene discovery tool on a large scale (genomic or subgenomic level) it is best used, given the variability of phenotypes, in the context of genetically sensitized systems to search for novel components in a given pathway. In this context it also serves as a suitable tool for target validation in *Drosophila*.

5.3   Chemical genetics: lead identification in *Drosophila*

As outlined in the previous sections, genetic and reverse genetic approaches are useful tools to identify or functionally validate drug targets. Whether these are *drugable*, however, is another question. Genetics selects for function, not *drugability*. If functionally relevant components of the disease-relevant signaling pathway can be identified by mutational inactivation of the corresponding gene, it should be possible to use this system to identify low-molecular-weight compounds that attenuate signaling by inhibiting the function of the same essential component. This approach, termed 'chemical genetics' by Schreiber (1998), relies on inhibitors to study the function of a protein within a cell. In the pre-RNAi era of mammalian cell culture studies, chemical genetics has contributed substantially to understanding the role of various proteins, including various protein kinases. Success stories are wortmannin, a PI3K inhibitor (Arcaro and Wymann, 1993), the MEK inhibitor PD098059 (Alessi *et al.*, 1995; Dudley *et al.*, 1995) and the p38 inhibitor SB203580 (Lee *et al.*, 1994; Cuenda *et al.*, 1995), to name just a few. Although wortmannin was first identified based on its inhibitory effect on the respiratory burst of neutrophils and its target was identified subsequently, both PD098059 and SB203580 were developed as specific inhibitors against the corresponding kinases. In the following, we will discuss how *Drosophila* can contribute to lead identification and characterization and we present an example of how the combined use of chemical genetics and classical genetics can provide targets and the corresponding leads.

Do leads that inhibit *Drosophila* proteins inhibit human proteins?

The usefulness of *Drosophila* as a lead discovery system depends obviously on the probability that a compound that inhibits a *Drosophila* protein will also inhibit its human homolog. Given the often relatively small degree of amino acid identity between homologs, this is indeed a real concern. Because there are no compounds identified in *Drosophila* on the market, this concern can be addressed only by asking how many of the inhibitors selected against human proteins also inhibit the *Drosophila* proteins. Here, the number is surprisingly high. Most of the inhibitors tested *in vivo* in *Drosophila* work in the expected manner (Table 5.2). For some of them (e.g. rapamycin and wortmannin; see below) it has been shown that they also function in embryos when injected and/or in larvae when delivered by feeding.

Therefore, it appears that there is a relatively high probability that biologically active substances in *Drosophila* will also possess a similar function in mammalian cells.

Advantages and disadvantages of *INVOSCREEN*™

The Genetics Company Inc. has developed an *in vivo* screening platform based on the administration of compounds during *Drosophila* development and on the evaluation of phenotypic readouts. In much the same way as genetic screens for novel components in disease-related pathways are performed in genetically sensitized systems, similar systems can be used in drug screening. In a genetically sensitized background, compounds will be detected even if they only partially inhibit the sensitized signaling pathway. As in the case of target identification in *Drosophila*, the most pertinent advantage of lead identification in an animal model is that the leads have been selected based on their biological activity and specificity. Compounds that are toxic because they lack specificity or because they affect basic cellular and metabolic processes will not be identified. Furthermore, this technology also offers the advantage of being able to select for orally available compounds and to find drugs that have to be metabolized to reach maximal activity. Finally, animal models permit the screening of complex biological phenotypes such as behavior, organ or body growth or neurodegenerative conditions, phenotypes that cannot be recapitulated in a tissue culture dish.

There are also obvious drawbacks to drug screening in animal models:

1. The throughput is relatively low compared with classical high-throughput screening. We estimate a throughput of ca. 10 000 compounds per month.

2. Many of the potential drugs go undetected because they are metabolized too rapidly or do not bind the fly target homolog. However, the problem of the high metabolic rate of *Drosophila* larvae can be partially overcome by using defined genetic backgrounds in which drug turnover is substantially reduced without significant impairment of the viability under laboratory conditions (M. Jünger, F. Rintelen and E. Hafen, in preparation).

Table 5.2   Examples of drug effects in humans and *Drosophila*

	Humans		*Drosophila*		
Compound	Therapeutic application	Oral available	Disease model	Genotype	Specific phenotype
Rapamycin[1]	Immunosuppres./cancer	Yes	Cancer	wt	Small flies
Wortmannin[1]	Cancer/PI3 kinase inhibitor	Yes	Cancer	wt	Small flies
PD098059[1]	MEK inhibitor	–	Cancer	wt embryo	Inhibition of terminal differentiation
UO126[1]	MEK inhibitor	–	Cancer	wt embryo	Inhibition of terminal differentiation
SB203580[2]	p38 kinase inhibitor	Yes	Cancer	tkv^{CA} mutants	Suppression of the tkv^{CA} wing phenotype
Doxurubicin[1]	Cancer	No	Cancer	wt	None
Amethopterin, busulfan, vinblastin[1]	Cancer	Yes	Cancer	wt	Developmental delay
L-Dopa, pergolide, bromo-criptine[3]	Parkinson's disease	Yes	Parkinson's disease	h-synuclein transgenic flies	Reestablishment of locomotor activity
Selegiline, benztropine, estrogen[4]	Parkinson's disease	Yes	Neuro-degenera-tion	*myb* transgenes	Delayed onset of neurological symptoms
Tacrine, nicotinamide, propento-phyllin[4]	Alzheimer's disease	Yes	Neurode-generative diseases	*myb* transgenes	Delayed onset of neurological symptoms
DAPT[5]	Alzheimer's disease	Yes	Notch pathway	wt	Notch mutant
Glyceryl trioleate oil[6]	Neuro-degenerative disorders	Yes	Neurode-generation	*bgm* mutants	Suppression of neuro-degeneration
Phenytoin, valproate	Epilepsy	Yes	Epilepsy	K-channel mutants	Leg shaking reduced[7]

[1] TGC, unpublished results.
[2] Adachi-Yamada *et al.* (1999); Han *et al.* (1998).
[3] Pendleton *et al.* (2002).
[4] Fogarty, P. and Lipstick, J. (2000). Patent application WO 00/55620A1. Palo Alto, CA: Leland Stanford Junior University.
[5] DAPT, N-[N-3,5-difluorophenacetyl)-L-alanyl]-S-phenylglycine t-butylester; Micchelli *et al.*, 2002.
[6] Min and Benzer (1999).
[7] Sharma, A. and Kumar, S. (2001). *Patent application US 6,291,739 BI*. New Delhi, India: Council of Scientific and Industrial Research.

3. A hit identifies a biologically active compound in a disease model but not directly its target. For further development it is therefore mandatory to have the right tools to identify the corresponding target (see later section on mechanism-of-action studies). For instance, the activity of a compound within a pathway can be narrowed down by testing its activity in different genetic backgrounds, or by the identification of resistance mutations by genetic screening.

In summary, considering the advantages and disadvantages of drug testing in *Drosophila*, *INVOSCREEN*™ is a powerful tool to screen for biologically active compounds for complex traits. It is most effectively used in combination with genetic screens because this increases the chance of identifying the golden triplet: the target, its lead and their function.

Search for compounds inhibiting cellular growth

Cellular growth is a prerequisite for tumor growth and involves more than just the control of the cell cycle machinery (Neufeld *et al.*, 1998; Stocker and Hafen, 2000). The elucidation of mechanisms underlying growth control will provide insight into the way to interfere with tumor growth, therefore one of the goals of The Genetics Company, Inc. is to identify the genes essential for cellular growth in *Drosophila* and to develop low-molecular-weight inhibitors against the corresponding proteins. For this purpose, in parallel to the genome-wide saturation screen for genes involved in cell and organ growth (see earlier section on screens for recessive mutations), we are performing a chemical genetic screen for compounds that inhibit cellular growth. The feasibility of this approach was demonstrated by the striking similarity of the phenotype obtained by genetic mutations of *dTOR* and *dPI3K* and the effects of the administration of the corresponding chemical inhibitors rapamycin and wortmannin (Stewart *et al.*, 1996; Weinkove *et al.*, 1999; Oldham *et al.*, 2000; Zhang *et al.*, 2000; H. Stocker, unpublished; Figure 5.5). It is worth noting that rapamycin, which has been used successfully as an immunosuppressant, is now also in clinical trials as an anticancer drug and for other cell-growth-related disorders such as restinosis (Hidalgo and Rowinsky, 2000).

Mechanism-of-action studies

For many drugs currently on the market the corresponding target is not known. *Drosophila* can be used to identify the target and the mechanism of action (MOA) of such drugs. Of course, the drug in question has to produce a specific, clearly detectable phenotype in *Drosophila* and this phenotype must be related to the action of the drug in humans. In many cases, the observed

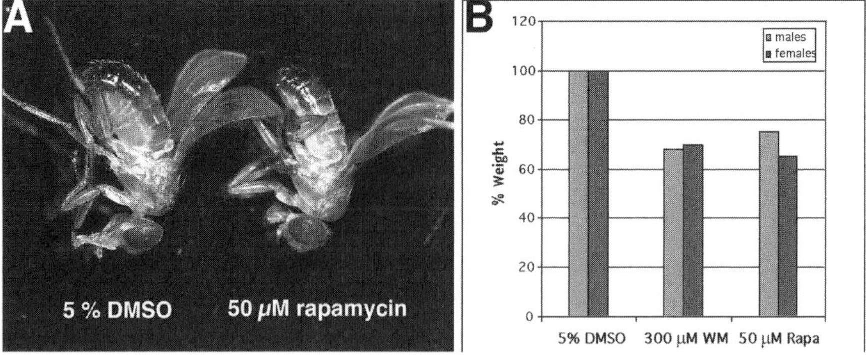

Figure 5.5 Effect of rapamycin and wortmannin on *Drosophila* development. Wild-type *Drosophila* were grown in the presence of 50 μM rapamycin (Rapa) 300 μM wortmannin (WM) or 5% DMSO as a control. Two days after eclosion, the weight of the flies was measured

phenotype in conjunction with the available information of the MOA in humans will provide information on which biological pathway is inhibited by the drug. Testing different mutations in genes encoding pathway components for resistance or hypersensitivity to the drug may further narrow down the target or identify it directly. When the target cannot be identified through existing mutations, mutations that render the flies resistant to the drug can be selected. Some of these mutations will identify the gene target or a closely associated protein. This approach has been used successfully for the identification of insecticide targets such as Methoprene and Ivermectin (Ashok *et al.*, 1998; Kane *et al.*, 2000).

5.4 Outlook

From all multicellular organisms, the unprecedented wealth of biological and genetic information makes *Drosophila* a very promising tool for target and lead discovery using classical and chemical genetics. The number of disease models in *Drosophila* increases rapidly. These include models for a variety of neurodegenerative diseases (Muqit and Feany, 2002), metabolic diseases such as type 2 diabetes and various cancers. The availability of genetically sensitized systems suitable for genetic and chemical screening offers a highly synergistic approach to target and lead discovery. Over the past 20 years, *Drosophila* research has revolutionized our understanding of developmental biology like no other organism. We are convinced that the next 20 years will see an equally strong impact of this small fly on pharmaceutical research. By delivering *in vivo* validated targets and leads rich in biological information about their function in a multicellular organism, we expect these targets and leads to have a much higher success rate in the validation process in other animal models and in clinical trials. The success of companies built on these model systems, including Exelixis, Inc., Develogen, AG and The Genetics Company, Inc., indicate that this is not only the view of passionate fly geneticists but is also perceived like this in the pharmaceutical and economic industries.

5.5 Acknowledgments

We thank S. Breuer, K. Nairz, S. Oldham, F. Rintelen, B. Schindelholz, H. Stocker and M. Vegh for discussions and E. Niederer and Greg Cole for expert technical help in setting up *INVOSCREEN*™.

5.5 References

Adachi-Yamada, T., Nakamura, M., Irie, K., Tomoyasu, Y., Sano, Y., Mori, E., Goto, S., *et al.* (1999). p38 mitogen-activated protein kinase can be involved in transforming growth factor beta superfamily signal transduction in *Drosophila* wing morphogenesis. *Mol. Cell. Biol.* **19**, 2322–2329.

Alessi, D. R., Cuenda, A., Cohen, P., Dudley, D. T. and Saltiel, A. R. (1995). PD 098059 is a specific inhibitor of the activation of mitogen-activated protein kinase kinase *in vitro* and *in vivo*. *J. Biol. Chem.* **270**, 27489–27494.

Arcaro, A. and Wymann, M. P. (1993). Wortmannin is a potent phosphatidylinositol 3-kinase inhibitor: the role of phosphatidylinositol 3,4,5-trisphosphate in neutrophil responses. *Biochem. J.* **296**, 297–301.

Ashok, M., Turner, C. and Wilson, T. G. (1998). Insect juvenile hormone resistance gene homology with the bHLH-PAS family of transcriptional regulators. *Proc. Natl. Acad. Sci. USA* **95**, 2761–2766.

Berg, C. A. and Spradling, A. C. (1991). Studies on the rate and site-specificity of P-element transposition. *Genetics* **127**, 515–524.

Berger, J., Suzuki, T., Senti, K. A., Stubbs, J., Schaffner, G. and Dickson, B. J. (2001). Genetic mapping with SNP markers in *Drosophila. Nat. Genet.* **29**, 475–481.

Bienz, M. and Clevers, H. (2000). Linking colorectal cancer to Wnt signaling. *Cell* **103**, 311–320.

Boettner, B. and Van Aelst, L. (2002). The RASputin effect. *Genes Dev.* **16**, 2033–2038.

Böhni, R., Riesgo-Escovar, J., Oldham, S., Brogiolo, W., Stocker, H., Andruss, B. F., Beckingham, K., *et al.* (1999). Autonomous control of cell and organ size by CHICO, a *Drosophila* homolog of vertebrate IRS1-4. *Cell* **97**, 865–875.

Bos, J. L. (1989). ras Oncogenes in human cancer: a review. [Erratum appears in *Cancer Res.* 1990; **50**, 1352.] *Cancer Res.* **49**, 4682–4689.

Brazil, D. P. and Hemmings, B. A. (2001). Ten years of protein kinase B signalling: a hard Akt to follow. *Trends Biochem. Sci.* **26**, 657–664.

Brogiolo, W., Stocker, H., Ikeya, T., Rintelen, F., Fernandez, R. and Hafen, E. (2001). An evolutionarily conserved function of the *Drosophila* insulin receptor and insulin-like peptides in growth control. *Curr. Biol.* **11**, 213–221.

Brunner, E., Peter, O., Schweizer, L. and Basler, K. (1997). Pangolin encodes a Lef-1 homologue that acts downstream of Armadillo to transduce the Wingless signal in *Drosophila*. *Nature* **385**, 829–833.

Burks, D. J., de Mora, J. F., Schubert, M., Withers, D. J., Myers, M. G., Towery, H. H., Altamuro, S. L., *et al.* (2000). IRS-2 pathways integrate female reproduction and energy homeostasis. *Nature* **407**, 377–382.

Butler, A. A. and Roith, D. L. (2001). Control of growth by the Somatropic axis growth hormone and the insulin-like growth factors have related and independent roles. *Annu. Rev. Physiol.* **63**, 141–164.

Cantley, L. C. and Neel, B. G. (1999). New insights into tumor suppression: PTEN suppresses tumor formation by restraining the phosphoinositide 3-kinase/AKT pathway. *Proc. Nat. Acad. Sci. USA* **96**, 4240–4245.

Capecchi, M. R. (1989). Altering the genome by homologous recombination. *Science* **244**, 1288–1292.

Chen, C., Jack, J. and Garofalo, R. S. (1996). The *Drosophila* insulin receptor is required for normal growth. *Endocrinology* **137**, 846–856.

Clark, S. G., Stern, M. J. and Horvitz, H. R. (1992). *C. elegans* cell-signalling gene *sem-5* encodes a protein with SH2 and SH3 domains. *Nature* **356**, 340–344.

Clemens, J. C., Worby, C. A., Simonson-Leff, N., Muda, M., Maehama, T., Hemmings, B. A. and Dixon, J. E. (2000). Use of double-stranded RNA interference in *Drosophila* cell lines to dissect signal transduction pathways. *Proc. Natl. Acad. Sci. USA* **97**, 6499–6503.

Cohen, P. and Frame, S. (2001). The renaissance of GSK3. *Nat. Rev. Mol. Cell Biol.* **2**, 769–776.

Cuenda, A., Rouse, J., Doza, Y. N., Meier, R., Cohen, P., Gallagher, T. F., Young, P. R., *et al.* (1995). SB 203580 is a specific inhibitor of a MAP kinase homologue which is stimulated by cellular stresses and interleukin-1. *FEBS Lett.* **364**, 229–233.

Dickson, B. J., van der Straten, A., Domínguez, M. and Hafen, E. (1996). Mutations modulating Raf signaling in *Drosophila* eye development. *Genetics* **142**, 163–171.

Dudley, D. T., Pang, L., Decker, S. J., Bridges, A. J. and Saltiel, A. R. (1995). A synthetic inhibitor of the mitogen-activated protein kinase cascade. *Proc. Natl. Acad. Sci. USA* **92**, 7686–7689.

Elbashir, S. M., Lendeckel, W. and Tuschl, T. (2001). RNA interference is mediated by 21- and 22-nucleotide RNAs. *Genes Dev.* **15**, 188–200.

Fire, A., Xu, S., Montgomery, M. K., Kostas, S. A., Driver, S. E. and Mello, C. C. (1998). Potent and specific genetic interference by double-stranded RNA in *Caenorhabditis elegans*. [See comments.]. *Nature* **391**, 806–811.

Gehring, W. J. (2002). The genetic control of eye development and its implications for the evolution of the various eye-types. *Int. J. Dev. Biol.* **46**, 65–73.

Hafen, E., Basler, K., Edstroem, J. E. and Rubin, G. M. (1987). *sevenless*, a cell-specific homeotic gene of *Drosophila*, encodes a putative transmembrane receptor with a tyrosine kinase domain. *Science* **236**, 55–63.

Halfar, K., Rommel, C., Stocker, H. and Hafen, E. (2001). Ras controls growth, survival and differentiation in the *Drosophila* eye by different thresholds of MAP kinase activity. *Dev. Suppl.* **128**, 1687–1696.

Han, Z. S., Enslen, H., Hu, X., Meng, X., Wu, I. H., Barrett, T., Davis, R. J., *et al.* (1998). A conserved p38 mitogen-activated protein kinase pathway regulates *Drosophila* immunity gene expression. *Mol. Cell. Biol.* **18**, 3527–3539.

Hanson, I. M. (2001). Mammalian homologues of the *Drosophila* eye specification genes. *Semin. Cell Dev. Biol.* **12**, 475–484.

Hidalgo, M. and Rowinsky, E. K. (2000). The rapamycin-sensitive signal transduction pathway as a target for cancer therapy. *Oncogene* **19**, 6680–6686.

Hunter, C. P. (2000). Gene silencing: shrinking the black box of RNAi. *Curr. Biol.* **10**, R137–140.

Ikeya, T., Galic, M., Belawat, P., Nairz, K. and Hafen, E. (2002). Nutrient-dependent expression of insulin-like peptides form neurosecretory cells in the CNS contribute to growth regulation in *Drosophila*. *Curr. Biol.* **12**, 1293–1300.

Kane, N. S., Hirschberg, B., Qian, S., Hunt, D., Thomas, B., Brochu, R., Ludmerer, S. W., *et al.* (2000). Drug-resistant *Drosophila* indicate glutamate-gated chloride channels are targets for the antiparasitics nodulisporic acid and ivermectin. *Proc. Natl. Acad. Sci. USA* **97**, 13949–13954.

Karim, F. D., Chang, H. C., Therrien, M., Wassarman, D. A., Laverty, T. and Rubin, G. M. (1996). A screen for genes that function downstream of Ras1 during *Drosophila* eye development. *Genetics* **143**, 315–329.

Kennerdell, J. R. and Carthew, R. W. (1998). Use of dsRNA-mediated genetic interference to demonstrate that frizzled and frizzled 2 act in the wingless pathway. *Cell* **95**, 1017–1026.

Kennerdell, J. R. and Carthew, R. W. (2000). Heritable gene silencing in *Drosophila* using double-stranded RNA. *Nat. Biotech.* **18**, 896–898.

Kramps, T., Peter, O., Brunner, E., Nellen, D., Froesch, B., Chatterjee, S., Murone, M., *et al.* (2002). Wnt/wingless signaling requires BCL9/legless-mediated recruitment of pygopus to the nuclear beta-catenin-TCF complex. *Cell* **109**, 47–60.

Lee, J. C., Laydon, J. T., McDonnell, P. C., Gallagher, T. F., Kumar, S., Green, D., McNulty, D., *et al.* (1994). A protein kinase involved in the regulation of inflammatory cytokine biosynthesis. *Nature* **372**, 739–746.

Lewis, E. B. and Bacher, F. (1968). Methods of feeding ethyl methane sulfonate (EMS) to *Drosophila* males. *Drosophila Inf. Serv.* **43**, 193.

Liu, J. P., Baker, J., Perkins, A. S., Robertson, E. J. and Efstratiadis, A. (1993). Mice carrying null mutations of the genes encoding insulin-like growth factor I (Igf-1) and type 1 IGF receptor (Igf1r). *Cell* **75**, 59–72.

Lowenstein, E. J., Daly, R. J., Batzer, A. G., Li, W., Margolis, B., Lammers, R., Ullrich, A., *et al.* (1992). The SH2 and SH3 domain-containing protein GRB2 links receptor tyrosine kinases to ras signaling. *Cell* **70**, 431–442.

Maconochie, M., Nonchev, S., Morrison, A. and Krumlauf, R. (1996). Paralogous Hox genes: function and regulation. *Annu. Rev. Genet.* **30**, 529–556.

McCormick, F. (1997). The superfamily of Ras-related GTPases. *Jpn. J. Cancer Res.* **88**, inside front cover.

Micchelli, C., Esler, W., Kimberly, W., Jack, C., Berezovska, O., Kornilova, A., Hyman, B., *et al.* (2002). g-Secretase/presenilin inhibitors for Alzheimer's disease phenocopy Notch mutations in *Drosophila*. *FASEB J.* **17**, 79–81.

Miklos, G. L. and Rubin, G. M. (1996). The role of the genome project in determining gene function: insights from model organisms. *Cell* **86**, 521–529.

Min, K. T. and Benzer, S. (1999). Preventing neurodegeneration in the *Drosophila* mutant bubblegum. [See comments.]. *Science* **284**, 1985–1988.

Misquitta, L. and Paterson, B. M. (1999). Targeted disruption of gene function in *Drosophila* by RNA interference (RNA-i): a role for nautilus in embryonic somatic muscle formation. *Proc. Natl. Acad. Sci. USA* **96**, 1451–1456.

Moon, R. T., Bowerman, B., Boutros, M. and Perrimon, N. (2002). The promise and perils of Wnt signaling through beta-catenin. *Science* **296**, 1644–1646.

Morin, P. J. (1999). beta-Catenin signaling and cancer. *Bioessays* **21**, 1021–1030.

Morrison, D. K. (2001). KSR: a MAPK scaffold of the Ras pathway? *J. Cell Sci.* **114**, 1609–1612.

Muqit, M. M. and Feany, M. B. (2002). Modelling neurodegenerative diseases in *Drosophila*: a fruitful approach? *Nat. Rev. Neurosci.* **3**, 237–243.

Nairz, K., Stocker, H., Schindelholz, B. and Hafen, E. (2002). High-resolution SNP mapping by denaturing HPLC. *Proc. Natl. Acad. Sci. USA* **99**, 10575–10580.

Nakae, J., Kido, Y. and Accili, D. (2001). Distinct and overlapping functions of insulin and IGF-I receptors. *Endocr. Rev.* **22**, 818–835.

Neufeld, T. P., Delacruz, A. F. A., Johnston, L. A. and Edgar, B. A. (1998). Coordination of growth and cell division in the *Drosophila* wing. *Cell* **93**, 1183–1193.

Newsome, T. P., Asling, B. and Dickson, B. J. (2000). Analysis of *Drosophila* photoreceptor axon guidance in eye-specific mosaics. *Development* **127**, 851–860.

Nüsslein, V. C. and Wieschaus, E. (1980). Mutations affecting segment number and polarity in *Drosophila*. *Nature* **287**, 795–801.

O'Kane, C. J. and Gehring, W. J. (1987). Detection *in situ* of genomic regulatory elements in *Drosophila*. *Proc. Natl. Acad. Sci. USA* **84**, 9123–9127.

Oldham, S. and Hafen, E. (2003). Insulin/IGF and target of rapamycin signaling: a TOR de force in growth control. *Trends Cell Biol.* **13**, 79–85.

Oldham, S., Montagne, J., Radimerski, T., Thomas, G. and Hafen, E. (2000). Genetic and biochemical characterization of dTOR, the *Drosophila* homolog of the target of rapamycin. *Genes Dev.* **14**, 2689–2694.

Olivier, J. P., Raabe, T., Henkemeyer, M., Dickson, B., Mbamalu, G., Margolis, B., Schlessinger, J., *et al.* (1993). A *Drosophila* SH2-SH3 adaptor protein implicated in coupling the sevenless tyrosine kinase to an activator of Ras guanine nucleotide exchange, Sos. *Cell* **73**, 179–191.

Ollmann, M., Young, L. M., Di Como, C. J., Karim, F., Belvin, M., Robertson, S., Whittaker, K., *et al.* (2000). *Drosophila* p53 is a structural and functional homolog of the tumor suppressor p53. *Cell* **101**, 91–101.

Parker, D. S., Jemison, J. and Cadigan, K. M. (2002). Pygopus, a nuclear PHD-finger protein required for Wingless signaling in *Drosophila*. *Dev. Suppl.* **129**, 2565–2576.

Peifer, M. and Polakis, P. (2000). Wnt signaling in oncogenesis and embryogenesis – a look outside the nucleus. *Science* **287**, 1606–1609.

Pendleton, R. G., Parvez, F., Sayed, M. and Hillman, R. (2002). Effects of pharmacological agents upon a transgenic model of Parkinson's disease in *Drosophila melanogaster*. [Erratum appears in *J. Pharmacol. Exp. Ther.* 2002; **300**, 1131.] *J. Pharmacol. Exp. Ther.* **300**, 91–96.

Polakis, P. (2000). Wnt signaling and cancer. *Genes Dev.* **14**, 1837–1851.

Poretsky, L., Cataldo, N. A., Rosenwaks, Z. and Giudice, L. C. (1999). The insulin-related ovarian regulatory system in health and disease. *Endocr. Rev.* **20**, 535–582.

Prober, D. A. and Edgar, B. A. (2002). Interactions between Ras1, dMyc, and PI3K signaling in the developing wing. *Genes Dev.* **16**, 2286–2299.

Reiter, L. T., Potocki, L., Chien, S., Gribskov, M. and Bier, E. (2001). A systematic analysis of human disease-associated gene sequences in *Drosophila melanogaster*. *Genome Res.* **11**, 1114–1125.

Rintelen, F., Stocker, H., Thomas, G. and Hafen, E. (2001). PDK1 regulates growth through PKB and S6K in *Drosophila. Proc. Natl. Acad. Sci. USA* **98**, 15020–15025.

Rodriguez-Viciana, P., Warne, P. H., Khwaja, A., Marte, B. M., Pappin, D., Das, P., Waterfield, M. D., *et al.* (1997). Role of phosphoinositide 3-OH kinase in cell transformation and control of the actin cytoskeleton by Ras. *Cell* **89**, 457–467.

Rommel, C., Clarke, B., Zimmermann, S., Nunez, L., Rossman, R., Reid, K., *et al.* (1999). Differentiation stage-specific inhibition of the Raf-MEK-ERK pathway by Akt. *Science* **286**, 1738–1741.

Rong, Y. S. and Golic, K. G. (2000). Gene targeting by homologous recombination in *Drosophila.* [See comments]. *Science* **288**, 2013–2018.

Rong, Y. S., Titen, S. W., Xie, H. B., Golic, M. M., Bastiani, M., Bandyopadhyay, P., Olivera, B. M., *et al.* (2002). Targeted mutagenesis by homologous recombination in *D. melanogaster. Genes Dev.* **16**, 1568–1581.

Rorth, P., Szabo, K., Bailey, A., Laverty, T., Rehm, J., Rubin, G. M., Weigmann, K., *et al.* (1998). Systematic gain-of-function genetics in *Drosophila. Development* **125**, 1049–1057.

Saltiel, A. R. and Kahn, C. R. (2001). Insulin signalling and the regulation of glucose and lipid metabolism. *Nature* **414**, 799–806.

Schreiber, S. L. (1998). Chemical genetics resulting from a passion for synthetic organic chemistry. *Bioorg Med. Chem.* **6**, 1127–1152.

Sharp, P. A. (1999). RNAi and double-strand RNA. *Genes Dev.* **13**, 139–141.

Simon, M. A., Bowtell, D., Dodson, G. S., Laverty, T. R. and Rubin, G. M. (1991). Ras1 and a putative guanine nucleotide exchange factor perform crucial steps in signaling by the sevenless protein tyrosine kinase. *Cell* **67**, 701–716.

Simon, M. A., Dodson, G. S. and Rubin, G. M. (1993). An SH3-SH2-SH3 protein is required for p21Ras1 activation and binds to sevenless and Sos proteins *in vitro. Cell* **73**, 169–177.

Spradling, A. C., Stern, D., Beaton, A., Rhem, E. J., Laverty, T., Mozden, N., Misra, S., *et al.* (1999). The Berkeley Drosophila Genome Project gene disruption project: single P-element insertions mutating 25% of vital *Drosophila* genes. *Genetics* **153**, 135–177.

St Johnston, D. (2002). The art and design of genetic screens: *Drosophila melanogaster. Nature Rev. Genet.* **3**, 176–188.

Sternberg, P. W. and Han, M. (1998). Genetics of RAS signaling in *C. elegans. Trends Genet.* **14**, 466–472.

Stewart, M. J., Berry, C. O., Zilberman, F., Thomas, G. and Kozma, S. C. (1996). The *Drosophila* p70s6k homolog exhibits conserved regulatory elements and rapamycin sensitivity. *Proc. Natl. Acad. Sci. USA* **93**, 10791–10796.

Stocker, H. and Hafen, E. (2000). Genetic control of cell size. *Curr. Opin. Genet. Dev.* **10**, 529–535.

Therrien, M., Chang, H. C., Solomon, N. M., Karim, F. D., Wassarman, D. A. and Rubin, G. M. (1995). KSR, a novel protein kinase required for RAS signal transduction. [See comments.] *Cell* **83**, 879–888.

Therrien, M., Wong, A. M. and Rubin, G. M. (1998). CNK, a RAF-binding multidomain protein required for RAS signaling. *Cell* **95**, 343–353.

Thissen, J. P., Underwood, L. E. and Ketelslegers, J. M. (1999). Regulation of insulin-like growth factor-I in starvation and injury. *Nutr. Rev.* **57**, 167–176.

Thompson, B., Townsley, F., Rosin-Arbesfeld, R., Musisi, H. and Bienz, M. (2002). A new nuclear component of the Wnt signalling pathway. *Nat. Cell Biol.* **4**, 367–373.

Tufts Center for the Study of Drug Development (2001). *Tufts Center for the Study of Drug Development Pegs Cost of a New Prescription Medicine at $802 Million.* Boston: Tufts Center for the Study of Drug Development, Tufts University.

Wassarman, D. A., Therrien, M. and Rubin, G. M. (1995). The Ras signaling pathway in *Drosophila*. *Curr. Opin. Genet. Dev.* **5**, 44–50.

Weinkove, D., Neufeld, T., Twardzik, T., Waterfield, M. and Leevers, S. (1999). Regulation of imaginal disc cell size, cell number and organ size by *Drosophila* class IA phosphoinositide 3-kinase and its adaptor. *Curr. Biol.* **9**, 1019–1029.

White, M. A., Nicolette, C., Minden, A., Polverino, A., Van Aelst, L., Karin, M. and Wigler, M. H. (1995). Multiple Ras functions can contribute to mammalian cell transformation. *Cell* **80**, 533–541.

Xu, T. and Rubin, G. M. (1993). Analysis of genetic mosaics in developing and adult *Drosophila* tissues. *Development* **117**, 1223–1237.

Yenush, L. and White, M. F. (1997). The IRS-signalling system during insulin and cytokine action. *Bioessays* **19**, 491–500.

Zhang, H., Stallock, J. P., Ng, J. C., Reinhard, C. and Neufeld, T. P. (2000). Regulation of cellular growth by the *Drosophila* target of rapamycin dTOR. *Genes Dev.* **14**, 2712–2724.

Zimmermann, S. and Moelling, K. (1999). Phosphorylation and regulation of Raf by Akt (protein kinase B). *Science* **286**, 1741–1744.

6
Mechanism of Action in Model Organisms: Interfacing Chemistry, Genetics and Genomics

Pamela M. Carroll, Kevin Fitzgerald and **Rachel Kindt**

Each year pharmaceutical companies lose billions of dollars on compounds that fall out of development due in part to an unknown mechanism of action. In addition, better versions of compounds that are currently on the market are, in some cases, not being pursued because the mechanism by which the compound functions is unclear. The ability of small molecules with known mechanisms of action to affect *Caenorhabditis elegans* and *Drosophila* through their therapeutically relevant targets demonstrates that these same systems might have utility in the identification of molecular targets for compounds with unknown mechanisms. Advances in *Drosophila* and *C. elegans* research allow the combination of genome sequence information, gene expression profiles and genome-wide mutations to be used in an unprecedented dissection of a complex organism. This chapter will focus on the technical and innovative advantages of model organisms in discovering compound mechanism of action, as well as providing detailed accounts of compound utility in simple organism 'disease models'.

Model Organisms in Drug Discovery. Edited by Pamela M. Carroll and Kevin Fitzgerald.
© 2003 John Wiley & Sons, Ltd. ISBN 0 470 84893 6

6.1 Introduction

How is it that drugs work? What is the mechanism by which they are able to alter disease processes inside the human body? Most drugs are composed of small organic compounds in pill form that are swallowed and absorbed through the stomach or small intestine. The molecules then permeate the body by riding along in the bloodstream until they find their targets and modulate its activity. The targets of most drugs are the cellular proteins that carry out most functions within our bodies. For instance, one particular subset of proteins, the G-protein-coupled receptors, are the targets of more than one-third of the drugs on the market today, representing 30% of the top-selling pharmaceuticals (Scussa, 2002). Given that drugs target proteins, some of which belong to the same subset, how is it that these compounds have specific effects? Generally speaking, the drugs presently available have undergone a rigorous selection process. This process aims to ensure that they target only the specific proteins or cellular function involved in a given disease and not others necessary for the normal functions of human cells. Occasionally, however, drugs may need to change the activity of more than one protein to be effective, or may cause unwanted effects because they are not specific enough. In addition, sometimes the proteins targeted by the drug are unknown. In these cases it is very important to identify the compound's target and therefore to understand the mechanism by which small molecules affect biological processes (Koh and Crews, 2002). Aptly, these types of studies are termed mechanism of action (MOA) and traditionally use a variety of biochemical assays to find the direct binding partners to a compound; however, model system genetics are proving useful for identifying the function of the compound's biology. Invertebrate models systems such as *Caenorhabditis elegans* and *Drosophila* are ideal for studying compounds and ultimately in identifying the protein target(s). In theory, MOA studies in living systems are only limited by the specificity and bioavailability of the drug in question. The objective of this chapter is to explore the need for MOA studies and the genetic systems in which they are effective, utilizing specific examples.

6.2 Introduction to compound development

In order to understand the need for MOA technologies in pharmaceutical companies, one must take a closer look at the process by which drugs have been developed. Some of the earliest records of drug discovery come from ancient times. Their methods of drug discovery were mostly *in vivo*, with *in vivo* in this case meaning not animal models but rather humans eating, drinking or topically applying crude extracts from one or several plants, animal venoms, secretions or parts. In what can only be classified as 'trial and error' clinical trials, the doctors (or shaman) would try different combinations of drug extracts. These extracts would have one of three different effects: nothing would happen, in which case a different combination of extract would be applied; the patient would get sicker, in which case the dosage would be stopped; or the patient would show some kind of improvement or effect that would be noted for the future. Often, the effects of plant extracts were hypothesized by acute observation of animal behavior or the effects of such plants or animal venom on wildlife. This type of drug discovery is still underway in large populations of people in the world, and over thousands of years through trial and error several combinations of plant extracts have been found to alleviate some human diseases and their symptoms. Most often in these cases the individual efficacious component of the complex plant extract has been (and often remains) unknown. Therapies that have been developed in this fashion continue to be sold today and are generally referred to as herbal remedies or 'natural products'.

During the 1970s a large portion of pharmaceutical drug discovery was devoted to separating the individual components of plant and animal extracts in the search for the active compound(s). These extracts would be applied to cell-based assays with readouts designed as surrogates for disease endpoints. Compounds that produced an effect were progressed forward into drug discovery and animal models. In most of these cases the target protein(s) whose activities were being modified by the compounds in the extract remained unknown. A good example of this type of drug discovery was the advent of the diabetes drugs thiazolidinediones, in which the target was unknown for many years (now known to be peroxisome proliferator-activated receptors, PPARs) (O'Moore-Sullivan and Prins, 2002). Another example in the same disease area occurs with the compound metformin, which has been used in the treatment of diabetes for decades and yet whose MOA remains undefined (Sirtori and Pasik, 1994). Additional compounds discovered in this way include parthenolide, digoxin, salicylic acid, opium, tropane alkaloids, galantamine and camptothecin, the active ingredients of Feverfew, Foxgloves, Willow bark, Poppy seeds, Snowdrops and Happy tree, respectively (Bynum, 1970; Heptinstall, 1988; Wall and Wani, 1995; Brune, 2002). The early methods of drug discovery had many shortcomings and were very inefficient. Natural products tend to be complex molecules that can be difficult to synthesize *de novo*. In addition, if a compound discovered through natural product separation has undesirable properties, such as solubility or pharmacokinetic issues, it is impossible, without knowing what the protein targets of the compound are, to find a new chemotype with similarly efficacious properties and the same MOA.

In the early to late 1980s with the advent of molecular biology, the understanding of basic biology exploded and recently has been combined with the knowledge of the human genome. In parallel to the innovations in biology, new methods in synthetic combinatorial chemistry arose that have allowed for the *de novo* synthesis of large amounts and numbers of small organic compounds. The combination of the two preceding trends has led to the current state of pharmaceutical development and the one-enzyme (or protein), one-compound hypothesis (Drews and Ryser, 1997). Using this paradigm, one begins with a protein of a particular biological activity and then sorts through compound decks containing millions of compounds that are tested *in vitro* for their ability to bind to and modify the activity of the protein. Compounds developed in this current mode of discovery include Viagra, Cox-2 inhibitors, Gleevac, cyclin-dependent kinase inhibitors and Ras farnesyltransferase inhibitors (Goldenberg, 1998; Brune, 2002; Capdeville *et al.*, 2002; Caponigro, 2002).

However, even within the 'one protein, one compound' discovery process, many compounds continue to fail in clinical development due to side-effects or 'off-target' activity. Even when a compound binds to one protein with high affinity, it does not necessarily mean that it does not bind to another protein with equal or better affinity, thus having 'off-target' activity. This becomes a problem especially with proteins that belong to large related families. Compounds also are sometimes metabolized into alternative compounds, which may take on novel properties and gain specificity for different protein targets than the parent compound. Many effective compounds previously mentioned, such as metformin, acetaminophen, lithium and amiodarone, developed through older methods of drug discovery, still have unknown MOAs. In short, there are older as well as new compounds that could benefit tremendously from a robust process capable of identifying all of their protein targets.

6.3   Model organisms arrive on the scene

During the early 1980s the molecular and genetic tools to analyze microorganisms were well established. The technical tools as well as the genetic techniques from these organisms were beginning to be exploited by researchers in other model systems, such as *C. elegans* and *Drosophila*. By 1990, *C. elegans* and *Drosophila* researchers were able to combine similar tools with the availability of some sequence information in order to analyze increasingly complex biology. It is only recently that *C. elegans* and *Drosophila* have been applied to drug discovery and proving useful in particular to the pharmacological modulation of specific biological pathways.

To be confident with any experimental system one must validate its usefulness for a given process. There are many cases of relevant compound effects in model organisms. For example, early studies with known acetylcholine receptor antagonists in *C. elegans* clearly demonstrated that these antagonists, developed in mammalian systems, were capable of inhibiting the cognate *C. elegans* enzyme (Walker *et al.*, 2000). In addition, these inhibitors were used in genetic screens and were crucial in the identification not only of the *C. elegans* acetylcholine receptor but also of several novel conserved components of acetylcholine signaling that had not yet been identified in mammalian systems (Weinshenker *et al.*, 1995). Phorbol esters and caffeine, which affect protein kinase C and phosphodiesterase, respectively, are very active in modulating their protein targets in both *C. elegans* and mammalian systems. Similarly caffeine, in *Drosophila* systems, acts as a stimulant and increases their motor activity and affects the time period in which they sleep (Hendricks *et al.*, 2000; Greenspan *et al.*, 2001). Clearly, many pharmacologically active compounds modulate the same protein families in model systems as in mammalian cells (Table 6.1). This may not be surprising, given that compounds generally bind to active sites in proteins and the modification of these sites is responsible for their effects.

The ability of small molecules with known MOAs to affect *C. elegans* and *Drosophila* through their related targets suggests that these same systems might have utility in the identification of molecular targets for compounds with unknown mechanisms. However, if model systems are truly to be utilized, then one must be confident that the underlying mode of drug action is similar enough for conclusions from one system to be informative in the other. One must ask the question: how similar is the underlying biology of *C. elegans* and *Drosophila* compared with humans? The answer is not quite so simple. From a genomic standpoint, on asking how highly conserved the human, *C. elegans* and *Drosophila* genomes are, the answer received is different for each. In a systematic analysis of human disease-associated genes for instance, 77% (i.e. 548 of 714 genes identified in the Online Mendelian Inheritance in Man, OMIM) are clearly related to genes in *Drosophila* (Reiter *et al.*, 2001) and a similar number in *C. elegans* (Culetto and Sattelle, 2000). These genes cover a range of diseases implicated in cancer and cardiovascular, renal, endocrine, innate immunity and metabolic disease. There are general physiological exceptions that are not preserved across species. For example, flies and worms have no erythrocytes, and so hemoglobin orthologs are absent. Also, genes specific to the rearrangement of immunoglobulin genes are absent in these organisms, which lack acquired immunity. It is likely that evolution saw no requirement for *Drosophila* and *C. elegans*, with their short lifespans, to evolve sophisticated long-acting immune systems. Because of this, there are some limitations as to which type of compounds one might wish to study in either the *C. elegans* or *Drosophila* system.

Evolution did not, however, reinvent the wheel unnecessarily. In analyzing the major pathways known to be involved in the development of human cancers, one is hard pressed to find a gene or signaling pathway in humans that is not represented in *C. elegans* and/or *Drosophila*. In fact, many proteins known to play the most crucial roles in human cancers were originally

Table 6.1 Partial list of compounds with common targets in invertebrate and vertebrate systems

Organism	Compound	Target	Reference
Yeast	Berberine	MAPKK	Jang *et al.*, 2002
Yeast	Fumagillin	Methionine aminopeptidase	Sin *et al.*, 1977
Yeast	Compactin	HMG-CoA reductase	Basson *et al.*, 1986
Yeast	Brefeldin A	Arf/Sec7 complex	Peyroche *et al.*, 1999
C. elegans	Levamisole	Acetyl choline receptor	Lewis *et al.*, 1987
C. elegans	Fluoxetine	Serotonin reuptake transporter	Choy and Thomas, 1999
C. elegans	Capsaicin	Vanilloid receptor VR1	Wittenburg and Baumeister, 1999
C. elegans	Carbachol	Muscarinic acetylcholine receptor	Hwang *et al.*, 1999
C. elegans	Olomoucine	Cyclin kinases	Abraham *et al.*, 1995
C. elegans	Phorbol esters	Phorbol ester receptor	Kazanietz, 1995
C. elegans	Nicotine	Nicotinic acetylcholine receptors	Fleming *et al.*, 1997
C. elegans	Oxotremorine	Muscarinic acetylcholine receptor	Hwang *et al.*, 1999
C. elegans	Ivermectin	GABA(A) receptor/glutamate gated chloride channels	Cully *et al.*, 1994
Drosophila	Adociasulfate-2	Kinesin motor	Sakowicz *et al.*, 1998
Drosophila	Cocaine	Monoamine transporter	McClung and Hirsh, 1999
Drosophila	Caffeine	Adenosine antagonists	Shaw *et al.*, 2000
Drosophila	Antihistamine	Histamine receptor	Shaw *et al.*, 2000
Drosophila/ *C. elegans*	Gamma-secretase inhibitor	Presenilin	This chapter and Micchelli *et al.*, 2002
Drosophila	Lithium	Glycogen synthase kinase-3 beta	Hedgepeth *et al.*, 1997
Drosophila/ *C. elegans*	Farnesyl protein transferase	Ras FTI	Hara and Han, 1995; Kauffmann *et al.*, 1995

identified and their activities detailed in model systems. A good example of this is the oncogene Ras. The Ras oncogene was discovered in mammals to be the homolog of the Harvey sarcoma virus ras gene in 1982 (Parada *et al.*, 1982). It was not until years later, when Ras was identified in yeast (Tatchell, 1986), *C. elegans* (Beitel *et al.*, 1990; Han and Sternberg, 1990) and *Drosophila* (Simon *et al.*, 1991), that details of the Ras cellular signaling pathway took form. For example, in *Drosophila*, a connection was made between activation of receptor tyrosine kinases and activation of the Ras-mitogen-activated protein kinase (MAPK) pathway (Simon, 2000). The human version of the K-Ras protein was found to be 84% identical to Ras of *C. elegans* and 87%

identical to Ras of *Drosophila*. Importantly, expression of an activated Ras gene similar to those found in human cancers resulted in excessive cell proliferation in both systems, suggesting functional as well as sequence conservation (Fortini *et al.*, 1992; Karim and Rubin, 1998). Discovery of a *C. elegans* and *Drosophila* 'oncogene' led to the discovery of many genes in the Ras-MAPK pathway and the subsequent isolation and validation of orthologous genes in mammalian systems (Matthews and Kopczynski, 2001). In the case of cancer, and the basic processes of cell proliferation and division, evolution was frugal indeed. It is this pathway conservation that is exploited when jumping from a compound's effects in humans to finding its protein target in model systems. Other notable examples include apoptosis, leading to the 2002 Nobel Prize in Medicine for *C. elegans* work, and Alzheimer's disease, which will be discussed later in this chapter.

6.4 Elucidating the mechanism of compound action

Choosing the correct highly conserved pathways in model systems leads to outcomes that are highly informative about the mammalian condition in question. So how does one go about finding the mechanism of drug action utilizing model systems? (Figure 6.1). The first step is to administer the compound to the animal, requiring mixing the compound in the fly or worm food or, in the case of *C. elegans*, bathing the animals in compound. Higher concentrations of compounds are often required in *C. elegans* and *Drosophila* than in mammalian cell-based and animal assays. The *C. elegans* nematode has a relatively impermeable cuticle, therefore a relatively large outside concentration of compound is often required to ensure a very small *in vivo* concentration. Typically 'wild-type' animals are tested, although other *C. elegans* or *Drosophila* strains may be tested as well, depending on what is known about the compound and its effects in mammalian cells. For instance, a compound with an unknown anticancer mechanism might be tested both on normal animals and in animals containing specific genetic mutations that modify the cell cycle or apoptosis (Kauffmann *et al.*, 1995). *Caenorhabtis elegans* or *Drosophila* disease models also may be tested (Jorgensen and Mango, 2002), as well as strains that lack the drug efflux pumps that may improve drug transport into cells.

The treated animals are observed closely for the compound effects on the animal's development, behavior and other abnormal characteristics. The observed effect, or 'phenotype', is then compared with the collective knowledge in the model system field. The importance of a robust phenotype cannot be overstated. Lack of a phenotype will effectively end the analysis – although it is interesting to note that over 60% of the pharmaceutical compounds that we tested show some effect in the worm (data not shown).

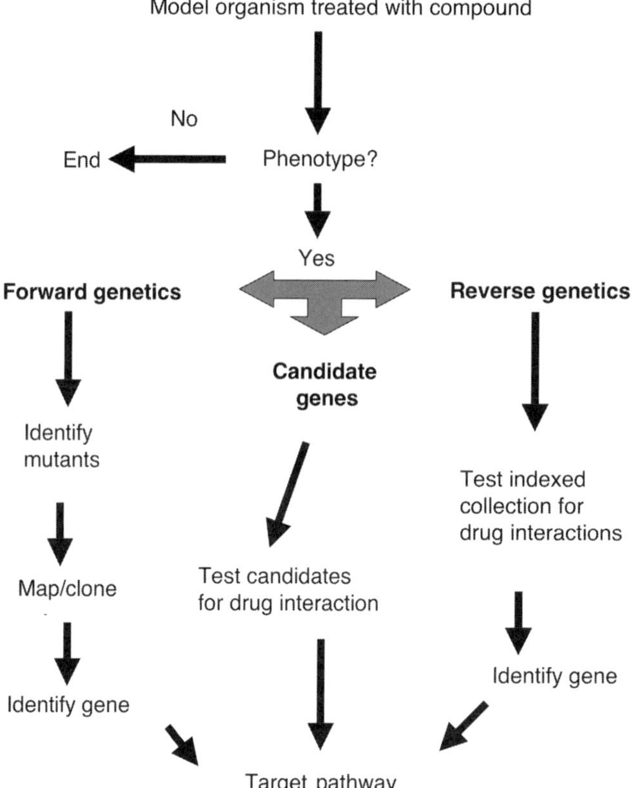

Figure 6.1   Mechanism of action process flow. First, animals are treated with compounds and analyzed for phenotypes. Once a robust compound-induced phenotype is identified, genetic modifiers (resistant or hypersensitive) of the compound are sought by a forward genetic screen (i.e. chemical mutagenesis), a reverse genetic screen (i.e. RNAi collection) and/or a candidate gene approach. In the forward genetic screen, mutations are mapped to a chromosome location and the mutated gene is identified. In the reverse genetic screen, the gene is known at the starting point. In the candidate gene approach, the starting phenotype hints at potential candidates that are tested directly (see text for details)

The more precise and well characterized the phenotype, the better the hypothesis. When an effect is observed from compound administration, then one can ask if the effect mimics that of a specific gene disruption phenotype. Often the phenotypic effect of a compound suggests a well-known signaling pathway that is being disrupted. This comparison is possible because the specific effect of individually removing many genes from *C. elegans* or *Drosophila* has been studied intensively, and because, in many cases, these genes have been integrated into pathways and genetic circuits. Follow-up experiments include the testing of sentinel strains that are compromised for components in the pathway, rendering the animals more sensitive or resistant to the compound. If a mutation in one gene makes animals resistant to the compound-induced phenotype, the compound is likely to target the same biological network as the wild-type version of the mutant gene product. However, compound-induced phenotypes often do not fall into a characterized pathway, and a mutagenesis screen is undertaken to identify genes that, when mutated, effect the compound-induced phenotype. Resistant and sensitive strains are identified and the underlying gene mutations and genes involved are isolated. The protein products produced from these genes then become candidates for the drug target of action. An important control is to test if mutants obtained are cross-resistant to other compounds, suggesting mutations in genes involved in non-specific drug transport.

Once potential protein candidates for a compound's target are identified in model systems, the identification of mammalian orthologs of those proteins may be complicated by the duplication of gene families in the evolution of the mammalian genomes. In these cases, identifying the true ortholog among a number of highly related proteins may be challenging. Once a candidate gene list is identified, experiments are conducted to see if the orthologous mammalian gene(s) is involved in the compound's activity. By analogy, one may expect that disruption of the mammalian ortholog may confer similar compound resistance or sensitivity. However, artifacts do arise because a compound may have unrelated effects in *C. elegans* and in mammalian cells due to differences in the species protein complement. In general, we have found that compounds have comparable effects. Once the list of candidate genes has been narrowed down and the signaling pathway has been implicated in both the mammalian and model systems, biochemistry is utilized to show a direct binding of the compound to the target protein. This type of approach may yield the mechanism of the compound's therapeutic effect but may also reveal off-target activities of the compound that may lead to potential toxicological effects.

A prototype example of MOA analysis in *C. elegans* is provided by studies on the drug Prozac. Prozac is well known to inhibit the mechanism by which neuronal cells recycle serotonin by interfering with a reuptake protein, but it was controversial as to whether all the therapeutic effects of Prozac could be accounted for by the serotonin pathway or whether the reported side-effects were due to unknown drug targets. Choy and Thomas (1999) tested the hypothesis that Prozac might have multiple protein targets and several mechanisms of action. They applied the compound to both wild-type animals and *C. elegans* that were mutant for the production of serotonin. In wild-type animals Prozac had effects consistent with the known mechanism of Prozac's action through effects on serotonin reuptake. Interestingly, however, they found that Prozac had effects on animals lacking in serotonin altogether, indicating a unique mechanism separate from the serotonin system. A genetic screen was carried out to investigate the biological pathways and proteins involved in Prozac's serotonin-independent phenotype. These screens revealed at least one novel class of proteins involved in Prozac's effects on *C. elegans*. Although it still remains to be determined if these proteins are involved in the therapeutic or side-effects of Prozac in humans, the studies act as a strong paradigm for the novel utility of model systems in MOA studies.

There are many published examples of pharmacological agents affecting similar pathways in *Drosophila* and mammals (Table 6.1); however, owing to technical challenges, there are fewer examples of *Drosophila* genetic screens for resistance to compound-induced phenotype than *C. elegans*. Technically, it can be difficult to obtain a consistent drug response amongst individual flies, even in the same vial (unpublished data). The two main problems appear to be ill-defined genetic backgrounds and drug delivery. First, laboratory strains are often highly polymorphic, making genetic mapping of drug-resistant mutations problematic. This is further confounded by the use of 'mapping chromosomes' that have accumulated mutations that may affect tracking phenotypes during genetic mapping. This problem can be addressed by rebuilding strain collections in an isogenic background, which is a large but feasible undertaking (unpublished data). Secondly, drug delivery is variable owing to the fly's aversion to high concentrations of drug in their food. Starving the *Drosophila* adults or larvae before drug exposure can minimize the variability of response. Testing different drug delivery methods, such as injection (Kauffmann *et al.*, 1995), vaporization as with ethanol (Moore *et al.*, 1998) and chronic versus short dosing, will help make genetic screens for MOA an increasingly fruitful undertaking in flies (Janssen *et al.*, 2000). Still there are success stories: Ffrench-Constant and colleagues have applied the MOA strategy towards understanding the molecular basis of insecticide resistance. In one case, a chloride ion channel gene was found to be mutated in *Drosophila* populations resistant to the insecticide cyclodiene. Further studies demonstrated that this ion channel was the direct target of cyclodiene (Ffrench-Constant *et al.*, 1993, 2000).

6.5   A case study for Alzheimer's disease drug discovery

The starting point for MOA analysis is often a compound with a desirable (or undesirable, e.g. toxic) biological activity and an unknown cognate target (direct binding) or pathway. The MOA analysis of a gamma-secretase inhibitor was an excellent proof of principle for this approach, demonstrating the power of phenotyping and genetic analysis in elucidating the compound mechanism.

The neuropathology of Alzheimer's disease (AD) is characterized by the presence of extracellular deposits called senile plaques, which are primarily composed of amyloid (Aβ) peptide. This peptide is generated by the sequential processing of amyloid precursor protein (APP). The N-terminus of Aβ is generated by the processing activity named beta-secretase, followed by proteolytic cleavage at the C-terminus by an activity called gamma-secretase. Inhibition of gamma secretase activity is an obvious therapeutic goal for the treatment of AD, but until recently the molecular components of gamma secretase were unknown (Esler and Wolfe, 2001). Presenilin proteins (PS1 and PS2 in mammals) were originally implicated in the etiology of AD based on the finding that patients with genetic mutations in these genes were predisposed to an early-onset form of AD (Levy-Lahad *et al.*, 1995; Sherrington *et al.*, 1995). Further studies linked PS proteins, and components of the membrane complex that they form, to gamma-secretase activity. The PS proteins are highly conserved through evolution and also have been implicated in the processing of Notch receptor (Levitan and Greenwald, 1995; Struhl and Greenwald, 1999).

Compound BMS AG6B was identified in a cell-based screen for compounds that altered the Aβ40/Aβ42 peptide ratio produced in Aβ processing (unpublished data). In order to identify candidate molecular targets for BMS AG6B, the drug was applied to *Drosophila* and *C. elegans* and the resulting phenotypes were analyzed. These experiments were carried out with the identity and biological activity unknown to the testers, such that the hypotheses generated were based solely on the information from the phenotypes.

Adult flies and their progeny were exposed to 10–40 mM BMS AG6B dissolved in dimethylsulfoxide (DMSO) and administered in the flies' food. Treated adults had grossly normal behavior and morphology at all the concentrations tested. However, exposure to 40 mM BMS AG6B was lethal to larvae (second instar stage). At lower doses (10 mM) some ($<10\%$) of the flies did survive to adulthood. When examined closely, the surviving adults flies were found to have morphological defects in a number of tissues, including notched wing margins, rough eyes, missing or fused leg segments and missing abdominal bristles (Figure 6.2). This combination of phenotypes is characteristic of mutations in members of the Notch signaling pathway (Shellenbarger and Mohler, 1978). When the phenotypes of treated flies were compared directly to those in a Notch hypomorph, they were found to be very similar. Overall, the *Drosophila* phenotypes suggested that the drug was disrupting the action of the Notch pathway. Similar Notch phenotypes in *Drosophila* were seen with other gamma-secretase inhibitors (Micchelli *et al.*, 2002).

Wild-type *C. elegans* were treated with 0.1–2.0 mM BMS AG6B (added to the bacterial lawn) throughout larval development and as adults. Worms treated as adults exhibited no discernable changes in morphology or behavior, and their progeny also appeared normal. All worms treated as larvae

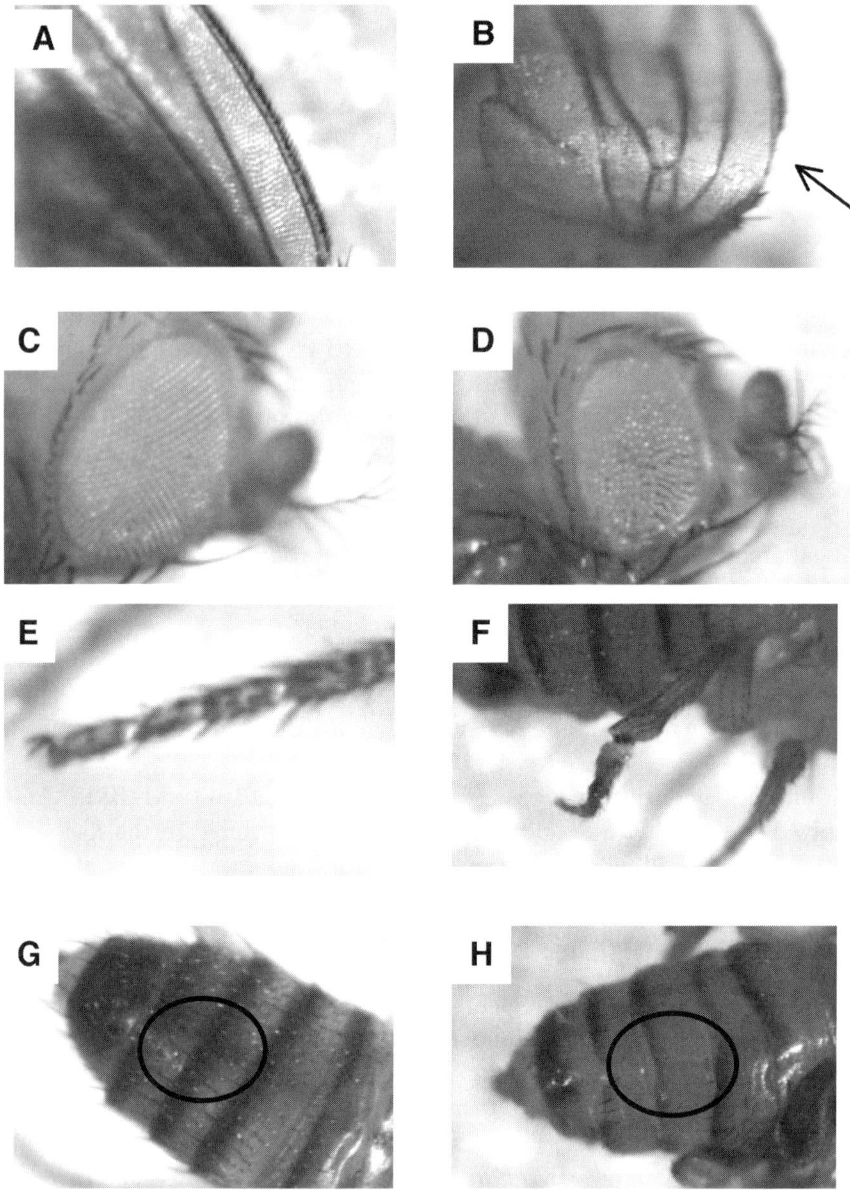

Figure 6.2 Compound BMS AG6B induces Notch-like phenotypes in *Drosophila*. Adult flies and their progeny were treated with 10 mM BMS AG6B. Micrographs show the representative control (DMSO)-treated (A, C, E, G) and BMS AG6B-treated (B, D, F, H) adult progeny: (A, B). The normally smooth wing margin is notched in B (indicated by an arrow); (C, D) 'Rough eye' phenotype is shown in D; (E, F) E shows a morphologically wild-type leg and in F the tarsal segment is missing or fused; (G, H) the wild-type abdomen shown in G is covered with dark bristles and an area in H lacking bristles after drug treatment is circled

developed to adulthood. However, some treated worms were egg-laying defective (Egl-d), as demonstrated by the presence of late-stage eggs in the uterus, which give the animals a bloated appearance that is apparent at low magnification. The phenotype was found to be dose dependent (Figure 6.3D).

There are many tissues and signaling pathways required for normal egg-laying in *C. elegans*, and the drug-induced phenotypes could potentially signify a disruption in any one or more of them. The observation that larval treatment was required for the effect suggested a developmental rather than a neuromuscular defect. When drug-treated worms were examined in late larval stages by high magnification for morphological or positional changes in important egg-laying tissues – vulva, somatic gonad, sex muscles and hermaphrodite-specific neuron – all of the structures appeared quite normal. However, close examination revealed that some of the treated animals harbored dead embryos (Figures 6.3A and 6.3B). Interestingly, some embryos were found to lack an anterior pharynx, known as an Aph phenotype (Figure 6.3C). This morphological signature also pointed to the Notch pathway. There are two Notch receptors in worms – *lin-12* and *glp-1* – that mediate many cell-fate decisions (Austin and Kimble, 1989; Yochem and Greenwald, 1989; Kimble and Simpson, 1997). Notch receptor *glp-1* is required for germline proliferation and embryonic development, particularly the signaling event that gives rise to the anterior pharyngeal lineage. Strong loss-of-function *glp-1* mutants are sterile and weaker alleles are maternal-effect embryonic lethal, giving rise to Aph (anterior pharynx missing) progeny. Notch receptor *lin-12* is required for many post-embryonic cell-fate decisions. *lin-12* null alleles are sterile and weaker alleles are Egl-d due to defects in the vulval precursor cells and sex muscles. The two Notch proteins together control zygotic viability because *lin-12; glp-1* double mutants die as larvae (Lambie and Kimble, 1991). Overall, the drug treatment phenotypes were consistent with decreased, if not eliminated, Notch activity in worms, although the phenotypic match was not exact. For example, BMS AG6B-treated worms had normal vulval induction and a single anchor cell (data not shown).

Because mutations in the pathway or target might be expected to enhance or suppress the phenotype, the compound was tested on a panel of mutants in the Notch pathway. To test the hypothesis that the drug could be decreasing Notch activity, worms carrying an activated Notch allele *lin-12(n137)* were treated with BMS AG6B. This *lin-12* mutant carries a missense mutation in the extracellular domain of the protein that acts to increase *lin-12* signaling (resulting in extra vulval structures) and is known as a Muv phenotype (Greenwald and Seydoux, 1990). The number and frequency of these extra vulval structures were significantly reduced in worms treated with BMS AG6B (Table 6.2). The simplest explanation for the results are that BMS AG6B inhibits Notch protein function directly, inhibits a positive regulator of Notch signaling or activates a negative regulator.

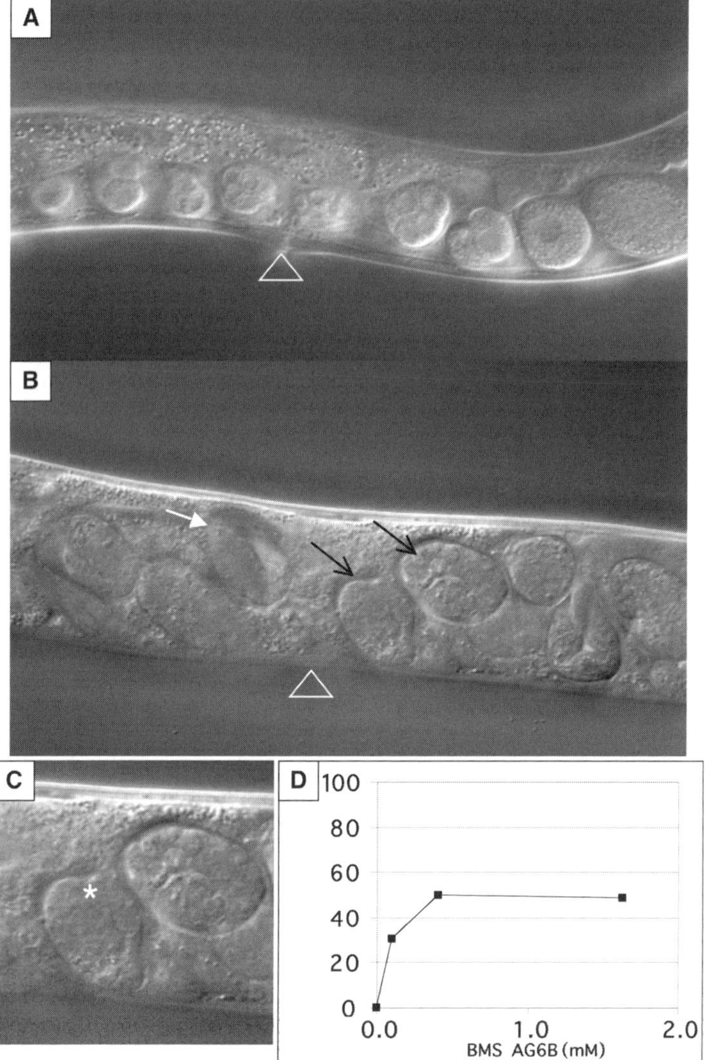

Figure 6.3   Egg-laying and embryogenesis defects in BMS AG6B-treated *C. elegans*. (A, B) Differential interference contrast photomicrographs focusing on the region of the hermaphrodite uterus. The location of the vulva is shown with a triangle. Photomicrograph A shows embryos retained by the control (DMSO)-treated animals. No embryos are older than comma stage, which is the wild-type phenotype. Photomicrograph B shows the same view of adults worms after larval treatment throughout larval stages with 1.6 mM BMS AG6B. Older embryos are retained in the uterus, one of which is indicated by a filled arrow. Two dead embryos are indicated with open arrows. (C) Close-up view of dead embryos from B. The embryo on the left is missing the anterior pharynx (area marked with an asterix). (D) Dose–response curve of the BMS AG6B Egl-d response. For each data point, 60–100 animals were scored for a bloated Egl-d appearance. A representative experiment is shown

Table 6.2   Compound BMS AG6B suppresses activated Notch/*lin-12(gf)* phenotypes

Genotype	BMS AG6B (mM)	Number of vulvae or pseudovulvae				
		1	2	3	4	5
Wild type	0	100	0	0	0	0
lin-12(gf)	0	0	0	0	44	56
	0.1	0	0	0	52	48
	0.4	0	17	33	50	0
	1.6	27	33	33	7	0

Caenorhabditis elegans larvae were treated with BMS AG6B or DMSO; >20 adult hermaphrodites were scored for vulvae or pseudovulvae using a dissecting microscope; the percent of scored animals in each category is reported; *lin-12(gf)* is *lin-12(n137);him-5 (e1467)*.

A number of positive regulators have been described for *C. elegans* Notch proteins that either alter downstream signaling or affect the processing of Notch (Levitan and Greenwald, 1998; Fares and Greenwald, 1999). The presenilins belong to the latter class; loss-of-function mutants in the presenilin ortholog *sel-12* were identified in a screen for suppressors of a Notch gain-of-function mutant. There are three presenilin proteins in the worm: *spe-4*, *sel-12* and *hop-1*. When *hop-1* and *sel-12* are both inactivated, *lin-12-* and *glp-1*-like Notch phenotypes are observed, suggesting that the genes act redundantly in Notch processing (Li and Greenwald, 1997; Westlund *et al.*, 1999). Presenilin *spe-4* is more divergent and appears to function separately in a distinct tissue. To test the model that BMS AG6B might act to antagonize presenilin function or a presenilin-modulated pathway, the drug was tested on *hop-1* and *sel-12* mutants.   When *hop-1* mutant worms were treated with BMS AG6B, a striking phenotype was observed: the worms were uniformly sterile. The effect was strong and specific: the observed sterility was fully penetrant as low as 0.1 mM, whereas wild-type drug-treated worms did not become sterile (Figure 6.4A). Examination of the sterile worms under high magnification revealed that these animals all lacked oocytes, and sperm were present distally in the gonad (Figures 6.4B and 6.4C). This specific sterile phenotype is characteristic of loss-of-function mutations in Notch/*glp-1* as well as *hop-1*; *sel-12* double mutants (Figure 6.4D). In contrast to the *hop-1*-treated animals, when *sel-12(ep6)* animals were treated with the drug no sterility was observed. The *sel-12(ep6)*-treated worms resembled wild-type drug-treated worms in that they displayed a mildly fertile and Egl-d phenotype, as well as dead embryos (Figures 6.5A and 6.5B). The fact that the drug treatment of wild-type animals resembles *sel-12(sp6)* and not the *sel-12(sp6);hop-1* double mutant animals is consistent with the compound acting on *sel-12* and not *hop-1*. The effects of drug treatment on the presenilin mutants are summarized in Table 6.3. These findings are consistent with a decrease in presenilin function (particularly *sel-12*) or in presenilin pathway signaling.

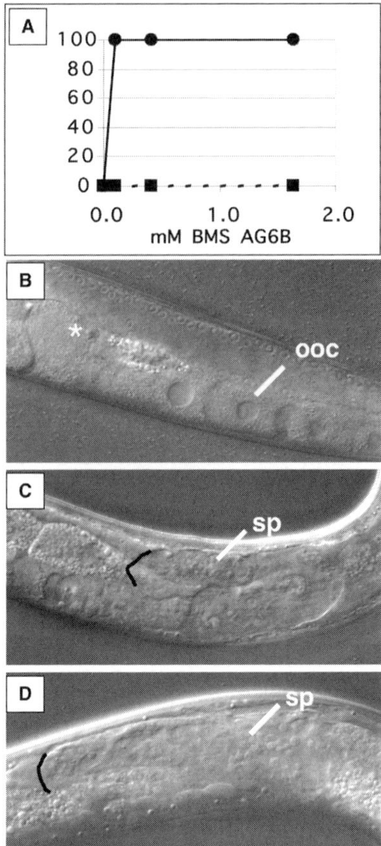

Figure 6.4 Compound BMS AG6B induces glp-1-like sterility in *hop-1* mutants. (A) Dose–response curve of wild-type and *hop-1* mutant worms treated with BMS AG6B as larvae; >99% of animals have the fertile, wild-type morphology. Worms were scored at adulthood under low magnification for a sterile appearance. A representative experiment is shown. (B, D) Differential interference contrast photomicrographs of adult hermaphrodites, focusing on one gonad arm. Photomicrograph B is the control (DMSO)-treated *hop-1*(ep171) worm. Oocytes are labeled 'ooc' and sperm are not visible in this focal plane but are found in the spermatheca (marked with an asterix). Photomicrograph D shows the *hop-1*(ep171);*sel-12*(ep6) double mutant, which shares the shortened gonad (most distal point outlined in black), lack of oocytes and distal sperm (sp) phenotypes. (C) The 0.8 mM BMS AG6B-treated *hop-1(ep171)* worm. Gonad extension is curtailed (the most distal point is outlined in black), no oocytes are found and sperm (sp) are found ectopically in the distal portion of the gonad arm

These genetic data are thus consistent with presenilin, Notch or, potentially, a presenilin complex component as the target of the gamma secretase inhibitor BMS AG6B. More recent work has solidified the hypothesis that presenilin is the gamma secretase: notably, the finding that transition-state inhibitors of gamma secretase inhibitors bind directly to heterodimeric forms

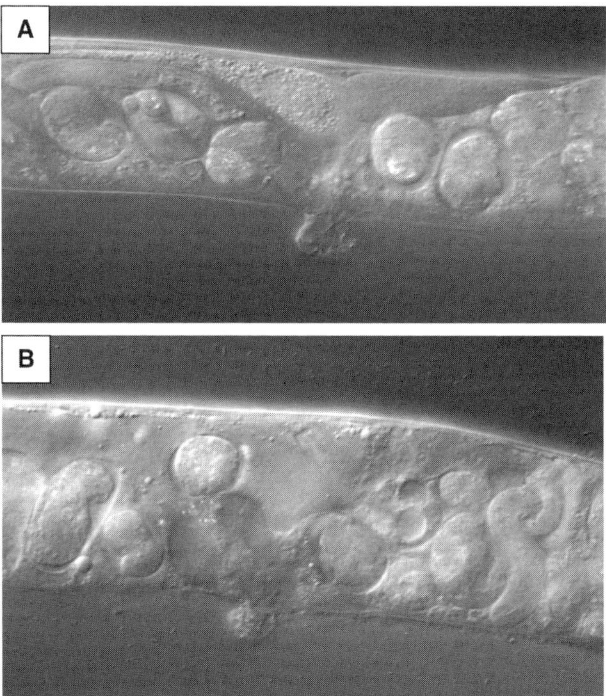

Figure 6.5 Compound BMS AG6B does not induce sterility in *sel-12* mutants. (A) Control (DMSO)-treated *sel-12(ep6)*. (B) A *sel*-12(ep6) hermaphrodite treated with 1.6 mM BMS AG6B. Animals are fertile but have dead eggs

Table 6.3 Summary of phenotypes of BMS AG6B-treated *C. elegans* presenilin mutants

Genotype	Drug	Egl-d	Dead embryos	Sterile
		Phenotypes		
Wild type	−	No	No	No
Wild type	+	Yes	Yes	No
sel-12	−	Yes	No	No
sel-12	+	Yes	Yes	No
hop-1	−	No	No	Yes (0.5%)
hop-1	+	N/A	N/A	Yes (100%)
hop-1;sel-12	−	N/A	N/A	Yes

of presenilin, suggesting that they contain the active site of the protease (Esler *et al.*, 2000; Li *et al.*, 2000). These data are also consistent with reports of worm and fly phenotypes induced by another gamma-secretase inhibitor, compound E (Francis *et al.*, 2002). The genetic data here also support the hypothesis that Notch processing is due to the same activity as APP processing, and this was further supported by the finding that BMS AG6B gamma secretase inhibited mammalian Notch processing *in vitro* (data not shown).

It is interesting that the genetic interactions with the two presenilins in worms differed. The finding that *hop-1* mutants were more sensitive than *sel-12* mutants to the effects of the drugs suggests that because the genes act redundantly the drug might inhibit *sel-12* or *sel-12*-dependent pathways more strongly. Presenilin *sel-12* is more homologous to human presenilin (50% identical to PS1 versus 33% identical for *hop-1*). Alternatively, differences in response to the two worm presenilins could be explained by different contributions of the two proteins to signaling in the affected tissues, supported by the observation that more severe phenotypes are observed in the absence of maternal *sel-12* than in the absence of maternal *hop-1* (Westlund *et al.*, 1999). In any case, the differential effect is an example of the high level of specificity that can be achieved by phenotyping and genetic analysis.

Not only do these experiments highlight the power of model organism genetics for target identification and analysis of disease genes, but they also point to the utility of compounds as pathway probes and screening tools. Because a compound effect on flies and worms is dose dependent, it can be used to generate an 'allelic series' of the inhibited genes for phenotyping and screening. Compound administration can be timed to mimic a temperature-sensitive mutant, which might avoid undesired lethality of a complete loss-of-function mutant. Compounds such as gamma secretase inhibitors are effective sensitizers of pathways and can be used as the entry point for genetic screens, not only for target identification but to generate pathway information. For example, it would be possible to screen for mutants that, say, enhanced the phenotype of BMS AG6B-treated worms such that they became *glp*-like sterile. In fact, a screen for enhancers of the *sel-12* mutant – similar in concept – identified two new regulators of presenilin signaling (Francis *et al.*, 2002).

6.6 New chemical genetic strategies: genome-wide cell-based genetic screens

The examples used above combine chemical-induced phenotypes and genetic mutagenesis screens to reveal the molecular basis of chemical action. Reverse genetic approaches utilizing RNA interference (RNAi) technology are becoming increasingly popular. The RNAi introduces double-stranded RNA (dsRNA) into a system (*C. elegans*, *Drosophila* and cell culture) as a post-transcriptional method of gene knock-down (Fraser *et al.*, 2000).The significance of this approach can be seen in *C. elegans*, where RNAi to every gene on chromosome I was systematically tested for gene function. Using

RNAi as a genetic screening tool in *C. elegans* is covered in Chapter 3. Another approach is to use RNAi technologies in MOA studies as a cell-based gene knock-out system in *Drosophila*-cultured cells to analyze systematically the function of the 14 000 predicted genes in the *Drosophila* genome. The simple addition of dsRNA to *Drosophila* cells in culture ablates the protein expression of target genes by RNAi mechanisms, thereby efficiently 'phenocopying' loss-of-function mutations (Caplen *et al.*, 2000, 2001; Clemens *et al.*, 2000). For example, the insulin signaling pathway was studied for RNAi efficiency in *Drosophila* S2 cells (Clemens *et al.*, 2000). As expected from knowledge of the insulin pathway, inhibiting the expression of MAPKK by dsRNA prevents human insulin-stimulated phosphorylation of MAPK. In another branch of the insulin pathway, dsRNA directed against PTEN (a negative regulator of insulin signaling), leads to constitutive activation of the insulin-responsive PI3K pathway. Therefore, RNAi combined with established biochemical reagents allows deeper characterization of complex signaling pathways.

In a drug discovery setting, RNAi in cell-based systems can be used to identify novel targets in compound-validated pathways. For example, modulation by an antagonist should 'phenocopy' cells treated with RNAi to the compound's target. Cell-based screening in *Drosophila* cells will be useful when compound activity can be correlated with phenotypic detection methods, such as using markers, functional assays or microscopic imaging of cells. In cell-based genetic screens RNAi is a rapid method for identifying MOA pathways but not all disease pathways can be represented in the limited cell lines available in *Drosophila*. Reasonable expectations can be made that S2 cells will be relevant in conserved cell-based functions such as apoptosis, cell division, cytoskeletal morphology and metabolism, or molecular readouts such as specific phosphorylation or gene expression changes.

The RNAi technologies for use in mammalian cell-based system are rapidly evolving but the ease, cost, efficiency and reproducibility using RNAi in S2 cells will allow for routine genome-wide functional analysis (Elbashir *et al.*, 2001; Tuschl, 2002). For example, RNAi was used to identify a cellular tyrosine kinase that acts upstream of the phosphorylation of Dscam, a protein found to be important in axonal pathfinding (Muda *et al.*, 2002). Only one of six RNAi treatments (Src42A) directed at suspected kinases was able to decrease tyrosine phosphorylation on Dscam in S2 cells. This suggests that Src42A acts upstream of Dscam and may be a candidate Src42A substrate. This approach could be scaled up in S2 cells to test all 200+ *Drosophila* kinases for changes in a phosphorylation event.

The RNAi in *Drosophila* is most effectively induced by dsRNAs of more than 80 nucleotides in length, which are easy to generate by polymerase chain reaction (PCR) (Clemens *et al.*, 2000). We routinely generate dsRNA to complementary (c)DNA clones using generic primers. This makes it functionally feasible to test the RNAi of all *Drosophila* genes by designing cell-based readouts in 96-well or 384-well formats. Conversely, mammalian cell experiments use 21–23 length oligonucleotides that are expensive to purchase and variably mediate an RNAi-like inhibition of gene expression in 'knock-down' efficiency. Presently, the expense of RNA oligonucleotides makes mammalian cell-based 'genetic' screens prohibitive, although oligonucleotide vectors for RNAi are being developed. Also, the smaller, less redundant genome of *Drosophila* may be more revealing. Notwithstanding, mammalian RNAi will be an important resource for rapid validation of *Drosophila* targets in vertebrate biology.

6.7   A case study for innate immunity and inflammation drug discovery

Drosophila S2 cells have macrophage-like properties and therefore should be informative in understanding cell-mediated innate immunity. Most components of innate immunity are conserved evolutionarily from *Drosophila* to humans, and only higher eukaryotes have acquired immunity (Silverman and Maniatis, 2001). Insects have a potent and rapid response to a broad spectrum of pathogens and the response discriminates between types of pathogens. Fungal and bacterial infections of *Drosophila* lead to transcriptional activation of sets of antimicrobial peptide (AMP) genes and eventually S2 cells will phagocytose the microbes. These responses are mediated by nuclear factor kappa B (NF-κB) family members, which are conserved transcription factors that also activate the expression of inflammation genes in mammals. Induction of each AMP gene is regulated by a balance of inputs that are manifested by combinations of the three Rel/NF-κB proteins Relish, Dorsal and Dif (Figure 6.6). Activation of Rel/NF-κB pathways is essential for the *Drosophila* innate immune response. For example, *Drosophila* carrying mutations in the Relish gene do not express certain classes of antimicrobial peptides, such as *Cecropins* and *Diptercin*, and are susceptible to Gram-negative (*Escherichia coli*) bacterial infection. Similarly, Dorsal or Dif is essential for activation of AMPs such as *Drosomycin*, involved in fungal and Gram-positive infections (Hedengren *et al.*, 1999). *Drosophila* Rel proteins, like mammalian Rels, are sequestered in the cytoplasm as a result of association with an IκB-like inhibitor protein such as Cactus. When cells are activated by pathogens, signaling pathways are activated leading to the release of IκB, nuclear translocation of Rel proteins and Rel-activated transcription. Cactus is the IκB protein that inhibits Dorsal and Dif. Like NF-κB, Relish is the mammalian homolog of p105, and contains both a Rel domain and an IκB inhibitory domain (Silverman and Maniatis, 2001).

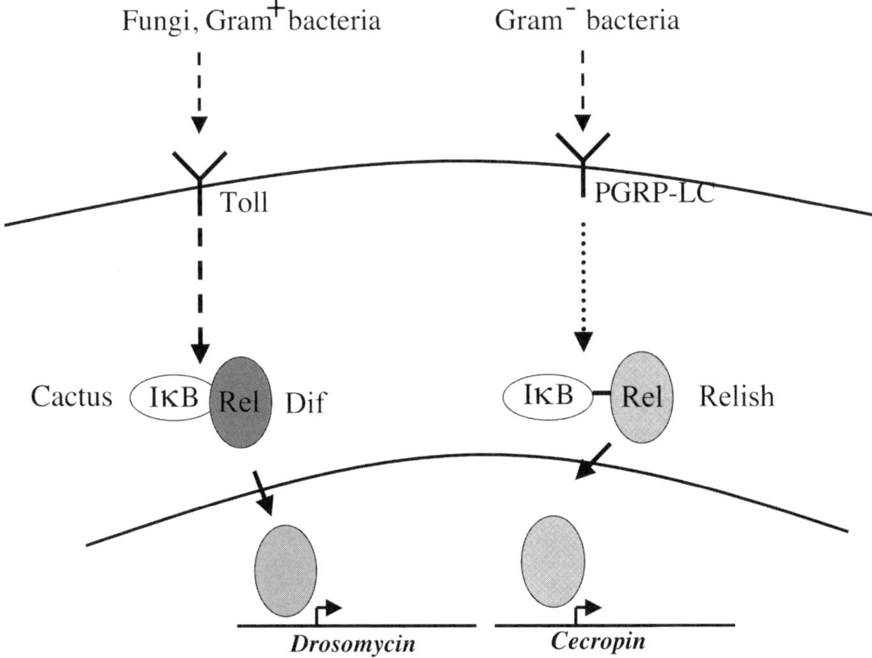

Figure 6.6 A 'simplified' model of Rel signal transduction in *Drosophila* innate immune responses. The Toll to Dif and the PGRP-LC to Relish transduction pathways are depicted. The Toll receptor is activated by fungi and Gram-positive bacteria, leading to degradation of Cactus, which is an IκB molecule that inhibits Dif by cytoplasmic sequestration. The activated Dif will induce gene expression of *Drosomycin*. The PGRP-LC receptor is activated by Gram-negative bacteria, which leads to endoproteolytic cleavage of the Relish protein between the IκB and Rel domains, thereby releasing the Rel domain to induce gene expression of *CecropinA1*

An immune response in S2 cells can be induced by lipopolysaccharide (LPS) – a Gram-negative bacterial cell wall component – to express a subset of AMPs (Han and Ip, 1999). In our experiments (Figure 6.7) and others (Silverman *et al.*, 2000; Sun *et al.*, 2002), the RNAi knock-down of Relish shows decreasing transcriptional activation of AMPs, in this case *CecropinA1*. This result demonstrates that LPS activation of *CecropinA1* requires the Rel protein Relish and parallels the *in vivo* response of Relish mutants. In contrast, RNAi of the IκB homolog Cactus, an inhibitor of Rel proteins Dif and Dorsal, causes significant upregulation of the fungal response gene *Drosomycin*, independent of activation. To compare the RNAi effect with the drug effect we used parthenolide, the active component of the anti-inflammatory medicinal herb Feverfew (*Tanacetun parthenium*), which is known to inhibit NF-κB signaling. Recent biochemical results suggest that the

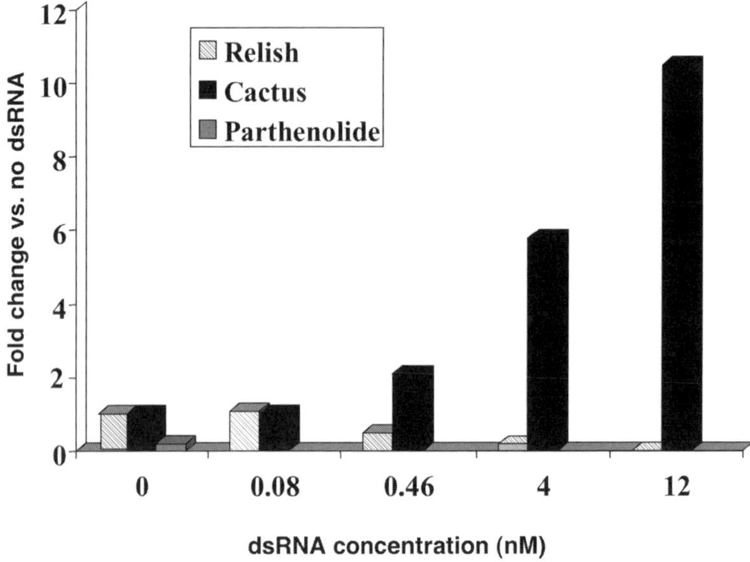

Figure 6.7   Antimicrobial gene expression is altered by the RNAi to NF-κB pathway and parthenolide in S2 cells. The hatch-shaded columns represent *CecropinA1* gene expression in samples treated with Relish dsRNA relative to no dsRNA treatment control. Relish RNAi treatment shows a dose-dependent inhibition of *CecropinA* gene expression. The black-shaded columns represent *Drosomycin* gene expression in samples treated with Cactus dsRNA relative to no dsRNA treatment control. Cactus RNAi treatment causes upregulation of *Drosomycin* gene expression. The no-treatment controls are represented by a 'one-fold' change in the zero dsRNA Relish and Cactus columns. All samples shown have been treated with LPS (20 μg/ml) for 1 h. The gray column represents S2 cell treatment with parthenolide (50 μm) for 30 min prior to LPS treatment.

probable MOA is that parthenolide binds IκK, a kinase that when activated inhibits IκB (Kwok *et al.*, 2001). We have found that pretreating S2 cells with parthenolide inhibits LPS-induced gene activation (Figure 6.7). This result indicates that compounds can target similar NF-κB pathways in *Drosophila* and mammals, and S2 cell-based experiments can model a high-content assay for the activity of the mammalian NF-κB pathway. For example, candidate novel components that function in NF-κB signaling could be determined in an RNAi-based or compound-based screen that tests for the disruption of LPS-inducible gene activation. In a related experiment, researchers using pools of random library generated dsRNAs identified 34 gene products as being involved in the phagocytosis of Gram-negative bacteria (Ramet *et al.*, 2002). One of these gene products was identified as PGRP-LC. Work by Ramet and others found PGRP-LC to be the elusive receptor for Gram-negative bacteria (Choe *et al.*, 2002; Gottar *et al.*, 2002; Ramet *et al.*, 2002).

6.8   Global gene expression studies in MOA

It is now realized that genes are regulated as networks and many genes are co-regulated in response to unique cellular conditions. Whole-genome expression profiling has been facilitated greatly by the development and standardization of DNA microarrays. Knowledge of global changes in gene expression will improve our ability to predict drug effects, both therapeutic and toxic side-effects. Changes in gene expression patterns that occur in response to the treatment of cells with small molecules may reveal specific patterns of gene expression that might reflect or explain the activity of the compound. The *Drosophila* genome has the two most common microarray technologies available: high-density oligonucleotide (Affymetrix GeneChips) or cDNA microarrays (Arbeitman *et al.*, 2002). *Caenorhabditis elegans* genome arrays are also available. The advantages of using *Drosophila* or *C. elegans* in genome profiling are that the complete genome is represented on a single array, there is a lack of protein function redundancy and the cost of the commercially available microarrays is significantly less than human microarrays. If a compound targets a model system ortholog then one can scan the compound activity with invertebrate microarrays. The conservation of pharmacology combined with the genetically tractable tools available in *Drosophila* will allow identification of the pathway of drug action. Substantive clues can be gained by monitoring levels of gene expression in normal and mutant conditions. For example, antagonist treatment and loss of its protein target should confer overlapping phenotypes and transcriptional profiles. In *Drosophila* or *C. elegans*, overlapping gene expression clusters in drug-treated animals or S2 cells may compare to a phenotype of a mutation or RNAi treatment of a gene involved in the drug action. In our unpublished data, a transcriptional profile pattern of LPS-stimulated S2 cells that are pretreated with parthenolide extensively overlaps with a profile from Relish RNAi pretreatment. By extension, examining gene clusters of parthenolide treatment would have predicted correctly that it was targeting the NF-κB pathway.

Establishment of a gene expression profile database will be important for MOA studies with microarrays. In *Saccharomyces cerevisiae* it is now possible to compare drug-induced profiles with the existing gene expression database, thereby identifying relevant biological pathway(s) or functions for the drug target. For example, Hughes *et al.* (2000) used a compendium approach to compare drug-induced expression profiles to reference profiles of known cellular pathways. The pattern of gene expression changes observed is treated as a molecular fingerprint for the compound. Pattern-matching algorithms are then used to determine whether a compound has a similar molecular signature to that of gene deletion. For example, a recent study found that the transcriptional profile pattern that occurred in response to the treatment of yeast cells with the anesthetic dyclonine had the same signature response as deletion of the ERG2 gene, revealing the protein target of dyclonine as Erg2 (see Chapter 2 for more details).

Ideally, there should be a public database that would include profiling results in all RNAi, mutant and gene overexpression experiments to compare with drug treatment profiles. To that end, Spellman and Rubin (2002) compiled transcriptional profiling data from 88 experimental conditions. One concerning note from this study is that over 20% of genes whose expression cluster together across a range of experimental conditions map also cluster as adjacent genes within a chromosome but are otherwise not functionally related. This suggests that there may be another order of gene expression that is related to regional chromatin accessibility and could complicate the interpretation of the profiling analysis.

6.9 Selecting and advancing compound leads using model systems

Given the success of studying drug action in model systems it is possible to utilize specific phenotypes as an application in identifying and prioritizing drug candidates. Model systems technologies can offer an understanding of the effects of compounds in a living system as well as help to characterize, evaluate and prioritize a compound. In MOA studies described earlier in the chapter, the only requirement to begin screening for genetic modifiers is an observable effect on the model system. Initiating a drug discovery program with model systems where compound libraries will be screened requires tailoring assay development to specific biological readouts, preferably with a highly validated target or target pathway. In most cases, compound screening with mammalian biochemical or cell-based assays will be preferable, but it is feasible to design a high-throughput chemical screen using worms and *Drosophila*. An intriguing possibility is to 'humanize' the model system target. Kaletta *et al.* in Chapter 3 refer to their efforts of expressing human ion channels in *C. elegans*.

Identifying high-quality lead compounds using model systems requires an experimental design built on extensive information around the disease or pathway. Pertinent information includes: convincing evidence that model systems are high-content mimics of mammalian models; for example, do worm/fly mutants exist that model the disease or a pathway conserved in the disease?; use of highly specific and easily scorable phenotypes or assays (such as a reporter construct) that are amenable to automated equipment; and 'disease' phenotypes accessible to drug action, e.g. compounds should be able to mimic the mutant phenotype.

As described above, the extensive information around *C. elegans* presenilin suggests a viable entry point for high-throughput compound screening for potential leads in Alzheimer's disease. The presenilin proteins sel-12 and hop-1 in *C. elegans* process Notch by similar mechanisms to those of mammalian Notch and APP processing. Lead compounds were tested that inhibit presenilin enzyme activity and these compounds behave as partial loss of Notch function, suggesting that affected tissues in *C. elegans* are accessible to compounds (Figure 6.2). Genetic mutations in *sel-12* are partial loss of presenilin function and in the *sel-12;hop-1* double mutant are complete loss of presenilin function. Because *C. elegans* presenilin mutant phenotypes are easy to score, a compound screen could be devised to screen for drug-induced *sel-12;hop1* phenotypes in a *sel-12* mutant background. To allow large numbers of compounds to be screened, automated sorting machines are available that dispense worms and *Drosophila* embryos or larvae in a multiwell format (Furlong *et al.*, 2001).

The S2 cell-based system is also amenable to compound screens and, when compared to whole-organism approaches, has the advantage of miniaturization and high-throughput formatting. In Figure 6.7 we show that parthenolide inhibits a NF-κB pathway very similar to that of humans and the drug's transcriptional profiling signature closely matches that of an NF-κB (Relish) RNAi treatment. One could consider automated screening of compounds that inhibit NF-κB transcriptional activation of a reporter construct. In some cases, the lack of pathway redundancy in *Drosophila* and *C. elegans* may work to their advantage in screening technologies.

Lead compound discovery may be aided by evaluating transcriptional profiling. The specificity of the candidate drug can be tested by matching drug treatment patterns to gene expression profiles of RNAi directed to the validated target. An antagonist that is specific to the intended target should produce an expression profile similar to that of target RNAi. Suboptimal drugs will interact with non-intended targets. Many off-site targets in model organisms will likely translate to many off-site targets in humans.

6.10 Future perspectives

It is clear from our work in the pharmaceutical industry that there will continue to be a strong demand to understand how drugs work at the molecular level. Only recently has the MOA of acetaminophen – one of the most widely used drugs available for decades – come to light (Chandrasekharan *et al.*, 2002). The massive information-driven growth in fields such as computational chemistry, structural biology and bioinformatics is leading to unparalleled opportunities in drug design and empowered drug discovery. The strength of the chemical genetic approach stems from the ability of mutations to alter the function of a single gene product within the context of a complex cellular environment. Once hooked into a pathway, many new genomics tools can be brought to bear on any given problem. Advances in *Drosophila* and *C. elegans* research allow the combination of genome sequence information, genome-wide cDNAs, mapping protein interactions, gene expression profiles and genome-wide mutations in an unprecedented dissection of a complex organism. In general, the challenge of a model system biology group in an industrial setting is to balance throughput with quality biological information. There is a significant amount of potential to enhance all target validation methodologies, including model systems. Improvement in automation, miniaturization and visualization of biological processes offers the most promise.

Studies with compounds can be integrated with many of the evolving genomics and proteomics tools. This chapter summarized the advantages of *C. elegans* and *Drosophila* as model systems in understanding a broad spectrum of MOA and lead compound identification issues. However, model organism approaches when combined with other methods, in parallel or circuit, can produce a complete biochemical and genetic profile of the drug target protein(s). There are many evolving approaches in chemical genomics, such as protein profiling and cell-based chemical screenings, that were beyond the scope of this review chapter (Zheng and Chan, 2002). The technologies developed for work in *S. cerevisiae* remain the model of researchers in the multicellular world (see Chapter 2).

6.11 Acknowledgments

Lisa Moore carried out the fly experiments shown in Figure 6.2. The authors would like to thank Jenny Kopczynski, Ross Francis, Garth McGrath, Steve Doberstein, Dan Curtis, Mark Cockett and Petra Ross-MacDonald for ideas and input, and Ben Burley for technical assistance. Hong Xiao, Bo Guan, Libeng Chen and Tiffany Vora conducted experiments in the S2 cell system. The authors would like to thank Becket Feierbach for her thoughtful manuscript review and helpful ideas.

6.12 References

Abraham, R. T., Acquarone, M., Andersen, A., Asensi, A., Belle, R., Berger, F., Bergounioux, C., *et al.* (1995). Cellular effects of olomoucine, an inhibitor of cyclin-dependent kinases. *Biol. Cell* **83**, 105–120.

Arbeitman, M. N., Furlong, E. E., Imam, F., Johnson, E., Null, B. H., Baker, B. S., Krasnow, M. A., *et al.* (2002). Gene expression during the life cycle of *Drosophila melanogaster*. *Science* **297**, 2270–2275.

Austin, J. and Kimble, J. (1989). Transcript analysis of *glp-1* and *lin-12*, homologous genes required for cell interactions during development of *C. elegans*. *Cell* **58**, 565–571.

Basson, M. E., Thorsness, M. and Rine, J. (1986). *Saccharomyces cerevisiae* contains two functional genes encoding 3-hydroxy-3-methylglutaryl-coenzyme A reductase. *Proc. Natl. Acad. Sci. USA* **83**, 5563–5567.

Beitel, G. J., Clark, S. G. and Horvitz, H. R. (1990). *Caenorhabditis elegans* ras gene *let-60* acts as a switch in the pathway of vulval induction. *Nature* **348**, 503–509.

Brune, K. (2002). Next generation of everyday analgesics. *Am. J. Ther.* **9**, 215–223.

Bynum, W. F. (1970). Chemical structure and pharmacological action: a chapter in the history of 19th century molecular pharmacology. *Bull. Hist. Med.* **44**, 518–538.

Capdeville, R., Buchdunger, E., Zimmermann, J. and Matter, A. (2002). Glivec (STI571, imatinib), a rationally developed, targeted anticancer drug. *Nat. Rev. Drug Discov.* **1**, 493–502.

Caplen, N. J., Fleenor, J., Fire, A. and Morgan, R. A. (2000). dsRNA-mediated gene silencing in cultured *Drosophila* cells: a tissue culture model for the analysis of RNA interference. *Gene* **252**, 95–105.

Caplen, N. J., Parrish, S., Imani, F., Fire, A. and Morgan, R. A. (2001). Specific inhibition of gene expression by small double-stranded RNAs in invertebrate and vertebrate systems. *Proc. Natl. Acad. Sci. USA* **98**, 9742–9747.

Caponigro, F. (2002). Farnesyl transferase inhibitors: a major breakthrough in anticancer therapy? Naples, 12 April 2002. *Anticancer Drugs* **13**, 891–897.

Chandrasekharan, N. V., Dai, H., Roos, K. L., Evanson, N. K., Tomsik, J., Elton, T. S. and Simmons, D. L. (2002). From the Cover: COX-3, a cyclooxygenase-1 variant inhibited by acetaminophen and other analgesic/antipyretic drugs: cloning, structure, and expression. *Proc. Natl. Acad. Sci. USA* **99**, 13926–13931.

Choe, K. M., Werner, T., Stoven, S., Hultmark, D. and Anderson, K. V. (2002). Requirement for a peptidoglycan recognition protein (PGRP) in Relish activation and antibacterial immune responses in *Drosophila*. *Science* **296**, 359–362.

Choy, R. K. and Thomas, J. H. (1999). Fluoxetine-resistant mutants in *C. elegans* define a novel family of transmembrane proteins. *Mol. Cell* **4**, 143–152.

Clemens, J. C., Worby, C. A., Simonson-Leff, N., Muda, M., Maehama, T., Hemmings, B. A. and Dixon, J. E. (2000). Use of double-stranded RNA interference in *Drosophila* cell lines to dissect signal transduction pathways. *Proc. Natl. Acad. Sci. USA* **97**, 6499–6503.

Culetto, E. and Sattelle, D. B. (2000). A role for *Caenorhabditis elegans* in understanding the function and interactions of human disease genes. *Hum. Mol. Genet.* **9**, 869–877.

Cully, D. F., Vassilatis, D. K., Liu, K. K., Paress, P. S., Van der Ploeg, L. H., Schaeffer, J. M. and Arena, J. P. (1994). Cloning of an avermectin-sensitive glutamate-gated chloride channel from *Caenorhabditis elegans*. *Nature* **371**, 707–711.

Drews, J. and Ryser, S. (1997). The role of innovation in drug development. *Nat. Biotechnol.* **15**, 1318–1319.

Elbashir, S. M., Harborth, J., Lendeckel, W., Yalcin, A., Weber, K. and Tuschl, T. (2001). Duplexes of 21-nucleotide RNAs mediate RNA interference in cultured mammalian cells. *Nature* **411**, 494–498.

Esler, W. P. and Wolfe, M. S. (2001). A portrait of Alzheimer secretases – new features and familiar faces. *Science* **293**, 1449–1454.

Esler, W. P., Kimberly, W. T., Ostaszewski, B. L., Diehl, T. S., Moore, C. L., Tsai, J. Y., Rahmati, T., *et al.* (2000). Transition-state analogue inhibitors of gamma-secretase bind directly to presenilin-1. *Nat. Cell Biol.* **2**, 428–434.

Fares, H. and Greenwald, I. (1999). SEL-5, a serine/threonine kinase that facilitates *lin-12* activity in *Caenorhabditis elegans*. *Genetics* **153**, 1641–1654.

Ffrench-Constant, R. H., Steichen, J. C., Rocheleau, T. A., Aronstein, K. and Roush, R. T. (1993). A single-amino acid substitution in a gamma-aminobutyric acid subtype A receptor locus is associated with cyclodiene insecticide resistance in *Drosophila* populations. *Proc. Natl. Acad. Sci. USA* **90**, 1957–1961.

Ffrench-Constant, R. H., Anthony, N., Aronstein, K., Rocheleau, T. and Stilwell, G. (2000). Cyclodiene insecticide resistance: from molecular to population genetics. *Annu. Rev. Entomol.* **45**, 449–466.

Fleming, J. T., Squire, M. D., Barnes, T. M., Tornoe, C., Matsuda K., Ahnn, J., Fire, A., *et al.* (1997) *Caenorhabditis elegans* levamisole resistance genes *lev-1*, *unc-29* and *unc-38* encode functional nicotinic acetylcholine receptor subunits. *J. Neurosci.* **17**, 5843–5857.

Fortini, M. E., Simon, M. A. and Rubin, G. M. (1992). Signalling by the sevenless protein tyrosine kinase is mimicked by Ras1 activation. *Nature* **355**, 559–561.

Francis, R., McGrath, G., Zhang, J., Ruddy, D. A., Sym, M., Apfeld, J., Nicoll, M., *et al.* (2002). aph-1 and pen-2 are required for Notch pathway signaling, gamma-secretase cleavage of betaAPP and presenilin protein accumulation. *Dev. Cell* **3**, 85–97.

Fraser, A. G., Kamath, R. S., Zipperlen, P., Martinez-Campos, M., Sohrmann, M. and Ahringer, J. (2000). Functional genomic analysis of *C. elegans* chromosome I by systematic RNA interference. *Nature* **408**, 325–330.

Furlong, E. E., Profitt, D. and Scott, M. P. (2001). Automated sorting of live transgenic embryos. *Nat. Biotechnol.* **19**, 153–156.

Goldenberg, M. M. (1998). Safety and efficacy of sildenafil citrate in the treatment of male erectile dysfunction. *Clin. Ther.* **20**, 1033–1048.

Gottar, M., Gobert, V., Michel, T., Belvin, M., Duyk, G., Hoffmann, J. A., Ferrandon, D., *et al.* (2002). The *Drosophila* immune response against Gram-negative bacteria is mediated by a peptidoglycan recognition protein. *Nature* **416**, 640–644.

Greenspan, R. J., Tononi, G., Cirelli, C. and Shaw, P. J. (2001). Sleep and the fruit fly. *Trends Neurosci.* **24**, 142–145.

Greenwald, I. and Seydoux, G. (1990). Analysis of gain-of-function mutations of the *lin-12* gene of *Caenorhabditis elegans*. *Nature* **346**, 197–199.

Han, M. and Sternberg, P. W. (1990). let-60, a gene that specifies cell fates during *C. elegans* vulval induction, encodes a ras protein. *Cell* **63**, 921–931.

Han, Z. S. and Ip, Y. T. (1999). Interaction and specificity of Rel-related proteins in regulating *Drosophila* immunity gene expression. *J. Biol. Chem.* **274**, 21 355–21 361.

Hara, M. and Han, M. (1995). Ras farnesyltransferase inhibitors suppress the phenotype resulting from an activated ras mutation in *Caenorhabditis elegans*. *Proc. Natl. Acad. Sci. USA* **92**, 3333–3337.

Hedengren, M., Asling, B., Dushay, M. S., Ando, I., Ekengren, S., Wihlborg, M. and Hultmark, D. (1999). Relish, a central factor in the control of humoral but not cellular immunity in *Drosophila*. *Mol. Cell* **4**, 827–837.

Hedgepeth, C. M., Conrad, L. J., Zhang, J., Huang, H. C., Lee, V. M. and Klein, P. S. (1997). Activation of the Wnt signaling pathway: a molecular mechanism for lithium action. *Dev. Biol.* **185**, 82–91.

Hendricks, J. C., Finn, S. M., Panckeri, K. A., Chavkin, J., Williams, J. A., Sehgal, A. and Pack, A. I. (2000). Rest in *Drosophila* is a sleep-like state. *Neuron* **25**, 129–138.

Heptinstall, S. (1988). Feverfew – an ancient remedy for modern times? *J. R. Soc. Med.* **81**, 373–374.

Hughes, T. R., Marton, M. J., Jones, A. R., Roberts, C. J., Stoughton, R., Armour, C. D., Bennett, H. A., *et al.* (2000). Functional discovery via a compendium of expression profiles. *Cell* **102**, 109–126.

Hwang, J. M., Chang, D. J., Kim, U. S., Lee, Y. S., Park, Y. S., Kaang, B. K. and Cho, N. J. (1999). Cloning and functional characterization of a *Caenorhabditis elegans* muscarinic acetylcholine receptor. *Recept. Channels* **6**, 415–424.

Jang, M. J., Jwa, M., Kim, J. H. and Song, K. (2002). Selective inhibition of MAPKK WisI in the stress-activated MAPK cascade of *Schizosaccharomyces pombe* by novel berberine derivatives. *J. Biol. Chem.* **277**, 12 388–12 395.

Janssen, S., Cuvier, O., Muller, M. and Laemmli, U. K. (2000). Specific gain- and loss-of-function phenotypes induced by satellite-specific DNA-binding drugs fed to *Drosophila melanogaster*. *Mol. Cell* **6**, 1013–1024.

Jorgensen, E. M. and Mango, S. E. (2002). The art and design of genetic screens: *Caenorhabditis elegans*. *Nat. Rev. Genet.* **3**, 356–369.

Karim, F. D. and Rubin, G. M. (1998). Ectopic expression of activated Ras1 induces hyperplastic growth and increased cell death in *Drosophila* imaginal tissues. *Development* **125**, 1–9.

Kauffmann, R. C., Qian, Y., Vogt, A., Sebti, S. M., Hamilton, A. D. and Carthew, R. W. (1995). Activated *Drosophila* Ras1 is selectively suppressed by isoprenyl transferase inhibitors. *Proc. Natl. Acad. Sci. USA* **92**, 10 919–10 923.

Kazanietz, M. G., Lewin, N. E., Bruns, J. D. and Blumberg, P. M. (1995). Characterization of the cysteine-rich region of the *Caenorhabditis elegans* protein Unc-13 as a high affinity phorbol ester receptor. Analysis of ligand-binding interactions, lipid cofactor requirements, and inhibitor sensitivity. *J. Biol. Chem.* **270**, 10 777–10 783.

Kimble, J. and Simpson, P. (1997). The LIN-12/Notch signaling pathway and its regulation. *Annu. Rev. Cell Dev. Biol.* **13**, 333–361.

Koh, B. and Crews, C. M. (2002). Chemical genetics. A small molecule approach to neurobiology. *Neuron* **36**, 563–566.

Kwok, B. H., Koh, B., Ndubuisi, M. I., Elofsson, M. and Crews, C. M. (2001). The anti-inflammatory natural product parthenolide from the medicinal herb Feverfew directly binds to and inhibits IkappaB kinase. *Chem. Biol.* **8**, 759–766.

Lambie, E. J. and Kimble, J. (1991). Two homologous regulatory genes, *lin-12* and *glp-1*, have overlapping functions. *Development* **112**, 231–240.

Levitan, D. and Greenwald, I. (1995). Facilitation of lin-12-mediated signalling by *sel-12*, a *Caenorhabditis elegans* S182 Alzheimer's disease gene. *Nature* **377**, 351–354.

Levitan, D. and Greenwald, I. (1998). LIN-12 protein expression and localization during vulval development in *C. elegans*. *Development* **125**, 3101–3109.

Levy-Lahad, E., Wasco, W., Poorkaj, P., Romano, D. M., Oshima, J., Pettingell, W. H., Yu, C. E., *et al.* (1995). Candidate gene for the chromosome 1 familial Alzheimer's disease locus. *Science* **269**, 973–977.

Lewis, J. A., Fleming, J. T., McLafferty, S., Murphy, H. and Wu, C. (1987). The levamisole receptor, a cholinergic receptor of the nematode *Caenorhabditis elegans*. *Mol. Pharmacol.* **31**, 185–193.

Li, X. and Greenwald, I. (1997). HOP-1, a *Caenorhabditis elegans* presenilin, appears to be functionally redundant with SEL-12 presenilin and to facilitate LIN-12 and GLP-1 signaling. *Proc. Natl. Acad. Sci. USA* **94**, 12 204–12 209.

Li, Y. M., Xu, M., Lai, M. T., Huang, Q., Castro, J. L., DiMuzio-Mower, J., Harrison, T., *et al.* (2000). Photoactivated gamma-secretase inhibitors directed to the active site covalently label presenilin 1. *Nature* **405**, 689–694.

Matthews, D. J. and Kopczynski, J. (2001). Using model-system genetics for drug-based target discovery. *Drug Discov. Today* **6**, 141–149.

McClung, C. and Hirsh, J. (1999). The trace amine tyramine is essential for sensitization to cocaine in *Drosophila*. *Curr. Biol.* **9**, 853–860.

Micchelli, C. A., Esler, W. P., Kimberly, W. T., Jack, C., Berezovska, O., Kornilova, A., Hyman, B. T., *et al.* (2002). g-Secretase/presenilin inhibitors for Alzheimer's disease phenocopy Notch mutations in *Drosophila*. *FASEB J.* **1**, 1.

Moore, M. S., DeZazzo, J., Luk, A. Y., Tully, T., Singh, C. M. and Heberlein, U. (1998). Ethanol intoxication in *Drosophila*: genetic and pharmacological evidence for regulation by the cAMP signaling pathway. *Cell* **93**, 997–1007.

Muda, M., Worby, C. A., Simonson-Leff, N., Clemens, J. C. and Dixon, J. E. (2002). Use of double-stranded RNA-mediated interference to determine the substrates of protein tyrosine kinases and phosphatases. *Biochem. J.* **366**, 73–77.

O'Moore-Sullivan, T. M. and Prins, J. B. (2002). Thiazolidinediones and type 2 diabetes: new drugs for an old disease. *Med. J. Aust.* **176**, 381–386.

Parada, L. F., Tabin, C. J., Shih, C. and Weinberg, R. A. (1982). Human EJ bladder carcinoma oncogene is homologue of Harvey sarcoma virus ras gene. *Nature* **297**, 474–478.

Peyroche, A., Antonny, B., Robineau, S., Acker, J., Cherfils, J. and Jackson, C. L. (1999). Brefeldin A acts to stabilize an abortive ARF-GDP-Sec7 domain protein complex: involvement of specific residues of the Sec7 domain. *Mol. Cell* **3**, 275–285.

Ramet, M., Manfruelli, P., Pearson, A., Mathey-Prevot, B. and Ezekowitz, R. A. (2002). Functional genomic analysis of phagocytosis and identification of a *Drosophila* receptor for *E. coli*. *Nature* **416**, 644–648.

Reiter, L. T., Potocki, L., Chien, S., Gribskov, M. and Bier, E. (2001). A systematic analysis of human disease-associated gene sequences in *Drosophila melanogaster*. *Genome Res.* **11**, 1114–1125.

Sakowicz, R., Berdelis, M. S., Ray, K., Blackburn, C. L., Hopmann, C., Faulkner, D. J. and Goldstein, L. S. (1998). A marine natural product inhibitor of kinesin motors. *Science* **280**, 292–295.

Scussa, F. (2002). World's best-selling drugs. *Med. Ad. News* **21**, 1–46.

Shaw, P. J., Cirelli, C. Greenspan, R. J. and Tononi, G. (2000). Correlates of sleep and waking in *Drosophila melanogaster*. *Science* **287**, 1834–1837.

Shellenbarger, D. L. and Mohler, J. D. (1978). Temperature-sensitive periods and autonomy of pleiotropic effects of l(1)Nts1, a conditional notch lethal in *Drosophila*. *Dev. Biol.* **62**, 432–446.

Sherrington, R., Rogaev, E. I., Liang, Y., Rogaeva, E. A., Levesque, G., Ikeda, M., Chi, H., *et al.* (1995). Cloning of a gene bearing missense mutations in early-onset familial Alzheimer's disease. *Nature* **375**, 754–760.

Silverman, N. and Maniatis, T. (2001). NF-kappaB signaling pathways in mammalian and insect innate immunity. *Genes Dev.* **15**, 2321–2342.

Silverman, N., Zhou, R., Stoven, S., Pandey, N., Hultmark, D. and Maniatis, T. (2000). A *Drosophila* IkappaB kinase complex required for Relish cleavage and antibacterial immunity. *Genes Dev.* **14**, 2461–2471.

Simon, M. A. (2000). Receptor tyrosine kinases: specific outcomes from general signals. *Cell* **103**, 13–15.

Simon, M. A., Bowtell, D. D., Dodson, G. S., Laverty, T. R. and Rubin, G. M. (1991). Ras1 and a putative guanine nucleotide exchange factor perform crucial steps in signaling by the sevenless protein tyrosine kinase. *Cell* **67**, 701–716.

Sin, N., Meng, L., Wang, M. Q., Wen, J. J., Bornmann, W. G. and Crews, C. M. (1997). The anti-angiogenic agent fumagillin covalently binds and inhibits the methionine aminopeptidase, MetAP-2. *Proc. Natl. Acad. Sci. USA* **94**, 6099–6103.

Sirtori, C. R. and Pasik, C. (1994). Re-evaluation of a biguanide, metformin: mechanism of action and tolerability. *Pharmacol. Res.* **30**, 187–228.

Spellman, P. T. and Rubin, G. M. (2002). Evidence for large domains of similarly expressed genes in the *Drosophila* genome. *J. Biol.* **1**, 5.

Struhl, G. and Greenwald, I. (1999). Presenilin is required for activity and nuclear access of Notch in *Drosophila*. *Nature* **398**, 522–525.

Sun, H., Bristow, B. N., Qu, G. and Wasserman, S. A. (2002). A heterotrimeric death domain complex in Toll signaling. *Proc. Natl. Acad. Sci. USA* **99**, 12871–12876.

Tatchell, K. (1986). RAS genes and growth control in *Saccharomyces cerevisiae*. *J. Bacteriol.* **166**, 364–367.

Tuschl, T. (2002). Expanding small RNA interference. *Nat. Biotechnol.* **20**, 446–448.

Walker, R. J., Franks, C. J., Pemberton, D., Rogers, C. and Holden-Dye, L. (2000). Physiological and pharmacological studies on nematodes. *Acta Biol. Hung.* **51**, 379–394.

Wall, M. E. and Wani, M. C. (1995). Camptothecin and taxol: discovery to clinic – thirteenth Bruce F. Cain Memorial Award Lecture. *Cancer Res* **55**, 753–760.

Weinshenker, D., Garriga, G. and Thomas, J. H. (1995). Genetic and pharmacological analysis of neurotransmitters controlling egg laying in *C. elegans*. *J. Neurosci.* **15**, 6975–6985.

Westlund, B., Parry, D., Clover, R., Basson, M. and Johnson, C. D. (1999). Reverse genetic analysis of *Caenorhabditis elegans* presenilins reveals redundant but unequal roles for sel-12 and hop-1 in Notch-pathway signaling. *Proc. Natl. Acad. Sci. USA* **96**, 2497–2502.

Wittenburg, N. and Baumeister, R. (1999). Thermal avoidance in *Caenorhabditis elegans*: an approach to the study of nociception. *Proc. Natl. Acad. Sci. USA* **96**, 10477–10482.

Yochem, J. and Greenwald, I. (1989). *glp-1* and *lin-12*, genes implicated in distinct cell–cell interactions in *C. elegans*, encode similar transmembrane proteins. *Cell* **58**, 553–563.

Zheng, X. F. and Chan, T. F. (2002). Chemical genomics: a systematic approach in biological research and drug discovery. *Curr. Issues Mol. Biol.* **4**, 33–43.

7

Genetics and Genomics in the Zebrafish – from Gene to Function and Back

Stefan Schulte-Merker

7.1 Zebrafish – a model system with utilities beyond the study of development

Ever since the pioneering efforts of G. Streisinger in the early 1980s (Streisinger *et al.*, 1981), increasing numbers of researchers have taken on zebrafish as their favorite system in which to address questions of developmental, physiological and medical biology. A great variety of zebrafish methods and techniques have been compiled over the years and, owing to its popularity, zebrafish is one of the vertebrates whose genome currently is being sequenced. The purpose of this chapter is to provide an introduction to some of the advantages and shortcomings of the zebrafish as a model organism. There is no attempt to cover all of the detailed zebrafish methodologies, instead this chapter is designed to highlight some of the principles and approaches that are being taken with zebrafish in order to address biological questions.

Initially, zebrafish were used primarily to study early developmental processes such as gastrulation and neuronal patterning. The embryos are transparent through the early phases of development, and many of the processes of interest to the developmental biologist are readily observable simply by focusing up and down a dissecting microscope. Moreover, fertilization is external, allowing embryos to develop synchronously in a

Model Organisms in Drug Discovery. Edited by Pamela M. Carroll and Kevin Fitzgerald.
© 2003 John Wiley & Sons, Ltd. ISBN 0 470 84893 6

simple salt solution within a petri-dish. There is no shortage of embryos to work with because a single pair of adult fish will spawn every week, producing a few hundred embryos per mating. Embryos develop quickly and reach the end of somatogenesis by 24 h post-fertilization. The heart starts to beat at 28 h and the first blood cells can be seen rushing through the vasculature by 30 h. At 72 h the intestine undergoes peristaltic movements and most cell types in the visceral tract have differentiated (Schilling, 2002). By day 5 larvae start to feed, whereas prior to that point they relied on their yolk supply.

Over the years it has become appreciated that the ease of manipulating embryos and zebrafish larvae opens up the opportunity to study organo-genesis in ways not previously possible. Researchers have developed novel methods to study their favorite fish organ systems and have developed genetic screens that previously were considered to be impossible in vertebrate systems. One impressive demonstration of the advantages of zebrafish in designing and carrying out genetic screens was carried out in retinal axons. A screen was designed where fixed larvae (5 days old) were mounted in agarose and two different lipophilic dyes (DiI and DiO) were injected into distinct positions, thereby labeling two separate populations of retinal ganglion axons within the eye (Baier *et al.*, 1996; Trowe *et al.*, 1996). The dyes travel along the corresponding neurons until they reach the respective areas of the contralateral optic tectum, outlining both the neuronal path from retina to tectum and the retinotectal projection. The method was so reproducible and scalable that it could be used as a basis for a genetic screen: overall processing of one larva, including mounting, dye injection and analysis, took just 1 min, and scoring 125 000 larvae resulted in the identification of 144 mutants in approximately 35 genes that exhibited defects in their retinotectal projections.

Although this example is a particularly impressive one, it merely highlights the versatility of zebrafish as a screening tool. Various laboratories are involved in looking at processes as diverse as thrombosis (Jagadeeswaran and Sheenan, 1999), angiogenesis (Weinstein *et al.*, 1995; Habeck *et al.*, 2002), hematopoiesis (Thisse and Zon, 2002) and many other areas that require studying recent medically relevant events. It is this versatility, combined with genetics and methods to manipulate both embryos and larvae alike, that has contributed to the success of zebrafish.

7.2 Pathway conservation between humans and fish: what difference do 400 million years make?

A common ancestor between humans and zebrafish lived roughly 400 million years ago, which at times has raised the question of whether the similarities between the two species are outnumbered by the differences. This is a question of particular relevance to those who use zebrafish as an entry point to learn about vertebrate physiology and human disease, but has less relevance to those who study fish development and biology in their own right. There are a number of themes surrounding the issue of conserved function between fish and humans, and we will try briefly to address the more relevant issues, namely genome duplication and synteny as well as functional conservation.

In zebrafish and other teleosts one finds, in 20–30% of cases, two homologous genes compared with the mammalian counterpart. Apparently, this stems from partial genome duplication or duplication of the entire genome with subsequent loss of much of the duplicated material. The resulting paralogs vary in function and expression pattern, which can complicate the comparison with mammalian equivalents. Eighty percent of the zebrafish and human genomes appear to be syntenic (Barbazuk *et al.*, 2000), which is very helpful in determining homology relationships in cases where members of the same protein family are to be compared. A reasonably precise assessment of the exact extent of genome duplication will have to await completion of the zebrafish genome sequencing and annotation effort, which is expected to be finished in 2005 (http://www.sanger.ac.uk).

A seemingly attractive way to address the question of conserved gene function is to compare fish mutants in a particular gene with mouse mutants in the corresponding gene. At present, there are roughly 150 zebrafish mutants that have been cloned (Frohnhöfer, 2002; Golling *et al.*, 2002) but this number is not nearly high enough to allow a meaningful comparison. Only about half of these mutants exhibit a well-described phenotype and there is not a mouse mutant counterpart for all of them.

Is zebrafish the perfect model of humans and human disease based on functional conservation between zebrafish and mammals? The answer is 'no' if one takes the question to be whether zebrafish is a model system for humans in each and every single case investigated. The answer is 'yes', however, if one considers individual cases (or genes), where it turns out that the genetic pathways between zebrafish and mammals have been conserved and the function of genes within those pathways has not changed. Examples of this are plentiful (see review by Dooley and Zon, 2000) and, as long as one is willing to 'embrace the differences and cherish the similarities' (phrase borrowed from G. Duyk) between zebrafish and humans, zebrafish offer a powerful experimental and genetic system for the understanding of vertebrate biology and disease.

7.3   The zebrafish tool kit

From function to gene: genetic screens

From its infancy as a model system until today, being able to identify mutants has been the driving force behind most people's interest in studying zebrafish.

The generation time of zebrafish is 3 months, which is short in vertebrate terms. Adult fish are 1 inch in size and the housing costs are very low once the initial tank system has been installed. The transparency of zebrafish until stages where organogenesis is well underway or completed makes zebrafish the vertebrate system of choice for forward genetic screens designed to investigate this process. Phenotypes are easily identified and the underlying gene may be subsequently cloned.

In addition to standard genetics there is quite an arsenal of genetic tricks that can be applied to zebrafish, including the generation of haploid and gynogenetic embryos (for review, see Kimmel, 1989), as well as novel methods to carry out maternal effect screens (Pelegri and Schulte-Merker, 1999). Still, the most common screening scenario still remains the induction of mutations in the parental generation and breeding the mutagenized individuals until two generations later. The F2 individuals are mated and the phenotypes can be examined in a homozygous situation (see below). Mutagenesis is carried out by utilizing gamma rays, retroviral insertions and, most commonly, the chemical mutagen ethyl–nitrosourea (ENU). These methods will be compared briefly below.

Irradiating post-meiotic sperm with x-rays or gamma-rays was the first attempt to generate fish mutants in a systematic fashion (Chakrabarti *et al.*, 1983) and it was successful in terms of very efficiently generating mutations. Mutation rates up to 2% have been reported (Chakrabarti *et al.*, 1983). However, many of the mutant lines have proved difficult to maintain and characterize molecularly, because irradiation tends to induce large deletions and chromosomal rearrangements. Other attempts to circumvent these problems and to establish protocols that induce small deletions while maintaining chromosomal stability have failed (Lekven *et al.*, 2000) and, unless one deliberately desires to induce deletions, other methods for generating mutant lines are preferable.

Insertional mutagenesis has proved extremely useful in the case of P elements in *Drosophila*. In this system, the mutagen consists of a transposable element that inserts into chromosomal DNA and compromises the expression or function of the gene and gene product. When successful, it is fairly straightforward to identify the underlying gene, because the P element serves as a tag that facilitates cloning. In zebrafish, the group led by Nancy Hopkins has established a protocol that makes use of a pseudotyped virus that is injected into blastula-stage embryos and inserts its genome into the genomic DNA of the fish embryo (Amsterdam and Hopkins, 1999). In those cases where the insertion happens to occur in a cell whose descendants become future germ cells, the insertion is passed through the germline and will, in a fraction of cases, mutate a gene to yield a detectable phenotype. The key features of this technology are producing a high-titer viral stock and genotyping the F1 fish in order to select fish with the highest number of independent insertions. Any phenotype of interest can be characterized molecularly with relative ease by testing which insertion tag co-segregates with the mutant phenotype, followed by cloning the flanking regions of the insertion.

Although it is reasonably straightforward with this approach to identify the mutated gene once a phenotype has been identified, the low efficiency of generating insertions has kept the zebrafish field from adopting this approach broadly. In comparison with to the widely used chemical mutagen ENU (see below) the frequency of generating mutations is only 5–10% (Pelegri, 2002), which means that in order to obtain the same number of mutants as with an ENU screen, one needs to maintain 10–20 times as many tanks and set up 10–20 times as many crosses. Because many investigators are not content with identifying just one mutant and, ideally, would rather collect all genes essential for the process under study, ENU mutagenesis has been favored.

The alkylating agent ENU has been used in many large- and small-scale screens and an estimate of well over 10 000 mutants have been generated in the three largest screens to date (Driever *et al.*, 1996; Haffter *et al.*, 1996; Odenthal *et al.*, Tübingen 2000 Screen, unpublished). Adult male fish are bathed in a solution of ENU, inducing mutations in premeiotic germ cells. These founder males are crossed with females to generate F1 fish that are heterozygous for the mutations induced in the previous generation. The F1 fish are crossed with unrelated F1 fish that stem from independent mutagenesis events. Brother–sister matings within the resulting F2 generation produce F3 egglays that are homozygous with respect to the mutation induced in the parental founder male. Naturally, there are many mutations per founder male and it is not uncommon to uncover more than one mutant phenotype within a single F3 egglay. Despite the fact that it can be cumbersome to clone an ENU-induced mutant, there are a number of reasons why ENU screens are popular: they require very little expertise (compared with insertional mutagenesis) and ENU is very efficient in generating single-locus mutations (compared with the low mutagenesis rate using retroviral insertions and the large size deletions that affect more than one gene). The high hit rate also opens up the opportunity to identify, even with a middle-sized screen, a number of mutants that affect the biological process under study and hence to identify a number of genes that result in identical or similar phenotypes.

From gene to function: reverse genetics using morpholinos

With the ever-increasing number of publicly available expressed sequence tags (ESTs) and the prospect of a fully sequenced and annotated genome, the lack of reliable techniques to perform reverse genetics has become more evident in the last few years. Approaches such as injecting antisense mRNAs made *in vitro*, or RNA interference, have proved less than satisfactory up to now, even though considerable effort has been invested into these technologies. The turn-around for reverse genetics in zebrafish arrived with a particular antisense chemical called a 'morpholino'. This technology was shown to work with remarkable efficiency in both frogs and fish (Heasman *et al.*, 2000; Nasevicius and Ekker, 2000). Morpholinos are uncharged oligomers made from subunits containing an adenine, cytosine, guanine or thymidine base that is linked to a six-membered morpholine ring. Non-ionic phosphorodiamidate intersubunits link the morpholine ring containing one of the respective bases together.

Morpholinos work by one of two mechanisms. If directed against the 5′ UTR (untranslated region) and the region of the gene equivalent to the first translated ATG, a morpholino oligomer will bind to the targeted mRNA and block translation by steric hindrance. This is an RNAse H-independent mechanism, which probably contributes to the specificity of morpholino activity because RNAse H-dependent mechanisms tend to affect other non-targeted mRNAs as well.

The second mechanism by which morpholinos show efficacy is to target them to exon–intron boundaries (Draper *et al.*, 2001). Here, they interfere with the splice machinery of the cell and, in the few cases where attempted, lead to missplicing or exon skipping (G. Stott, unpublished observation).

Morpholinos are delivered to the zebrafish embryo through injection at the 1–4 cell stage. This is done manually with the aid of a simple dissecting scope and an injection set-up. An experienced person can inject around 1500 embryos in the course of a morning. Morpholinos are not charged, and embryos seem to tolerate nanogram amounts of most morpholinos without any apparent adverse reactions such as gastrulation abnormalities, retardation or necrosis, all of which are undesired side-effects often encountered when using alternative antisense strategies. The high degree of tolerance that zebrafish embryos and larvae exhibit when challenged with morpholinos might well be the reason why morpholinos are superior to other chemistries. There is no obvious reason why morpholinos should bind better to their target mRNA compared with other antisense technologies but morpholinos might turn out to be one of very few chemicals enabling sufficiently high amounts of reagent per cell to enable a blocking effect. The amount of RNA in an early zebrafish embryo equals roughly 1 μg, 50 ng of which can be estimated to be mRNA. Injecting nanogram amounts of a particular morpholino directed against one specific mRNA into the early embryo is therefore a vast excess concentration of blocking agent versus target molecule. Even when diluted out over time through cell cleavages and some degradation, there are plenty of morpholino molecules left to accomplish inhibition of translation.

After it was discovered that morpholinos were efficacious in frogs (Heasman *et al.*, 2000) and zebrafish (Nasevicius and Ekker, 2000), it was readily appreciated that they were useful not only in verifying gene identity at the end of a positional cloning effort (see below) but also by paving the way for systematic reverse genetics in these organisms. It was suddenly feasible to study the function of a large number of vertebrate genes on the level of the whole vertebrate organism.

From gene to mutant to function: targeted mutagenesis

One of the obvious shortcomings of zebrafish has always been the lack of a specific technology that has made the mouse so useful: the knock-out (removal) of genes via homologous recombination in embryonic stem (ES) cells. It is of small comfort that zebrafish are in good company in this respect, but it would be highly desirable to be able to eliminate genes at will and study the resulting phenotype in a loss-of-function situation. The use of morpholinos (see above) is helpful in those cases where an early-acting gene is of interest, but the knock-down caused by morpholinos is transient (it lasts up to 5 days) and does not generate stable mutant lines.

Establishing ES cells and keeping them in culture in order to be able to attempt homologous recombination *in vitro* has been the bottleneck in zebrafish and many other systems (there might be other bottlenecks down the road, but for the time being this is the most eminent problem). Only very recently was it reported that a primary spleen cell line from rainbow trout (Ganassin and Bols, 1999) is able to support the growth of zebrafish blastomeres in culture and to keep most of the blastomeres in an undifferentiated state (Ma *et al.*, 2001). Blastomeres were transplanted into host zebrafish embryos and were able to populate the germline (Ma *et al.*, 2001). Thus, they fulfill one important requirement for ES cells. Further experiments are underway to determine whether these blastomeres can undergo homologous recombination *in vitro* (Paul Collodi, personal communication), which would satisfy another important criterion. Interestingly, it has been shown recently that injecting morpholinos directed against the 'dead end' gene renders the injected embryos void of pregonial germ cells (Ciruna *et al.*, 2002). Such embryos would be ideal recipients for *in vitro* manipulated zebrafish ES cells, because if the ES cells were to populate the germline, the whole germline would consist of manipulated cells of the desired genotype, thereby circumventing the nuisance of mosaic germlines.

In the absence of ES cell-mediated knock-out technologies, other means were found to create stable mutant lines in genes of interest. Wienholds *et al.* (2002) have reported a way of generating multiple ENU-induced alleles in a gene of interest. They have mutagenized zebrafish males using standard protocols (Pelegri, 2002) and generated a library of F1 males. Sperm samples were taken and stored frozen, whereas DNA was prepared from the remainder of the fish. Over 2700 DNA samples were used as templates for polymerase chain reactions (PCRs), amplifying 2.7 kbp of a gene of interest, in this case *rag-1*. Subsequent sequencing revealed 15 point mutations, one of which resulted in a premature stop codon. Going back to the corresponding sperm sample, Wienholds and colleagues established a stable *rag-1* mutant line.

The method outlined above is the only one at present that allows a mutant zebrafish line to be defined in a preselected gene. In contrast to the knock-out technology in mice, it is impossible to predetermine which nucleotide will be mutated, let alone the possibility of deleting whole exons. On the other hand, the method will provide the investigator with a number of mutant alleles per gene to analyze, which is often very useful. The method is scalable and, depending on the number of sequencing lanes one is willing to run, there is no *a priori* reason why particular genes should be untractable by this approach. Importantly, the frozen sperm and the DNA constitute a resource that can be used over and over again, making it necessary to generate this resource only once.

7.4   Drug screening in zebrafish

There is yet another interesting twist to screens and phenotypes in zebrafish. In recent years, an increasing number of laboratories have caught on to the idea of testing the effects of pharmacological drugs on zebrafish embryos. In hindsight, the idea makes perfect sense. There is a high degree of conservation between vertebrate genes and, consequently, the physiological effect that a particular drug causes in mammals should have a high chance of affecting the orthologous target protein in zebrafish. This notion has been put to the test in a number of cases and has been found to work in some instances. Interfering with nitric oxide levels by nitroprusside or N (G)-nitro-L-arginine methyl ester (L-NAME), for example, results in changes in vessel diameter when applied to zebrafish larvae (Fritsche *et al.*, 2000). A complete loss of all vessels was reported by Chan *et al.* (2002), who used the tyrosine kinase inhibitor PTK787/ZK222584 to block the activity of vascular endothelial growth factor receptors. Warfarin, an inhibitor of hemostatic proteins in mammals, induces bleeding in zebrafish (Jagadeeswaran and Sheenan, 1999), which is consistent with the notion of warfarin inhibiting the process of thrombosis and coagulation in both mammals and fish.

A particularly elegant example of the possible uses of drugs in zebrafish was provided by Langheinrich *et al.* (2002), who studied the function of p53, a protein known to cause cell cycle arrest and apoptosis in cells that are severely stressed or have undergone DNA damage. Using morpholinos, they demonstrated that the lack of p53, as such, has no detectable morphological effect in zebrafish embryos, a scenario very comparable to mouse embryos mutant in p53. However, when exposed to UV light (inducing DNA fragmentation) or when challenged with the anticancer drug camptothecin, zebrafish embryos devoid of p53 exhibited a far lesser degree of apoptosis than control embryos. This experiment shows that p53 function is conserved across species boundaries and, at the same time, that camptothecin acts through p53.

The application of chemicals to zebrafish embryos is easily accomplished by bathing the embryos in the respective chemical or, in those cases where penetration turns out to be problematic, injecting the compounds into the embryo. Because this can even be done in a 96-well format (Peterson *et al.*, 2000), scenarios of screening chemicals in zebrafish become feasible. There is one elegant example of this approach in which a cell cycle arrest zebrafish mutant was challenged with thousands of compounds in order to identify successfully the small number of compounds that were able to revert and rescue the mutant phenotype (Len Zon, personal communication).

Clearly, the zebrafish has potential as a screening tool and assay system for testing compounds and drugs. How far that potential can reach will, in large part, be determined by the degree of automation that can be integrated into the screening process.

7.5 Organs in color: transgenic zebrafish

In addition to forward and reverse genetics, zebrafish offer the opportunity to interfere with gene activity by overexpressing genes, either through injecting *in vitro* synthesized mRNAs or through transgenesis. The former method applies to genes and processes that have an early effect on development or organ formation. The half-life of the injected mRNA and the corresponding protein determines how late a process can be interfered with. Usually, this is a matter of hours or a couple of days at best.

The latter method, transgenesis, is employed in those cases where stable expression of a particular gene is desired, either ubiquitously or in a time- and tissue-specific manner. Transgenes in zebrafish are commonly generated via injection of DNA into the zygote (Gilmour *et al.*, 2002). By a poorly understood process, the DNA is amplified by the embryo and DNA concatamers are integrated at random positions (Stuart *et al.*, 1998). Integration only happens occasionally at the one cell stage, and as a consequence the founder animal (i.e. the fish that initially got injected) more often than not is mosaic, with some cells carrying the transgene and others not. Consequently, it is necessary to test whether the germline of any founder fish carries the transgene. This is accomplished by crossing the founder fish and examining the resulting progeny via PCR or, alternatively, by visual inspection of the animals in cases where a fluorescent gene product results from the transgene. Transgenesis rates in the range 0–20% using this method (Higashijima *et al.*, 1997; Gilmour *et al.*, 2002; Langenau *et al.*, 2003; N.

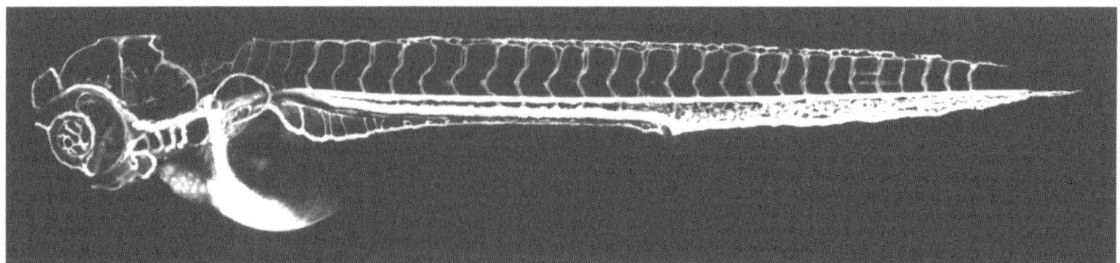

Figure 7.1   Angiography of a live zebrafish larva at 3 days of age. Anterior is to the left. Note the high resolution of individual vessels, which are fluorescently labeled

Scheer, personal communication). Once established, the transgenic line can be maintained by conventional breeding and the transgene is passed onto the next generations in a strictly Mendelian fashion.

It is the transparency of zebrafish that makes using transgenes attractive to researchers. Although transgenic fish have been put to use in a number of cases before, it is the elegant combination of transparency and fluorescently labeled proteins such as green fluorescent protein (GFP) that offers advantages peculiar to the zebrafish (see Figure 7.1). Fluorescent proteins under the control of specific promoters allow the generation of transgenic lines that display fluorescently marked blood (Long *et al.*, 1997), blood vessels (Lawson and Weinstein, 2002) and labeled lymphoid cells (Langenau *et al.*, 2003), to name a few examples. Such lines are useful for cell sorting specific populations but, more importantly, they offer the opportunity to observe biological processes over time *in vivo* with minimal interference. A beautiful example of this can be viewed under http://dir.nichd.nih.gov/lmg/uvo/weinslab.html where a rare chance to observe sprouting blood vessels *in vivo* is offered.

Several GFP-labeled lines also have been utilized for screens, where they provide the added advantage of screening the same embryo with more than one assay. For instance, a transgenic line that expresses GFP under the control of a vessel-specific promoter can be analyzed in a screen for mutants lacking vessels, the same embryos can be checked for motility defects a day later and yet another day later they can be fixed and scored for defects in ossification.

7.6   Genomic technologies

With all genetic model systems, the development of genomic tools goes hand in hand with genetics, because every interesting phenotype raises an immediate question: which gene has been mutated to cause the phenotypic alteration?

Developing genomic tools such as libraries, meiotic mapping panels and large-insert libraries for zebrafish has been slow initially. The work has sped up considerably, however, with the success of the first two large-scale forward genetic screens (Driever *et al.*, 1996; Haffter *et al.*, 1996). These ground-breaking screens succeeded in demonstrating that there were plenty of mutants to work on, and since that time genomic technologies have advanced rapidly. The zebrafish genome is about 1.7 Gbp in size, which is a little more than half the genome size of humans and mice. There are 25 chromosomes (haploid set) and approximately 2700 cM (Postlethwait *et al.*, 1994, and references therein).

What steps are involved in cloning a mutant of choice? Basically, there are five steps: mapping the mutant to a linkage group (chromosome); identifying flanking markers that define a chromosomal interval in which the mutated gene is located; generating markers within the interval that allow narrowing down of the interval size; sequencing the region of interest; and identifying the gene in question among the coding units within the region. There are detailed descriptions for how to carry out all of these steps elsewhere (Geisler, 2002, and references therein), therefore it will suffice here to give a tour-de-raison through the process, highlighting the existing public resources and pointing out the time-lines involved in all of these steps.

First, once a mutant has been identified, it is necessary to determine which chromosome the mutated gene resides on. To that end, a heterozygous carrier is outcrossed with a wild-type fish from a polymorphic strain, and carriers are identified from the resulting filial generation. These fish are used to produce homozygous mutant as well as sibling embryos, both of which are collected separately. Sorting of homozygous embryos is done phenotypically. The DNA from both mutant and sibling pools is then used to carry out a number of PCRs with primers amplifying so-called CA-repeats (microsatellites) – short DNA fragments that differ in length between polymorphic strains. By comparing whether particular CA-markers are co-segregating with the homozygous mutant embryos, it is possible in most cases to establish a linkage of the mutant gene with one or more of the polymorphic markers. This candidate linkage is then confirmed by testing individual embryos with such markers, which confirms and establishes the number of recombination events between the markers and the mutant locus. Because the PCR products have been mapped previously, both meiotically and on a radiation hybrid map, the position of the PCR products is known with respect to the chromosome (Knapik *et al.*, 1996).

Commonly, a marker set of roughly 200–250 polymorphic markers is used. Given the genome size of 2700 cM, the average resolution that can be achieved with this method is of the order of 10 cM. Agarose gels are used to resolve the polymorphic markers (Geisler, 2002) or, alternatively, acrylamide gels can be employed, allowing the use of 96-well capillary systems such as the ABI 3700 or MegaBACE (T. Wagner, personal communication) and a higher throughput. Once a mapping pipeline has been set up (which involves considerable work initially), one person can put two to three mutants on the linkage map per week (P. Beeckmann and T. Wagner, personal communication).

The information that one obtains from this initial mapping is very useful. In cases where a lot of mutants are to be mapped, binning the mutants into 'chromosomal groups' tremendously reduces the amount of complementation work that needs to be done to determine the number of genes, because only mutants mapping to the same linkage group need be considered for complementation crosses. Also, getting information about the rough position of the mutated gene of interest opens the door for a possible candidate gene approach, where candidate genes in the vicinity of the mutant locus can be considered for further linkage analysis.

The second step in a positional cloning exercise consists of defining the closest markers left and right of the locus of interest. To that end, all available markers in the region determined in step one are tested for linkage on a single embryo basis. This ideally identifies the two flanking markers that show the fewest recombination events with the mutant locus. The first map provided for the zebrafish anchoring CA-repeats (simple sequence-length polymorphisms) on the map consisted of 102 markers (Knapik *et al.*, 1996), but now over 10 000 CA-repeat markers are available (Zebrafish Webserver, http://zebrafish.mgh.harvard.edu), and more markers are added onto the map at a regular pace. Testing an additional 10 markers on a panel of 96 embryos usually will take only a few days. Not all of these markers may turn out to be polymorphic in the two strains that are being used in a particular experiment, but in many cases investigators have been able to limit the interval size to a couple of centimorgans (one centimorgan equals roughly 660 kbp) or less.

During the third step, the markers defining the interval are used to inititate a chromosomal walk. Genomic libraries of high quality have been made available very recently. From every new BAC, PAC or YAC, new SNPs (single-nucleotide polymorphisms) can be generated and tested for recombi-nation events. Collecting mutant embryos from a particular strain is not limiting in fish, and usually more than 2500 embryos (equaling 5000 meioses) are used for fine mapping, resulting in a resolution of 0.02 cM (or 13 kbp). Once the interval has been narrowed down sufficiently, the whole region is sequenced. Sequencing is the fourth step and takes about 4 weeks, depending on the expertise and the number of sequencing lanes available. From the genomic sequence, enough coding information can be retrieved to make predictions about the genes within the region.

The final step is to prove which one of the genes, if mutated, is responsible for the phenotype. There are a number of ways to accomplish this, and in most cases a combination of approaches is taken. The candidate gene is sequenced in both its wild type and mutant allelic form. Moreover, if the injection of a phospho-morpholino against the candidate gene can phenocopy the mutant phenotype, then this is a strong indication that the correct gene has been found. Also, expression of the mRNA of the respective gene should be detectable at or before the stage where the phenotype becomes apparent and ideally is restricted to the tissue affected by the phenotype. This final step can take anywhere from 2 weeks (in those cases where multiple mutant alleles are available and all of them carry convincing molecular lesions) to 2 months (in those cases where a phospho-morpholino needs to be ordered and the mutations are difficult to identify on the molecular level).

Although none of the technologies necessary for the positional cloning approach outlined above are unique to zebrafish, there are a couple of specifics that should be borne in mind. Unlike in other vertebrate systems, it is comparatively easy to collect a few thousand mutant embryos. Consequently, it is possible to let the fish do much of the 'genetic work', such that fine mapping with a very high degree of resolution allows a quick narrowing down of the interval in question. The downside to this approach is that one needs to wait for an entire generation time until one is in the position to start collecting homozygous mutant mapping embryos. Therefore, with any positional cloning project one will never be able to push the time-lines below the biological limits of generation time. However, the molecular work will, in years to come, become more efficient and will be supported by more complete resources such as libraries, expanded marker sets and the zebrafish genome sequence. This will considerably decrease the time-lines for positional cloning projects.

7.7 Outlook: the future has stripes

Zebrafish have evolved rapidly from a pet-shop inhabitant to a widely used genetic and experimental system. The times are long past when zebrafish researchers unvaryingly started their seminars by explaining why they work on zebrafish. The available resources and technologies that have been developed in zebrafish over the last few years are truly impressive. More development, however, is still needed. For example, setting up large-scale genetic screens where thousands of embryos or larvae are scored on a daily basis for 6 months remains very difficult on the screeners. In this area any sort of automated screening would be highly desirable. Semi-automated image capturing can be envisaged for at least a number of assays and would be a step forward in terms of time-lines and labor costs for a screen. Another area that would benefit from shorter time-lines is positional cloning. Starting with a mapping panel (48 or 96 mutant and sibling embryos each from a mapping cross) in hand, positional cloning of a mutant can take anywhere from 3 months to 1 year.

Here, the steps of assembling a physical contiguity is often rate limiting, however, with a fully annotated genome sequence well on its way this will become much less of an issue.

The versatility of zebrafish will undoubtedly continue to excite scientists. There will be more forward genetic screens using increasingly sophisticated assays and endpoints that will allow the identification of novel gene functions in increasingly complex assay systems (e.g. Farber *et al.*, 2001). There will be large-scale reverse genetic screens in which whole classes of proteins will be scanned for their role in a biological process of interest. Targeted mutagenesis will be used to generate stable mutant lines that do not exhibit a lethal phenotype on their own and can therefore be used as the basis for screens in genetically sensitized backgrounds. The number of transgenic lines that express fluorescent proteins under the control of a cell-type specific promoter will increase, and some of these will constitute the basis for screens utilizing cameras instead of the human eye as a first filter. Sensitized genetic backgrounds and the possibility for semi-automated readouts can be combined with compound screens, where thousands of chemicals are being tested for their effect on a whole organism level. Although this technology is unlikely to reach ultrahigh-throughput screening levels where millions of compounds are being tested, compound screens in fish could be useful to test those compounds that stem from a cell-based high-throughput screen and that need to be screened for further efficacy, toxicity or teratogenic side-effects (Nagel, 2002).

Finally, for those whose foremost interest is studying human diseases, it will be an interesting challenge to create human disease models that can be utilized in combination with the technologies listed above. One recent interesting example of this has been reported by Langenau *et al.* (2003), who described the induction of clonally derived T-cell acute lymphoblastic leukemia in zebrafish transgenic for the mouse *c-myc* gene. Suppressor screens using disease models such as this offer an exciting avenue for understanding better the genes contributing to human disease states, thereby defining future potential drug targets. Here, and in other areas of developmental, physiological and medical relevance, the zebrafish system will continue to make valuable contributions.

7.8 Acknowledgments

I would like to thank P. Beeckmann, T. Kidd, U. Langheinrich, N. Scheer and G. Stott for discussions and reading of the manuscript. H. Habeck provided the figure. Owing to space limitations, in many cases reviews are cited rather than original publications and I apologize to those whose original work I was not able to cite.

7.9 References

Amsterdam, A. and Hopkins, N. (1999). Retrovirus-mediated insertional mutagenesis in zebrafish. *Methods Cell Biol.* **60**, 87–98.

Baier, H., Klostermann, S., Trowe, T., Karlstrom, R. O., Nüsslein-Volhard, C. and Bonhoeffer, F. (1996). Genetic dissection of the retinotectal projection. *Development* **126**, 415–425.

Barbazuk, W. B., Korf, I., Kadavi, C., Heyen, J., Tate, S., Wun, E., Bedell, J.A., *et al.* (2000). The syntenic relationship of the zebrafish and human genomes. *Genome Res.* **10**, 1351–1358.

Chan, J., Bayliss, P. E., Wood, J. M. and Roberts, T. M. (2002). Dissection of angiogenic signaling in zebrafish using a chemical genetic approach. *Cancer Cell* **1**, 257–265.

Chakrabarti, S., Streisinger, G., Singer, F. and Walker, C. (1983). Frequency of gamma-ray induced specific locus and recessive lethal mutations in mature germ cells of the zebrafish, *Brachydanio rerio*. *Genetics* **103**, 109–123.

Ciruna, B., Weidinger, G., Knaut, H., Thisse, B., Thisse, C., Raz, E. and Schier, A. (2002). Production of maternal-zygotic mutant zebrafish by germ-line replacement. *Proc. Natl. Acad. Sci. USA* **99**, 14919–14924.

Dooley, K. and Zon, L. I. (2000). Zebrafish: a model system for the study of human disease. *Curr. Opin. Genet. Dev.* **10**, 252–256.

Draper, B., Morcos, P. A. and Kimmel, C. B. (2001). Inhibition of zebrafish fgf8 pre-mRNA splicing with morpholino oligos: a quantifiable method for gene knockdown. *Genesis* **30**, 154–1566.

Driever, W., Solnica-Krezel, L., Schier, A. F., Neuhauss, S. C. F., Malicki, J., Stemple, D. L., Stainier, D. Y. R., *et al.* (1996). A genetic screen for mutations affecting embryogenesis in zebrafish. *Development* **123**, 37–46.

Farber, S. A., Pack, M., Ho, S. Y., Johnson, I. D., Wagner, D. S., Dosch, R., Mullins, M. C., *et al.* (2001). Genetic analysis of digestive physiology using fluorescent phospholipid reporters. *Science* **292**, 1385–1388.

Frohnhöfer, H. G. (2002). Table of zebrafish mutants. In *Zebrafish*, C. Nüsslein-Volhard and R. Dahm (eds), pp. 237–292. Oxford: Oxford University Press.

Fritsche, R., Schwerte, T. and Pelster, B. (2000). Nitric oxide and vascular reactivity in developing zebrafish, *Danio rerio*. *Am. J. Physiol. Reg. Integr. Comp. Physiol.* **279**, 2200–2207.

Ganassin, R. C. and Bols, N. C. (1999). A stromal cell line from rainbow trout spleen, RTS34ST, that supports the growth of rainbow trout macrophages and produces conditioned medium with mitogenic effects on leukocytes. *In Vitro Cell Dev. Biol. Anim.* **35**, 80–86.

Geisler, R. (2002). Mapping and cloning. In *Zebrafish*, C. Nüsslein-Volhard and R. Dahm (eds), pp. 175–212. Oxford: Oxford University Press.

Gilmour, D. T., Jessen, J. R. and Lin, S. (2002). Transgenesis. In *Zebrafish*, C. Nüsslein-Volhard and R. Dahm (eds), pp. 121–143. Oxford: Oxford University Press.

Golling, G., Amsterdam, A., Sun, Z., Antonelli, M., Maldonado, E., Chen, W., Burgess, S., *et al.* (2002). Insertional mutagenesis in zebrafish rapidly identifies genes essential for early vertebrate development. *Nat. Genet.* **31**, 135–140.

Habeck, H., Walderich, B., Odenthal, J., Maischein, H.-M., Tübingen 2000 Screen Consortium and Schulte-Merker, S. (2002). Analysis of a zebrafish VEGF receptor mutant reveals specific disruption of angiogenesis. *Curr. Biol.* **12**, 1405–1412.

Haffter, P., Granato, M., Brand, M., Mullins, M. C., Hammerschmidt, M., Kane, D. A., Odenthal, J., *et al.* (1996). The identification of genes with unique and essential functions in the development of the zebrafish, *Danio rerio. Development* **123**, 1–36.

Heasman, J., Kofron, M. and Wylie, C. (2000). Beta-catenin signaling activity dissected in the early *Xenopus* embryo: a novel antisense approach. *Dev. Biol.* **222**, 124–134.

Higashijima, S., Okamoto, H., Ueno, N., Hotta, Y. and Eguchi, G. (1997). High-frequency generation of transgenic zebrafish which reliably express gfp in whole muscles or the whole body by using promoters of zebrafish origin. *Dev. Biol.* **192**, 289–299.

Jagadeeswaran, P. and Sheenan, J. P. (1999). Analysis of blood coagulation in the zebrafish. *Blood Cells Mol. Dis.* **25**, 239–249.

Kimmel, C. B. (1989). Genetics and early development of zebrafish. *Trends Genet.* **5**, 283–288.

Knapik, E. W., Goodman, A., Atkinson, O. S., Roberts, C. T., Shiozawa, M., Sim, C. U., Weksler-Zangen, S., *et al.* (1996). A reference cross DNA panel for zebrafish (*Danio rerio*) anchored with simple sequence length polymorphisms. *Development* **123**, 451–460.

Langenau, D. M., Traver, D., Ferrando, A. A., Kutok, J. L., Aster, J. C., Kanki, J. P., Lin, S., *et al.* (2003). Myc-induced T cell leukemia in transgenic zebrafish. *Science* **299**, 887–890.

Langheinrich, U., Hennen, E., Stott, G. and Vacun, G. (2002). Zebrafish as a model organsim for the identification and characterization of drugs and genes affecting p53 signaling. *Curr. Biol.* **12**, 2023–2028.

Lawson, N. D. and Weinstein, B. M. (2002). *In vivo* imaging of embryonic vascular development using transgenic zebrafish. *Dev Biol.* **248**, 307–318.

Lekven, A. C., Helde, K. A., Thorpe, C. J., Rooke, R. and Moon, R. T. (2000). Reverse genetics in zebrafish. *Physiol. Genom.* **2**, 37–48.

Long, Q., Meng, A., Wang, H., Jessen, J. R., Farrell, M. J. and Lin, S. (1997). GATA-1 expression pattern can be recapitulated in living transgenic zebrafish using GFP reporter gene. *Development* **124**, 4105–4111.

Ma, C., Fan, L., Ganassin, R., Bols, N. and Collodi, P. (2001). Production of zebrafish germ-line chimeras from embryo cell cultures. *Proc. Natl. Acad. Sci. USA* **98**, 2461–2466.

Nagel, R. (2002). DarT: the embryo test with the zebrafish *Danio rerio* – a general model in ecotoxicology and toxicology. *ALTEX* **19** (Suppl. 1), 38–48.

Nasevicius, A. and Ekker, S. C. (2000). Effective targeted gene 'knockdown' in zebrafish. *Nat. Genet.* **26**, 216–220.

Pelegri, F. (2002). Mutagenesis. In *Zebrafish*, C. Nüsslein-Volhard and R. Dahm (eds), pp. 145–174. Oxford: Oxford University Press.

Pelegri, F. and Schulte-Merker, S. (1999). A gynogenesis-based screen for maternal-effect genes in the zebrafish, *Danio rerio. Methods Cell Biol.* **60**, 1–20.

Peterson, R. T., Link, B. A., Dowling, J. E. and Schreiber, S. L. (2000). Small molecule developmental screens reveal the logic and timing of vertebrate development. *Proc. Natl. Acad. Sci. USA* **97**, 12965–12969.

Postlethwait, J. H., Johnson, S. L., Midson, C. N., Talbot, W. S., Gates, E., Ballinger, E. W., Africa, D., *et al.* (1994). A genetic linkage map for the zebrafish. *Science* **264**, 699–703.

Schilling, T. F. (2002). The morphology of larval and adult zebrafish. In *Zebrafish*, C. Nüsslein-Volhard and R. Dahm (eds), pp. 59–94. Oxford: Oxford University Press.

Streisinger, G., Walker, C., Dower, N., Knauber, D. and Singer, F. (1981). Production of clones of homozygous diploid zebrafish (*Brachydanio rerio* I). *Nature* **291**, 293–296.

Stuart, G. W., McMurray, J. V. and Westerfield, M. (1998). Replication, integration and stable germ-line transmission of foreign sequences injected into early zebrafish embryos. *Development* **103**, 403–412.

Thisse, C. and Zon, L. I. (2002). Organogenesis – heart and blood formation from the zebrafish point of view. *Science* **295**, 457–462.

Trowe, T., Klostermann, S., Baier, H., Granato, M., Crawford, A. D., Grunewald, B., Hoffmann, H., *et al.* (1996). Mutations disrupting the ordering and topographic mapping of axons in the retinotectal projection of the zebrafish, *Danio rerio*. *Development* **123**, 439–450.

Weinstein, B. M., Stemple, D. L., Driever, W. and Fishman, M. C. (1995). Gridlock, a localized heritable vascular patterning defect in the zebrafish. *Nat. Med.* **1**, 1143–1147.

Wienholds, E., Schulte-Merker, S., Walderich, B. and Plasterk, R. (2002). Target-selected inactivation of the zebrafish *rag1* gene. *Science* **297**, 99–102.

8

Lipid Metabolism and Signaling in Zebrafish

Shiu-Ying Ho, Steven A. Farber and **Michael Pack**

Although best known as a model organism used in developmental studies, the zebrafish is also suited to physiological analysis. Zebrafish process dietary lipids in a manner that closely resembles humans, and lipid metabolism can be inhibited by drugs used to treat human lipid disorders. Zebrafish also utilize prostanoid lipid signaling molecules, such as the prostaglandins and thromboxanes, and their synthesis can be inhibited by commonly prescribed non-steroidal antiinflammatory drugs. This chapter reviews studies devoted to lipid metabolism in zebrafish and identifies screening strategies for the identification of novel regulators of dietary lipid processing and prostanoid synthesis.

8.1 Introduction

As components of cell membranes, mediators of cell signaling and an energy source, lipids play an essential role in the physiology of all vertebrate cells. Given such diverse roles, it is not surprising that lipids also are important modulators of human disease. Perturbation of lipid metabolism is associated with heritable and acquired disease syndromes that predispose affected individuals to diabetes mellitus and atherosclerosis (Garg, 1998; Pajukanta and Porkka, 1999; Joffe *et al.*, 2001; McNeely *et al.*, 2001). Lipid mediators also regulate the activation of immune cells associated with these conditions

Model Organisms in Drug Discovery. Edited by Pamela M. Carroll and Kevin Fitzgerald.
© 2003 John Wiley & Sons, Ltd. ISBN 0 470 84893 6

and other disorders such as cancer and autoimmune diseases (Calder, 2001; Gupta and Dubois, 2001; Tilley *et al.*, 2001; Vivanco and Sawyers, 2002). Although classical studies have defined how lipids are absorbed, transported, deposited and mobilized, our knowledge of the genetic regulation of these and other aspects of 'lipomics' is far from complete. For these reasons, the analysis of lipid metabolism remains an active area of biomedical research.

In this chapter, we describe our experience with the zebrafish as a model system to study mammalian lipid metabolism and signaling. We have shown that zebrafish process dietary phospholipid and cholesterol in a manner analogous to humans and other mammals (Farber *et al.*, 2001). We also have shown that zebrafish and mammals utilize a conserved pathway to regulate the synthesis of prostanoids, an important class of lipid signaling molecules that are generated by the action of cyclooxygenases (Grosser *et al.*, 2002). These similarities of teleost and mammalian physiology are noteworthy because pharmacological inhibitors of cholesterol synthesis and cyclooxygenases are among the most commonly prescribed drugs used for the treatment and prevention of human diseases (Knopp, 1999; Crofford, 2001; Hennekens, 2001; Chau and Cunningham, 2002). Together, these studies confirm the utility of the zebrafish as a model system for drug discovery in areas related to the absorption and processing of lipids and their cellular metabolites. Such studies may have an impact on the development of new strategies for the treatment and prevention of common human diseases.

8.2 Fish as a model organism to study human physiology and disease

Through the pioneering work of Streisinger *et al.* (1981) the zebrafish, *Danio rerio*, has developed as an important model system to study vertebrate development (Haffter *et al.*, 1996). As outlined by Schulte-Merker in Chapter 7, advantages of the zebrafish include its short generation time, external fertilization, optically clear embryos and the large number of offspring produced from a single female. Although advantageous for embryological studies, these features also have facilitated the performance of large-scale forward genetic studies using chemical mutagenesis, gamma irradiation and, most recently, retroviral insertions (Driever *et al.*, 1996; Haffter *et al.*, 1996; Fisher *et al.*, 1997; Chen *et al.*, 2002). Such studies have led to the identification of diverse mutant phenotypes that affect embryo-genesis at various developmental stages, including axis formation, gastrulation and organogenesis. These studies have led to the recognition that genetic analyses in zebrafish are relevant to biomedical research, given that most mutants are predicted to derive from single gene defects and that most of these genes will be orthologs of mammalian genes whose function in development, cell signaling or organ physiology has been conserved evolutionarily (Postlethwait *et al.*, 1998).

A number of laboratories have utilized the zebrafish as a model to study human diseases (Barut and Zon, 2000; Amatruda *et al.*, 2002; Ward and Lieschke, 2002). Recent work from several zebrafish laboratories has identified important aspects of vertebrate physiology that are shared between zebrafish and mammals. Examples include the biosynthetic pathways of iron absorption and heme metabolism, which are essential to red blood cell production (Donovan *et al.*, 2000), and the biology of contractile proteins that regulate the function of cardiac and skeletal muscle (Sehnert *et al.*, 2002). Genetic analyses of neural and behavioral physiology, angiogenesis and cancer biology have been initiated using zebrafish and it is anticipated that genes discovered in these and other novel mutagenesis screens will identify genes that play a role in a diverse group of human diseases.

Although relatively few studies devoted specifically to the analysis of zebrafish physiology have been reported in the past, related teleosts (such as the carp) and other fish have served for many years as valuable models for the analysis of mammalian organ function. Most recently, direct analyses of zebrafish physiology have been performed using pharmacological agents. These compounds, whose mechanisms of action in humans are well characterized, show striking conservation of established effects on vascular tone, behavior, thyroid metabolism and blood coagulation. For example, vasoconstrictors active in humans, such as phenylephrin and N(G)-nitro-L-arginine methyl ester (L-NAME), cause a reduction of vascular flow through selected arterial beds of zebrafish larvae (Fritsche *et al.*, 2000; Schwerte and Pelster, 2000). Similarly, sodium nitroprusside, a vasodilator used to treat severe hypertension, causes arterial and venous dilatation in zebrafish larvae, as is observed in humans (Fritsche *et al.*, 2000). Studies also have demonstrated that diazepam, pentobarbitol and melatonin can induce a hypnotic-like state in zebrafish, akin to their effects in mammals (Zhdanova *et al.*, 2001). Importantly, co-administration of specific pharmacological inhibitors for these compounds prevents their effect with zebrafish. Finally warfarin, a well-known anticoagulant, exhibits similar effects in zebrafish, and amiodarone, an important cardiac drug that can inhibit thyroid hormone metabolism in humans, causes hypothyroidism when administered to zebrafish larvae (Jagadeeswaran and Sheehan, 1999; Liu and Chan, 2002).

Recently, we have begun genetic analyses of dietary lipid metabolism and lipid signaling mediators (prostanoids) using the zebrafish. These studies were born, in part, from our observation that zebrafish larvae digest and process cholesterol and phospholipids in a manner that is highly analogous to humans and other mammals. Subsequently, we showed that drugs used to inhibit cholesterol metabolism in humans have related effects in zebrafish.

Concomitantly, we identified the zebrafish orthologs of the mammalian cyclooxygenases-1 and -2 genes and showed that they are metabolically active, and have related pharmacological specificities and physiological roles to those of their mammalian counterparts. In the following sections we describe these studies in detail and address the design of screens for genes that contribute to the regulation of these essential aspects of human physiology.

8.3   Lipid metabolism screen

Many genes known to play important roles in mammalian lipid metabolism are conserved in the optically clear zebrafish larvae. We have initiated a large-scale *N*-ethyl-*N*-nitrosourea (ENU) screening using fluorescent lipid analogs to identify mutations with perturbed lipid metabolism. In this section, lipid metabolism in fish, optical biosensors and drug screening are discussed.

Lipid metabolism in fish

Numerous researchers have studied the major components of lipid metabolism in teleost fish: absorption, transport, storage and mobilization. It is now clear that lipid transport and mobilization in fish are similar to those observed in mammals but absorption and storage in fish are slightly different (Sheridan, 1988). In fish, both non-esterified fatty acids and triacylglycerol-enriched chylomicrons are transported to the liver via the blood circulation. The lipolysis processes in fish are accomplished by various lipases and hormones that are similar to those of mammals (Sheridan, 1988, 1994). In addition, the plasma lipoproteins, including apolipoprotein A- and B-like proteins, are comparable to mammals (Babin and Vernier, 1989). In mammals, absorption is accomplished by hydrolyzing lipids such as fatty acids and mono-acylglycerol, re-esterifying them into triacylglycerol and then lipoprotein loading in enterocytes. These newly made chylomicrons are subsequently secreted into the lymphatic system for transport to the liver (Tso and Fujimoto, 1991). Fish, however, not only contain this slow triacylglycerol delivery system (Sire *et al.*, 1981) but also absorb and deliver fatty acids directly into the peripheral tissues via the blood circulation (Sheridan *et al.*, 1985). Mammals deposit lipids primarily in adipose tissue; in contrast, fish store lipids not only in mesenteric adipose tissue but also in muscle and liver (Sheridan, 1994). The major stored lipids in fish are triacylglycerol and polyunsaturated fatty acids, with some minor lipid classes such as glycerylether analogs and alkoxydiacylglycerol (Sheridan, 1994).

Data from our laboratories have shown that when zebrafish larvae begin feeding at 5 days post-fertilization, they process dietary lipids in a similar manner to mammals. This includes lipid hydrolysis in the intestine, lipid transport from the intestine to the liver and hepatic secretion of bile necessary for emulsification and absorption of hydrophobic lipids within the intestine (Farber *et al.*, 2001). These data were obtained using the fluorescent optical biosensors that we developed, which can visualize lipid processing in living zebrafish larvae.

Optical biosensors to visualize lipid metabolism in live larvae

Phospholipase A_2 (PLA$_2$) activity is important for lipid signaling, host defenses, lipid absorption and cancer (MacPhee *et al.*, 1995; Dennis, 1997). In order to visualize PLA$_2$ enzymatic activity in live larvae, we have developed a family of fluorescent lipid biosensors (Farber *et al.*, 1999; Hendrickson *et al.*, 1999). One substrate, PED6 (*N*-{[6-(2,4-dinitrophenyl)amino]hexanoyl}-1-palmitoyl-2-BODIPY-FL-pentanoyl-*sn*-glycerol-3-phosphoethanolamine) (Figure 8.1A), exhibits a spectral change upon cleavage by PLA$_2$ by releasing

A

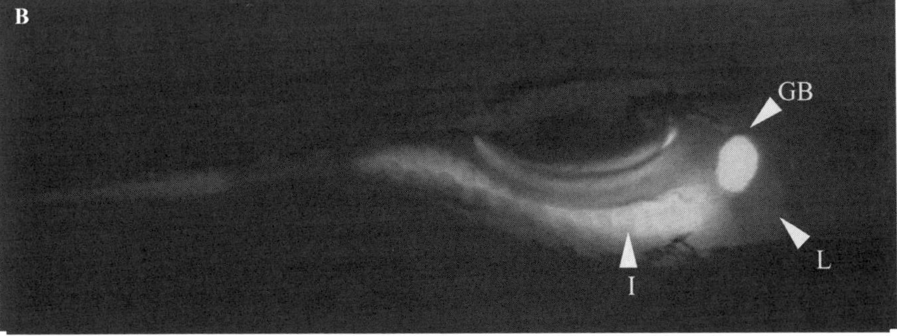

Figure 8.1 Substrate PED6 can visualize lipid metabolism in live zebrafish larvae. (A) The structure of PED6. The intact PED6 has no fluorescence because the emission of fluorophore at the *sn*-2 position is quenched by the dinitrophenyl group at the *sn*-3 position. Upon PLA$_2$ cleavage of the *sn*-2 BODIPY-labeled acyl chain a green fluorescence is observed. (B) Zebrafish larva 5 days post-fertilization labeled with PED6 (0.3 µg/ml, 6 h of incubation). Arrows show the liver (L), gall bladder (GB) and intestine (I)

a fluorescent BODIPY acyl chain; this event is organ-specific (Hendrickson *et al.*, 1999). We utilized PED6 to visualize PLA$_2$ activity in zebrafish larvae 5 days post-fertilization. As shown in Figure 8.1B, the intestine and the gall bladder are labeled by cleaved PED6 metabolites. Based on this observation and time-course studies we hypothesized that quenched PED6 is cleaved by PLA$_2$ in the intestine following PED6 ingestion, and the cleaved products – unquenched green fluorescent PED6 metabolites – are rapidly transported to the liver. These fluorescent metabolites are then secreted into newly formed bile and stored in the gall bladder. Following extrusion from the gall bladder, the fluorescent bile enters the intestine, where it is easily visualized.

To test our hypothesis, another fluorescent lipid reporter, BODIPY FR-PC (Figure 8.2A), was generated (Farber *et al.*, 2001). This fluorophore has two BODIPY acyl chains that exhibit fluorescence resonance energy transfer (FRET) to emit different spectra upon PLA$_2$ cleavage. When excited (505 nm), the intact substrate emits orange (568 nm). Upon PLA$_2$ cleavage, the same excitation results in a green emission (515 nm). Such a molecule can be used to localize PLA$_2$ activity. As shown in Figure 8.2B, only green fluorescence (cleaved product of PLA$_2$) was observed in the gall bladder and liver, where intact substrate (orange fluorescence) was located only in the intestinal epithelium. In conclusion, these studies suggest that lipid digestion and absorption systems in zebrafish larvae are similar to those in mammals.

We have initiated a physiological genetic screen *in vivo* with ENU mutagenized zebrafish using these biosensors because they provide a rapid readout of lipid metabolism and digestive organ morphology in living zebrafish larvae. So far, we have identified eight mutants. Among the mutants is one recessive lethal mutant, *fat-free*, that fails to accumulate fluorescently labeled lipids in the gall bladder following PED6 and NBD-cholesterol (22-[*N*-(7-nitronbenz-2-oxa-1,3-diazol-4-yl) amino]-23,24-bisnor-5-cholen-3-ol) ingestion, but its digestive system appears morphologically normal. Phenotypic analysis of this mutant indicated that the PLA$_2$ activity and swallowing are normal (Farber *et al.*, 2001). In contrast, *fat-free* had nearly normal fluorescence in the digestive organ after BODIPY FL-C5 ingestion. Because BODIPY FL-C5 is a short-chain fatty acid analog that is less hydrophobic and more soluble in aqueous solution, emulsifiers (such as bile) are not critical for its absorption. Instead, PED6 and NBD-cholesterol, the more hydrophobic molecules, require biliary emulsification in order to be processed and absorbed. Because the absorption of short-chain fatty acids is nearly normal in *fat-free*, we hypothesized that the *fat-free* mutation may attenuate bile synthesis or secretion.

Additional evidence that the *fat-free* mutant might be a potential animal model to study biliary synthesis or secretion are the results of a statin drug treatment study. As we have shown previously, when wild-type zebrafish larvae are treated with the statin drug atorvastatin (Lipitor), PED6 processing

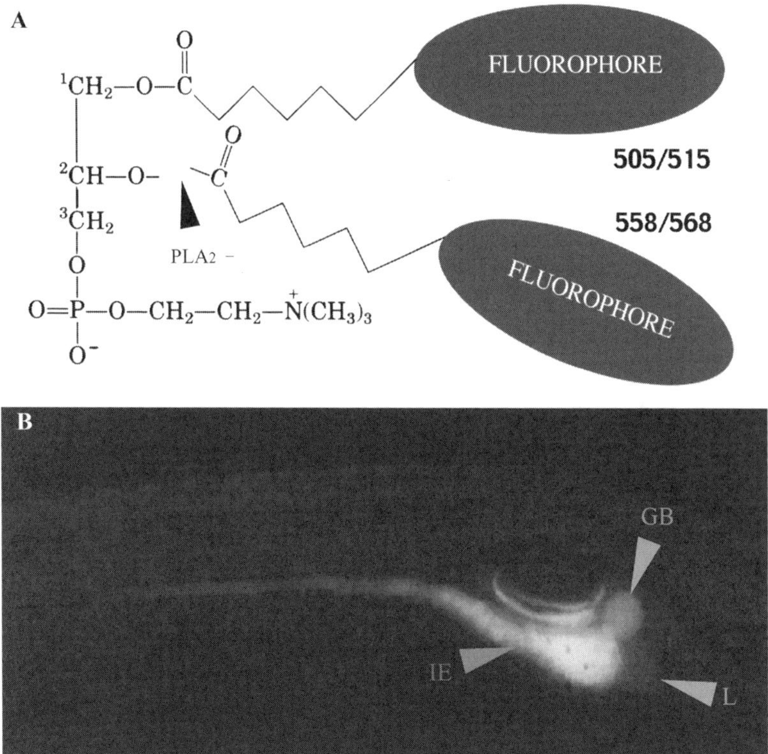

Figure 8.2 Labeling with BODIPY FR-PC. (A) The structure of BODIPY FR-PC. When the molecule is intact in the cell, excitation at 505 nm results in orange (568 nm) emission due to fluorescence resonance energy transfer (FRET) between the two BODIPY-labeled moieties. Upon PLA_2 cleavage at the *sn*-2 position, the BODIPY moiety at the *sn*-1 position results in green (515 nm) emission when excited (505 nm). (B) BODIPY FR-PC (5 µg/ml)-labeled zebrafish larva 5 days post-fertilization. The liver (L) and gall bladder (GB) showed green fluorescence (green arrow), indicating the accumulation of cleaved products. Uncleaved orange BODIPY FR-PC (orange arrow) is observed only in the intestinal epithelium (IE)

is profoundly attenuated in a similar manner to that observed in *fat-free* larvae (Farber *et al.*, 2001). Addition of exogenous fish bile reversed the blocking effect of Lipitor, suggesting that Lipitor blocks the synthesis of the cholesterol-derived biliary emulsifiers that are required for lipid absorption. However, the effect of Lipitor on NBD-cholesterol processing in wild-type larvae was slightly different from that observed in *fat-free* mutant larvae. Wild-type larvae had markedly reduced NBD-cholesterol fluorescence in the intestinal lumen following Lipitor treatment but gall bladder fluorescence was preserved (Figure 8.3A). In contrast, *fat-free* failed to accumulate NBD-cholesterol either in the intestine or in the gall bladder (Figure 8.3B). The

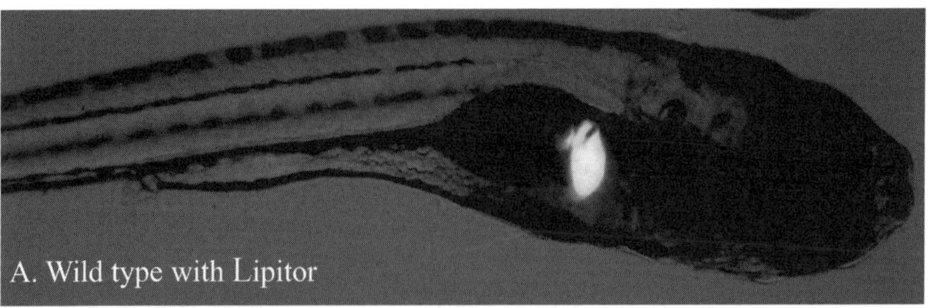

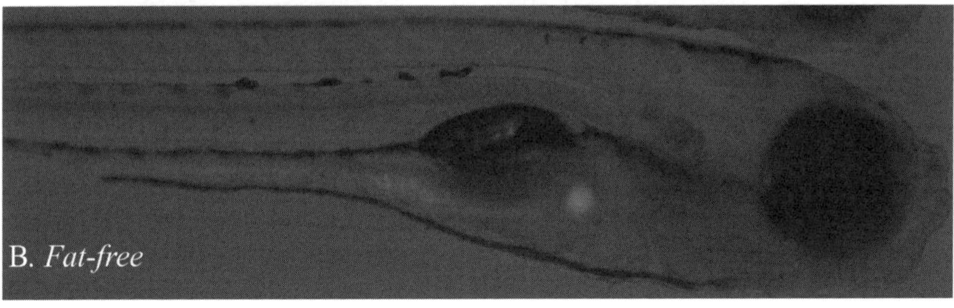

Figure 8.3   Labeling with NBD-cholesterol. (A) Wild-type larva 5 days post-fertilization incubated with both NBD-cholesterol and atorvastatin (Lipitor), had reduced fluorescence in the intestinal lumen but gall bladder fluorescence was preserved. (B) The *fat-free* mutant larva 5 days post-fertilization exhibits no fluorescence in the intestine and significantly reduced fluorescence in the gall bladder

fluorescence seen in the gall bladder of Lipitor-treated wild-type larvae is presumably due to NBD-cholesterol absorption with pre-existing bile. Because *fat-free* mutants have impaired bile synthesis or secretion, NBD-cholesterol is almost non-absorbable. Recently, numerous studies have shown that nuclear receptors regulate bile synthesis and processing (Chawla *et al.*, 2001; Goodwin amd Kliewer, 2002; Makishima *et al.*, 2002), but the mechanisms of bile homeostasis are not yet fully understood. For this reason, identification of additional genes that regulate bile synthesis and/or secretion, such as the zebrafish *fat-free* gene, is important.

Screen drugs with radioactive lipid precursors

We have successfully applied isotopic labeling techniques to study lipid profiles (lipomics) in a single larva. Briefly, we labeled zebrafish larvae with radioactive lipid precursors, followed by lipid extraction and thin-layer chromatography (TLC). We analyzed lipid fractions on the TLC plate using a radioactivity scanner. Using this technique, we immersed zebrafish larvae in

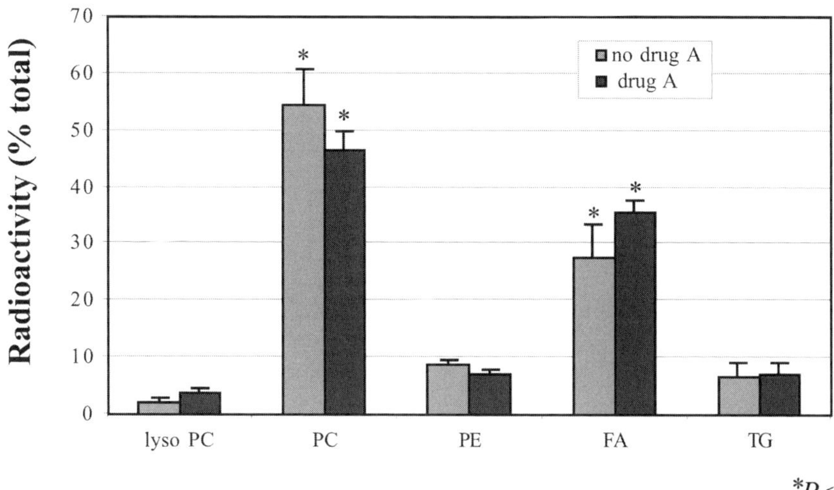

Figure 8.4 Lipomics analysis. Larvae (5 days post-fertilization) were incubated with radioactive oleic acid for 20 h, followed by lipid extraction and thin-layer chromatography (TLC). The solvent chloroform–ethanol–triethylamine–water (30:34:30:8) was used to develop the TLC plate. The radioactivities were scanned. The major metabolites derived from oleic acid (FA) are phosphatidylcholine (PC), phosphatidylethanolamine (PE), triacylglycerol (TG) and lysophosphatidylcholine (lysoPC). Data are means $\pm$ SD ($n = 3$).

the embryo media containing radioactive lipids with or without statins. Here, we show one example of our lipomics study using ^{14}C-oleic acid labeling with Drug A (20 h of treatment). We first found that Drug A interferes with fluorescent reporters in live zebrafish larvae as described above, then we went on to study ^{14}C-oleic acid labeling. The results showed that Drug A significantly decreased phosphatidylcholine synthesis and that most radio-activities remained in the fatty acid fraction for the Drug A treatment group (Figure 8.4). This suggests that Drug A may interfere with the phospholipid synthesis pathway. By combining these two high-throughput techniques we can perform large-scale screening of the chemical compounds that perturb lipid metabolism in zebrafish larvae and gain some information about the pathway by which these compounds interfere.

Screening strategies

The zebrafish system can be used to screen angiogenic drugs (Chan *et al.*, 2002) – compounds that affect embryogenesis (Peterson *et al.*, 2000) – and is suggested to be utilized for the screening of anticancer drugs (Amatruda *et al.*, 2002). As mentioned, fluorescent lipid analogs such as PED6 and NBD-cholesterol provide an easy readout for drug screening. Zebrafish larvae arrayed into multiwell plates that contain different chemical compounds and fluorescent lipid reporters can be screened for changes in gall bladder fluorescence. Automation of this process for drug screening is possible: a robotic fish sorter can distribute zebrafish larvae into multiwell plates; the fluorescent lipid reporters and different chemical compounds can be added using a robotic sample processor; and the intensity of fluorescence can be assayed using a multiwell spectrometer. Once the compound that alters the fluorescence intensity in the zebrafish larvae is identified, isotopic lipid labeling studies may help define where these compounds act.

8.4   Zebrafish as a model system to study prostanoid metabolism

Prostanoids are autocoid lipid signaling molecules that regulate important aspects of vertebrate cellular and organ physiology, such as immunity, renal function, cell proliferation, hemostasis and angiogenesis. Perhaps the best studied prostanoids are the prostaglandins and thromboxanes, whose synthesis is dependent upon cyclooxygenases (COXs), the enzymes targeted by aspirin and other non-steroid antiinflammatory drugs (NSAIDs) (reviewed in Serhan and Oliw, 2001). Humans and other mammals synthesize prostanoids via the actions of two COX paralogs (Figure 8.5) that are encoded on separate genes (Smith and Langenbach, 2001). The COX-2 isoform is largely induced in response to pathological and physiological stimuli, whereas COX-1 is constitutively expressed and therefore predicted to have a homeostatic role in most cell types. The presence of COX proteins in non-mammalian vertebrates suggested that orthologs of the COXs and other enzymes required for prostanoid biosynthesis and metabolism were encoded within the zebrafish genome. Given the important role of COX inhibition in the treatment of human disease (reviewed by Patrono *et al.*, 2001), we sought to identify the zebrafish orthologs of the mammalian COX genes and to determine whether the pharmacological characteristics of the zebrafish COX proteins were comparable to their human counterparts.

Prostanoid synthesis and signaling

In mammals, prostaglandins (PGs) and thromboxanes (TXs) are derived from the action of a family of synthases that convert PGH2 to bioactive PGs and

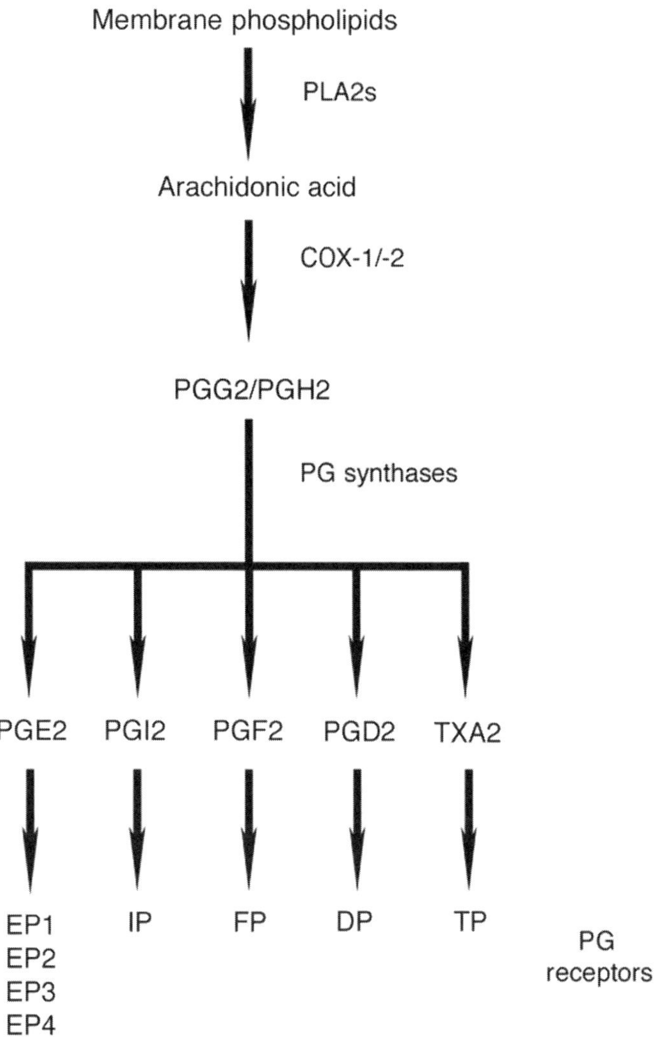

Figure 8.5 Prostaglandin (PG) biosynthesis. Schematic outlining the principal pathway for PG synthesis. Activation of cellular phospholipase A2 isozymes (PLA2s) liberates membrane-bound arachidonic acid, which is converted enzymatically by cyclooxygenases-1 or -2 (COX-1/-2) to PGG and PGH. These short-lived intermediaries are converted to specific PGs through the activity of specific PG synthases. Targeted disruption of specific PG receptors identifies the following physiological roles: EP1, cell proliferation/ transformation; EP2, ovulation/fertilization, salt retention, vascular tone, bronchodilation; EP-3, febrile response, mucosal secretion and integrity, vascular tone, renal water excretion; EP-4, maturation of fetal circulation, vascular tone, bone resorption; IP, hemostasis, vascular tone, inflammatory response and nocioception; FP, parturition; DP, response to inhaled allergens; TP, hemostasis (reviewed in Narumiya and FitzGerald, 2001).

TXs. PGH2 itself is derived from COX-mediated modification of arachidonic acid, the principal fatty acid present at the *sn2* position of membrane phospholipids (PL). Arachidonic acid liberation from membrane PLs is considered to be the rate-limiting step in prostanoid biosynthesis, and arachidonic acid bioavailability appears to be regulated by distinct PLA2 genes that are activated in response to different physiological stimuli (reviewed by Fitzpatrick and Soberman, 2001). The COX-2 isoform appears more active than COX-1 at low arachidonic acid levels and because of this COX-2 is believed to be the principal source of PG production during periods of sustained arachidonic acid release, such as occurs in arthritis, cancer and other chronic diseases.

Both PGs and TXs function as autocoid signaling molecules. They are secreted by cells and activate their own PG and TX receptors. To date, eight such receptors have been identified and splice variants of three have been discovered (Narumiya and FitzGerald, 2001). The PG/TX receptors are rhodopsin-type G-coupled transmembrane receptors that effect cytosolic calcium or cAMP levels. Restricted expression of PG/TX synthase and the prostanoid receptor genes provides the tissue specificity of prostanoid bioactivity. However, receptor fidelity is by no means absolute, and individual PGs can activate more than one PG receptor. The PGs may also function transcellularly to activate PG production in neighboring cells (reviewed by Serhan and Oliw, 2001). There is also now considerable evidence that PGs may activate the peroxisome proliferator-activated receptor (PPAR) family of nuclear hormone receptors in various cell types, such as adipocytes (Kliewer *et al.*, 1995), monocytes (Jiang *et al.*, 1998) and macrophages (Ricote *et al.*, 1998).

Functional analyses of vertebrate COX proteins

Analyses of mammalian COX proteins have defined conserved regions of both COX paralogs that are required for the conversion of arachidonic acid to PGH2. However, although the two COX isoforms share high sequence homology, important differences in their tertiary structure have been identified (FitzGerald and Loll, 2001). Such differences have enabled the design of selective COX inhibitors, such as the COXIBs (Patrono *et al.*, 2001). These differences also account for the differential response of the two COX proteins to non-selective inhibitors. For example, it has been observed recently that the acetylation of COX-2 by aspirin does not completely inactivate this isoform, as occurs with COX-1 (Claria and Serhan, 1995; Mancini *et al.*, 1997). Residual acetylated COX-2 activity is postulated to lead to the production of eicosanoids with novel actions that may play an important role in aspirin's antiinflammatory effects.

Regulatory regions of both COX genes also have been defined. Far more is known about this aspect of the COX-2 than the COX-1 gene, which may be expected given that COX-1 is constitutively expressed at high levels in most cell types. Knock-outs of both COX isoforms also have been generated. In contrast to COX-1 mutant mice, which are viable (Langenbach *et al.*, 1995), COX-2 mutants are infertile and commonly develop progressive renal disease that affects longevity (Morham *et al.*, 1995). As a result, COX-2 mutant mice are of limited use for functional analyses of the COX-2 protein.

Molecular, biochemical, pharmacological and functional analyses of zebrafish COXs

We chose to use the zebrafish model system to study COX pharmacology and biology because of the potential to apply large-scale forward genetic analysis and gene targeting studies to questions relevant to vertebrate prostanoid biology. The optical clarity, rapid development and accessibility of the zebrafish embryo and larva suggested that the zebrafish might serve as a useful model to identify genes that regulate or modify COX activity. Furthermore, the results of large-scale sequencing efforts suggest that many other enzymes involved in eicosanoid metabolism are active in zebrafish. This suggests that related studies designed to identify regulators of PG synthases, PG receptors and other important genes will be feasible in zebrafish.

Our original studies of zebrafish prostanoid biology were designed to address several simple questions, such as whether zebrafish cDNAs with sequence homology to mammalian COXs were in fact transcripts of functional orthologs of the mammalian COX genes whether the putative zebrafish COX orthologs share isoform-specific properties with their mammalian counterparts, and whether the physiological role of either zebrafish COX paralog had been conserved during vertebrate evolution. Answers to these questions would likely determine the suitability of the zebrafish model system for detailed analysis of prostanoid biology.

To address these questions, full-length zebrafish cDNAs with high sequence homology to the mammalian COXs were obtained and their presence in various cell types was assayed using reverse-transcription polymerase chain reaction (Grosser *et al.*, 2002). These data revealed that the two COX isoforms were expressed in adult tissues and at embryonic and larval stages. Expression of COX-2, and to a lesser degree COX-1, was prominent in the developing vasculature, suggesting a role for COX activity during zebrafish blood vessel development. This finding is of great interest because COX-2 is believed to play an important role in mammalian tumor angiogenesis (Masferrer *et al.*, 2000).

Sequence analysis of the zebrafish COXs revealed a high degree of conservation with their mammalian orthologs. Particularly noteworthy was the conservation of amino acids critical for catalysis, aspirin acetylation, heme coordination and the presence of multiple N-glycosylation sites. Further, the zebrafish COX-1 and COX-2 orthologs had characteristic N- and C-terminal and 3'-UTR (untranslated region) insertions, respectively. Sequence comparison of the amino acid residues within the arachidonate-binding channel of each enzyme was surprising. Between the two zebrafish COX isoforms, only one amino acid substitution is present within this region (Ile-434-Val), whereas the mammalian COXs differ in the identity of three critical residues. This finding was noteworthy because differences in the volume of this channel between the two COX isoforms is thought to be responsible for the pharmacological specificity of COX inhibitors. This raised the question as to whether such pharmacological specificity was also a feature of non-mammalian vertebrate COX proteins (discussed below).

Chromosomal localization studies provided additional evidence that the zebrafish cDNAs were orthologs of the mammalian COXs. Both genes reside in regions of the zebrafish genome where gene synteny has been conserved. Zebrafish and human COX-1 reside in close proximity to the *RXRG* and *Notch1B* genes (Grosser *et al.*, 2002). Similarly, the zebrafish and human COX-2 genes are in close proximity to their respective *CPLA2* orthologs.

Functional analysis of the zebrafish COXs was first addressed in transient transfections assays (Grosser *et al.*, 2002). These studies revealed that both COX isoforms drove PG production when introduced into COS-7 cells, which lack endogenous COX activity. Following stimulation with arachidonic acid, PG synthesis was measured using mass spectrometry. Introduction of either zebrafish COX gene led to the production of PGE2, whereas there was minimal PGE2 production in COS cells transfected with vector alone. Using mass spectrometry it was also shown that adult zebrafish produce PGE2, PGI2 and TXB2. Most importantly, prostanoid synthesis was inhibited in a dose-dependent manner in transfected COS cells and in live fish by both non-selective and selective COX inhibitors (indomethacin and NS-398, respectively). Furthermore, 50% inhibition of the zebrafish and mammalian COX proteins was achieved using similar doses of both inhibitors. Finally, it was shown that the selective COX inhibitors have similar pharmacological specificities against zebrafish and mammalian COX proteins.

Functional assays of zebrafish COXs suggested that prostanoid-mediated mechanisms of hemostasis and cell motility/proliferation have been conserved in non-mammalian vertebrates. In adult fish, thrombocyte aggregation (*ex vivo*) was inhibited by indomethacin (a non-selective COX inhibitor) but not by NS-398 (a selective COX-2 inhibitor) (Grosser *et al.*, 2002). This finding is noteworthy because restricted expression of COX-1 in mammalian platelets is, in large measure, responsible for the cardioprotective effects of aspirin, which has potent inhibitory effects on the aggregation of human platelets (reviewed by Patrono *et al.*, 2001). The role of zebrafish COXs during embryonic development was also analyzed. In mammals, zygotic transcription of both COX genes appears to be dispensable during embryonic development, although postnatal renal dysplasia develops in COX-2-deficient mice (Langenbach *et al.*, 1995; Morham *et al.*, 1995). Knock-down of zebrafish COX-2 protein also had no discernable effect on embryonic development. However, knock-down of zebrafish COX-1 caused a significant delay in epiboly, a developmental process dependent upon cell proliferation and cell migration. The discordant embryonic phenotypes produced by inhibition of teleost versus mammalian COX-1 may be explained by the fact that antisense morpholinos are capable of inhibiting the translation of both maternal and zygotic COX transcripts in zebrafish, whereas gene targeting in mammals perturbs only zygotic gene expression.

8.5   Future directions

Elucidation of the regulatory mechanisms that control prostanoid production and bioactivity remains an active area of research. Given the high degree of structural and functional conservation between zebrafish and humans COX genes, studies directed toward these questions seem feasible using this model system. High-throughput genetic analyses are particularly attractive to questions of gene regulation. For example, mutagenesis strategies that assay COX protein levels immunohistochemically, or via reporter genes in transgenic fish, may identify mutations that perturb COX RNA or protein expression and/or stabilization. Such mutants could lead to the identification of novel COX-1 regulators, which to date have largely eluded detection. Similarly, such screens may also define motifs within either COX protein that are pharmacologically relevant. The COX-deficient mutants recovered in this manner, which would be predicted to be fully viable, could be used to generate compound mutants by matings with fish that carry established mutations. Such compound mutants then could be assayed for a variety of prostanoid-related biochemical or physiological defects.

Biochemical-based mutagenesis screens are also feasible using the zebrafish. High-throughput assays of prostanoid production using mass spectrometry is one example. A physiological mutagenesis screen such as this would identify not only mutations that perturb COX activity directly but also mutations that perturb the function of upstream and downstream COX regulators, such as the genes predicted to couple COXs to PLA2s or PG synthases. The zebrafish also provides a convenient means to assay the role of known genes in prostanoid biosynthesis using the aforementioned antisense techniques. Finally, recently devised techniques for directly identifying specific gene mutations from mutagenized sperm offer the promise of generating libraries of mutant alleles that can be assayed in live fish generated through *in vitro* fertilization (Draper *et al.*, 2001; Wienholds *et al.*, 2002). This methodology, commonly referred to as 'TILLING' (McCallum *et al.*, 2000), offers the chance to perform a comprehensive analysis of genes regulating prostanoid synthesis and activity.

8.6 Summary

Recent work has shown that it is possible to assay phospholipid metabolism and prostanoid synthesis in zebrafish (Farber *et al.*, 1991; Grosser *et al.*, 2002). These preliminary studies suggest that important questions of lipid biology are amenable to large-scale, high-throughput analyses in this model system. Lipid metabolism now can be added to the growing list of vertebrate developmental and physiological processes that can be assayed in zebrafish. The potential to identify novel genes (or novel functions of known genes) that regulate the metabolism of dietary lipids or the generation of lipid signaling molecules has important pharmacological implications. By using this strategy, ultimately it may be possible to devise combined biochemical and physiological assays of small-molecule modulators of lipid metabolism. Such studies may provide a rapid and accurate screening methodology of great pharmacological value. As an example, a recent pilot screen of 640 bioavailable compounds from a chemical library (Prestwick Chemicals) identified several compounds that inhibit the accumulation of gall bladder fluorescence in zebrafish larvae fed the quenched lipid reporter PED6 (A. Rubinstein, Zygogen, Inc., personal communication). Multiple developmental and physiological pathways are predicted to have an impact on PED6 processing. Some of these, such as lipid absorption and transport, have important clinical implications and their analysis may prove to be tractable using zebrafish-based assays.

8.7 References

Amatruda, J. F., Shepard, J. L., Stern, H. M. and Zon, L. I. (2002). Zebrafish as a cancer model system. *Cancer Cell* **1**, 229–231.

Babin, P. J. and Vernier, J. M. (1989). Plasma lipoproteins in fish. *J. Lipid Res.* **30**, 467–489.

Barut, B. A. and Zon, L. I. (2000). Realizing the potential of zebrafish as a model for human disease. *Physiol. Genom.* **2**, 49–51.

Calder, P. C. (2001). Polyunsaturated fatty acids, inflammation, and immunity. *Lipids* **36**, 1007–1024.

Chan, J., Bayliss, P. E., Wood, J. M. and Roberts, T. M. (2002). Dissection of angiogenic signaling in zebrafish using a chemical genetic approach. *Cancer Cell* **1**, 257–267.

Chau, I. and Cunningham, D. (2002). Cyclooxygenase inhibition in cancer – a blind alley or a new therapeutic reality? *N. Engl. J. Med.* **346**, 1085–1087.

Chawla, A., Repa, J. J., Evans, R. M. and Mangelsdorf, D. J. (2001). Nuclear receptors and lipid physiology: opening the X-files. *Science* **294**, 1866–1870.

Chen, W., Burgess, S., Golling, G., Amsterdam, A. and Hopkins, N. (2002). High-throughput selection of retrovirus producer cell lines leads to markedly improved efficiency of germ line-transmissible insertions in zebra fish. *J. Virol.* **76**, 2192–2198.

Claria, J. and Serhan, C. N. (1995). Aspirin triggers previously undescribed bioactive eicosanoids by human endothelial cell–leukocyte interactions. *Proc. Natl. Acad. Sci. USA* **92**, 9475–9479.

Crofford, L. J. (2001). Rational use of analgesic and antiinflammatory drugs. *N. Engl. J. Med.* **345**, 1844–1846.

Dennis, E. A. (1997). The growing phospholipase A2 superfamily of signal transduction enzymes. *Trends Biochem. Sci.* **22**, 1–2.

Donovan, A., Brownlie, A., Zhou, Y., Shepard, J., Pratt, S. J., Moynihan, J., Paw, B. H., *et al.* (2000). Positional cloning of zebrafish ferroportin1 identifies a conserved vertebrate iron exporter. *Nature* **403**, 776–781.

Draper, B. W., Morcos, P. A. and Kimmel, C. B. (2001). Inhibition of zebrafish fgf8 pre-mRNA splicing with morpholino oligos: a quantifiable method for gene knockdown. *Genesis* **30**, 154–156.

Driever, W., Solnica-Krezel, L., Schier, A. F., Neuhauss, S. C., Malicki, J., Stemple, D. L., Stainier, D. Y., *et al.* (1996). A genetic screen for mutations affecting embryogenesis in zebrafish. *Development* **123**, 37–46.

Farber, S. A., Buyukuysal, R. L. and Wurtman, R. J. (1991). Why do phospholipid levels decrease with repeated stimulation? A study of choline-containing compounds in rat striatum following electrical stimulation. *Ann. NY Acad. Sci.* **640**, 114–117.

Farber, S. A., Olson, E. S., Clark, J. D. and Halpern, M. E. (1999). Characterization of Ca^{2+}-dependent phospholipase A2 activity during zebrafish embryogenesis. *J. Biol. Chem.* **274**, 19338–19346.

Farber, S. A., Pack, M., Ho, S. Y., Johnson, I. D., Wagner, D. S., Dosch, R., Mullins, M. C., *et al.* (2001). Genetic analysis of digestive physiology using fluorescent phospholipid reporters. *Science* **292**, 1385–1388.

Fisher, S., Amacher, S. L. and Halpern, M. E. (1997). Loss of cerebum function ventralizes the zebrafish embryo. *Development* **124**, 1301–1311.

FitzGerald, G. A. and Loll, P. (2001). COX in a crystal ball: current status and future promise of prostaglandin research. *J. Clin. Invest.* **107**, 1335–1337.

Fitzpatrick, F. A. and Soberman, R. (2001). Regulated formation of eicosanoids. *J. Clin. Invest.* **107**, 1347–1351.

Fritsche, R., Schwerte, T. and Pelster, B. (2000). Nitric oxide and vascular reactivity in developing zebrafish, *Danio rerio. Am. J. Physiol. Regul. Integr. Comp. Physiol.* **279**, R2200–2207.

Garg, A. (1998). Dyslipoproteinemia and diabetes. *Endocrinol. Metab. Clin. North Am.* **27**, 613–625, ix–x.

Goodwin, B. and Kliewer, S. A. (2002). Nuclear receptors. I. Nuclear receptors and bile acid homeostasis. *Am. J. Physiol. Gastrointest. Liver Physiol.* **282**, G926–931.

Grosser, T., Yusuff, S., Cheskis, E., Pack, M. A. and FitzGerald, G. A. (2002). Developmental expression of functional cyclooxygenases in zebrafish. *Proc. Natl. Acad. Sci. USA* **99**, 8418–8423.

Gupta, R. A. and Dubois, R. N. (2001). Colorectal cancer prevention and treatment by inhibition of cyclooxygenase-2. *Nat. Rev. Cancer* **1**, 11–21.

Haffter, P., Granato, M., Brand, M., Mullins, M. C., Hammerschmidt, M., Kane, D. A., Odenthal, J., *et al.* (1996). The identification of genes with unique and essential functions in the development of the zebrafish, *Danio rerio*. *Development* **123**, 1–36.

Hendrickson, H. S., Hendrickson, E. K., Johnson, I. D. and Farber, S. A. (1999). Intramolecularly quenched BODIPY-labeled phospholipid analogs in phospholipase A(2) and platelet-activating factor acetylhydrolase assays and *in vivo* fluorescence imaging. *Anal. Biochem.* **276**, 27–35.

Hennekens, C. H. (2001). Current perspectives on lipid lowering with statins to decrease risk of cardiovascular disease. *Clin. Cardiol.* **24**(7 Suppl), II-2–5.

Jagadeeswaran, P. and Sheehan, J. P. (1999). Analysis of blood coagulation in the zebrafish. *Blood Cells Mol. Dis.* **25**, 239–249.

Jiang, C., Ting, A. T. and Seed, B. (1998). PPAR-gamma agonists inhibit production of monocyte inflammatory cytokines. *Nature* **391**, 82–86.

Joffe, B. I., Panz, V. R. and Raal, F. J. (2001). From lipodystrophy syndromes to diabetes mellitus. *Lancet* **357**, 1379–1381.

Kliewer, S. A., Lenhard, J. M., Willson, T. M., Patel, I., Morris, D. C. and Lehmann, J. M. (1995). A prostaglandin J2 metabolite binds peroxisome proliferator-activated receptor gamma and promotes adipocyte differentiation. *Cell* **83**, 813–819.

Knopp, R. H. (1999). Drug treatment of lipid disorders. *N. Engl. J. Med.* **341**, 498–511.

Langenbach, R., Morham, S. G., Tiano, H. F., Loftin, C. D., Ghanayem, B. I., Chulada, P. C., Mahler, J. F., *et al.* (1995). Prostaglandin synthase 1 gene disruption in mice reduces arachidonic acid-induced inflammation and indomethacin-induced gastric ulceration. *Cell* **83**, 483–492.

Liu, Y. W. and Chan, W. K. (2002). Thyroid hormones are important for embryonic to larval transitory phase in zebrafish. *Differentiation* **70**, 36–45.

MacPhee, M., Chepenik, K. P., Liddell, R. A., Nelson, K. K., Siracusa, L. D. and Buchberg, A. M. (1995). The secretory phospholipase A2 gene is a candidate for the Mom1 locus, a major modifier of ApcMin-induced intestinal neoplasia. *Cell* **81**, 957–966.

Makishima, M., Lu, T. T., Xie, W., Whitfield, G. K., Domoto, H., Evans, R. M., Haussler, M. R., *et al.* (2002). Vitamin D receptor as an intestinal bile acid sensor. *Science* **296**, 1313–1316.

Mancini, J. A., Vickers, P. J., O'Neill, G. P., Boily, C., Falgueyret, J. P. and Riendeau, D. (1997). Altered sensitivity of aspirin-acetylated prostaglandin G/H synthase-2 to inhibition by nonsteroidal anti-inflammatory drugs. *Mol. Pharmacol.* **51**, 52–60.

Masferrer, J. L., Leahy, K. M., Koki, A. T., Zweifel, B. S., Settle, S. L., Woerner, B. M., Edwards, D. A., *et al.* (2000). Antiangiogenic and antitumor activities of cyclo-oxygenase-2 inhibitors. *Cancer Res.* **60**, 1306–1311.

McCallum, C. M., Comai, L., Greene, E. A. and Henikoff, S. (2000). Targeting induced local lesions IN genomes (TILLING) for plant functional genomics. *Plant Physiol.* **123**, 439–442.

McNeely, M. J., Edwards, K. L., Marcovina, S. M., Brunzell, J. D., Motulsky, A. G. and Austin, M. A. (2001). Lipoprotein and apolipoprotein abnormalities in familial combined hyperlipidemia: a 20-year prospective study. *Atherosclerosis* **159**, 471–481.

Morham, S. G., Langenbach, R., Loftin, C. D., Tiano, H. F., Vouloumanos, N., Jennette, J. C., Mahler, J. F., *et al.* (1995). Prostaglandin synthase 2 gene disruption causes severe renal pathology in the mouse. *Cell* **83**, 473–482.

Narumiya, S. and FitzGerald, G. A. (2001). Genetic and pharmacological analysis of prostanoid receptor function. *J. Clin. Invest.* **108**, 25–30.

Pajukanta, P. and Porkka, K. V. (1999). Genetics of familial combined hyperlipidemia. *Curr. Atheroscler. Rep.* **1**, 79–86.

Patrono, C., Patrignani, P. and Garcia Rodriguez, L. A. (2001). Cyclooxygenase-selective inhibition of prostanoid formation: transducing biochemical selectivity into clinical read-outs. *J. Clin. Invest.* **108**, 7–13.

Peterson, R. T., Link, B. A., Dowling, J. E. and Schreiber, S. L. (2000). Small molecule developmental screens reveal the logic and timing of vertebrate development. *Proc. Natl. Acad. Sci. USA* **97**, 12965–12969.

Postlethwait, J. H., Yan, Y. L., Gates, M. A., Horne, S., Amores, A., Brownlie, A., Donovan, A., *et al.* (1998). Vertebrate genome evolution and the zebrafish gene map [see Comments]. *Nat. Genet.* **18**, 345–349.

Ricote, M., Li, A. C., Willson, T. M., Kelly, C. J. and Glass, C. K. (1998). The peroxisome proliferator-activated receptor-gamma is a negative regulator of macrophage activation. *Nature* **391**, 79–82.

Schwerte, T. and Pelster, B. (2000). Digital motion analysis as a tool for analysing the shape and performance of the circulatory system in transparent animals. *J. Exp. Biol.* **203**, 1659–1669.

Sehnert, A. J., Huq, A., Weinstein, B. M., Walker, C., Fishman, M. and Stainier, D. Y. (2002). Cardiac troponin T is essential in sarcomere assembly and cardiac contractility. *Nat. Genet.* **31**, 106–110.

Serhan, C. N. and Oliw, E. (2001). Unorthodox routes to prostanoid formation: new twists in cyclooxygenase-initiated pathways. *J. Clin. Invest.* **107**, 1481–1489.

Sheridan, M. A. (1988). Lipid dynamics in fish: aspects of absorption, transportation, deposition and mobilization. *Comp. Biochem. Physiol. B* **90**, 679–690.

Sheridan, M. A. (1994). Regulation of lipid metabolism in poikilothermic vertebrates. *Comp. Biochem. Physiol. B* **107**, 495–508.

Sheridan, M. A., Allen, W. V. and Kerstetter, T. H. (1985). Changes in the fatty acid composition of steelhead trout, *Salmo gairdnerii* Richardson, associated with parr-smolt transformation. *Comp. Biochem. Physiol. B* **80**, 671–676.

Sire, M. F., Lutton, C. and Vernier, J. M. (1981). New views on intestinal absorption of lipids in teleostean fishes: an ultrastructural and biochemical study in the rainbow trout. *J. Lipid Res.* **22**, 81–94.

Smith, W. L. and Langenbach, R. (2001). Why there are two cyclooxygenase isozymes. *J. Clin. Invest.* **107**, 1491–1495.

Streisinger, G., Walker, C., Dower, N., Knauber, D. and Singer, F. (1981). Production of clones of homozygous diploid zebra fish (*Brachydanio rerio*). *Nature* **291**, 293–296.

Tilley, S. L., Coffman, T. M. and Koller, B. H. (2001). Mixed messages: modulation of inflammation and immune responses by prostaglandins and thromboxanes. *J. Clin. Invest.* **108**, 15–23.

Tso, P. and Fujimoto, K. (1991). The absorption and transport of lipids by the small intestine. *Brain Res. Bull.* **27**, 477–482.

Vivanco, I. and Sawyers, C. L. (2002). The phosphatidylinositol 3-kinase AKT pathway in human cancer. *Nat. Rev. Cancer* **2**, 489–501.

Ward, A. C. and Lieschke, G. J. (2002). The zebrafish as a model system for human disease. *Front Biosci.* **7**, d827–833.

Wienholds, E., Schulte-Merker, S., Walderich, B. and Plasterk, R. H. (2002). Target-selected inactivation of the zebrafish *rag1* gene. *Science* **297**, 99–102.

Zhdanova, I. V., Wang, S. Y., Leclair, O. U. and Danilova, N. P. (2001). Melatonin promotes sleep-like state in zebrafish. *Brain Res.* **903**, 263–268.

9

Chemical Mutagenesis in the Mouse: a Powerful Tool in Drug Target Identification and Validation

Andreas Russ, Neil Dear, Geert Mudde, Gabriele Stumm, Johannes Grosse, Andreas Schröder, Reinhard Sedlmeier, Sigrid Wattler and **Michael Nehls**

In the search for innovative therapeutic approaches, high-throughput *in vitro* technologies such as genome sequencing, DNA microarrays and proteomics have opened unprecedented opportunities, but they have also created new bottlenecks in the drug discovery process because newly identified candidate drug targets have to be linked to a physiological function *in vivo*. The genetic analysis of gene function in a mammalian model organism, typically the laboratory mouse, is one of the cornerstones in the elucidation of new molecular pathways. In addition to the standard tools of transgenesis and targeted mutagenesis in the mouse, chemical mutagenesis strategies have been established recently. They can be applied to the scalable gene-driven validation of potential targets *in vivo*, as well as the discovery of new therapeutic opportunities by phenotype-driven screens for new physiological pathways.

9.1   Introduction

Starting with the discovery of recombinant DNA, and accelerating with the genomics revolution, drug discovery strategies have undergone a transition

Model Organisms in Drug Discovery. Edited by Pamela M. Carroll and Kevin Fitzgerald.
© 2003 John Wiley & Sons, Ltd. ISBN 0 470 84893 6

from being driven by experimental pharmacology and physiology into a process dominated by the molecular characterization of potential drug targets. The classical approach started with prototypical compounds showing physiological effects with therapeutic potential, usually identified by investigations in physiology, pharmacology or endocrinology. Many key insights into mammalian biology are inseparable from the experimental pharmacology that was instrumental in their discovery. Importantly, experimental pharmacology is relying heavily on the use of whole organisms or isolated organs to provide information about physiology (Black, 1989).

Although the classical approach delivered both a lead compound and a defined physiological pathway with therapeutic potential, the modern molecular approach requires consecutive steps of drug target identification, validation and chemical drug discovery. With the complete sequence of the human genome at hand, a comprehensive catalogue of most or all potential protein targets is now available (Lander *et al.*, 2001; Venter *et al.*, 2001). Because experiments *in silico* and *in vitro* provide information about biochemical and cellular function, rather than whole-organism physiology, there is an additional requirement for evidence linking the newly identified candidate target to a physiological function likely to provide the desired therapeutic effect (Harris, 2001; Sanseau, 2001). The genetic analysis of animal models can provide this crucial link between molecular target and physiological function. In this sense, genetics partly fulfils the role that experimental pharmacology and physiology had in the classic era.

In drug discovery, there is not only the frequently cited need for rapid and reliable 'target validation' (the confirmation of an already existing therapeutic hypothesis for a given drug target); the need to discover new pathways of potential therapeutic use and to link new molecular targets to well-known physiological processes is as urgent.

This chapter discusses the strategies and applications of forward and reverse genetics in the murine model for the discovery and validation of candidate drug targets, and the analysis of the associated physiological pathways in the context of a complex system.

Forward and reverse genetics: complementary genetic approaches to target discovery and validation

In its classical definition, genetics investigates the patterns of inheritance of phenotypic variation. Since the advent of transgenic organisms, it is necessary to distinguish two fundamental strategies of genetic analysis.

Classical genetics, starting with the observation of phenotypic variation in a given population and working towards a molecular understanding of the underlying genetic factors, is termed forward genetics. Forward genetics is driven by phenotypic analysis, and looks for the 'phenotype first', and then the molecular basis of a given trait (Figure 9.1A).

In contrast, approaches involving the direct manipulation of specific genes, either by transgenesis or targeted mutagenesis, are summarized as reverse genetic strategies. This 'gene first' strategy is driven by the manipulation of DNA, rather than the observation of phenotypes, and investigates the functional consequences of a specific mutation in the context of the whole organism (Figure 9.1B).

Both strategies are complementary and have been used widely in all model organisms. The strength of reverse genetic technologies is the fine dissection of defined pathways and the testing of specific hypotheses about gene function, frequently applied in the analysis of complex gene families (Harris and Foord, 2000; Harris, 2001). In contrast, the realm of forward genetics is the discovery of the molecular basis of physiological pathways where no previous information exists (Hrabe and Balling, 1998; Justice *et al.*, 1999; Justice 2000; Balling, 2001; Nelms and Goodnow, 2001). Thus, reverse genetics is well suited for target validation, because it can test the therapeutic hypothesis for a given drug target. In contrast, forward genetics is the primary tool to put new molecular signposts into the 'white spots' of the functional map of pathways, and to discover innovative targets *de novo*.

Mouse genetics in target discovery and validation

When discussing the use of murine models in drug discovery, it is very important to distinguish three typical classes of experimental concepts, designed to answer fundamentally very different questions: efficacy testing, target validation and target discovery *de novo*. Although this text focuses on the latter two, it is essential to discuss the differences between the approaches to avoid misconceptions.

For efficacy testing of novel compounds, disease models are needed that reflect the course of the human disease as closely as possible. These models are frequently generated by non-genetic experiments using exogenous challenges to induce disease phenotypes. Typical examples are xenograft models for antitumor activity, or the induction of autoimmune diseases in collagen-induced arthritis or experimental autoimmune encephalitis. Some models rely on genetically altered animals, such as Apo-E knock-out mouse displaying increased susceptibility to atherosclerosis.

Although generally useful and widely accepted as standard tools, these applications are limited by factors other than the evolutionary conservation of the primary physiological pathway the target is acting in, i.e. drug administration, metabolism, excretion, etc. In addition, many of the

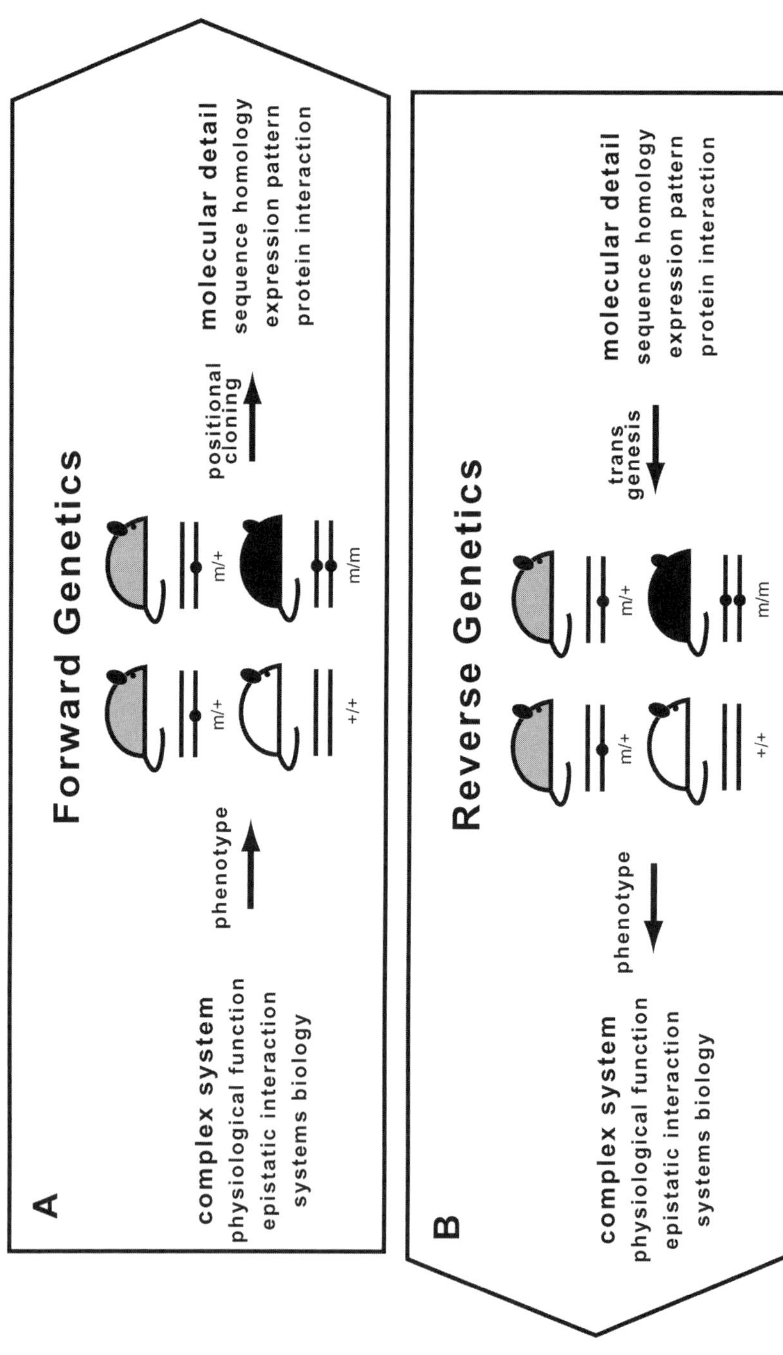

Figure 9.1 Forward genetics starts with the observation of a phenotype in the context of the complex system, and derives information about molecular detail by positional cloning (phenotype-driven strategy). In contrast, reverse genetics investigates the hypothesis about the phenotypic consequences of specific molecular changes by transgenic techniques, leading to the observation of a phenotype in the complex *in vivo* system (genotype-driven strategy)

established models show disease phenotypes similar in result to the human condition, but of very different etiology and dynamics, thus limiting their predictive power.

In genetic target validation, which is used increasingly in genomics-driven drug discovery, the *in vivo* function of a postulated drug target is tested by the phenotypic analysis of a mouse line mutant for the gene encoding the target protein. This concept relies on the evolutionary conservation of the physiological pathway between mammalian species. The mutant can be a classical spontaneous allele but most common is the use of a reverse genetic model, be it a transgenic animal or a targeted constitutive or conditional mutant (Harris and Foord, 2000; Harris, 2001). Although observation of the spontaneous phenotype of the mutant can give important clues for target validation, it is desirable to show the amelioration of a disease phenotype in an accepted model. This can be done by showing the resistance of the mutant to certain phenotypic challenges (Langenbach *et al.*, 1999; Morteau *et al.*, 2000; McPherron and Lee 2002), or by epistatic analysis. In this case, the phenotype of a disease causing mutation is fully or partially normalized by the introduction of a second mutation (Erickson *et al.*, 1996; Cohen *et al.*, 2002).

It is very important to note that the spontaneous phenotype of a mouse mutant for a certain drug target does not necessarily provide the relevant validation of the target. This is best exemplified by mutant mice carrying a targeted mutation of cyclooxygenase genes. Although there is no impressive spontaneous phenotype, these mice show resistance to inflammatory challenges, thus validating cyclooxygenase as the target for the widely used non-steroidal antiinflammatory drugs (NSAIDs) (Langenbach *et al.*, 1999; Morteau *et al.*, 2000).

In contrast to the hypothesis-driven validation of candidate drug targets, the discovery of potential novel targets and their physiological pathways requires a different experimental design. In target validation a well-defined hypothesis needs to be verified or falsified, whereas target discovery requires a search strategy that is open enough to uncover completely unpredicted findings but has powerful filtering functions to enrich for the desired signal in the vast amount of data generated by a genome-wide search. Although it would be possible in theory to reduce target discovery to the identification of novel genes from the genomic sequence, to be followed directly by target validation experiments, testing genetically engineered mutations in every gene in all relevant validation settings would not be economically viable in practice.

It is this combination of unbiased search with efficient filtering functions that makes forward genetics a powerful strategy for discovery biology. In mice, forward genetics can use different substrates of spontaneous or induced genetic diversity to achieve its goal (Brown and Balling, 2001).

9.2   Chemical mutagenesis in forward and reverse genetics

In the past two decades, reverse genetics approaches have dominated the use of the mouse model in drug discovery. The common denominator of the powerful techniques of transgenesis and gene targeting by homologous recombination is that specific DNA sequences are manipulated *in vitro* and introduced into the mouse germline by embryo manipulation. These requirements restrict the scalability of these approaches, especially in the context of forward genetics where the availability of several alleles for every gene is desired.

These limitations can be overcome by random mutagenesis applied *in vivo*. Although the basic techniques have been established for many years (Russell *et al.*, 1979; Hitotsumachi *et al.*, 1985), they did not find wide application until recently. Random mutagenesis in the mouse was 'rediscovered' in the mid-1990s and its potential was harnessed in context with the recent progress in genome mapping and sequencing (Hrabe de Angelis *et al.*, 2000; Nolan *et al.*, 2000). Its main advantage over DNA-based technologies is scalability; large numbers of mutants can be generated *in vivo* and analyzed in forward or reverse genetic screens.

To generate informative allelic series, including gain-of-function, loss-of-function and hypomorphic mutations, the induction of point mutations is the desired mode of action. Many mutagens, such as ionizing radiation or chlorambucil, frequently generate DNA deletions or inversions that might involve more that one gene and complicate the molecular characterization of the mutated locus. By far the most popular mutagenic agent is therefore *N*-ethyl-*N*-nitrosourea (ENU), a supermutagen inducing point mutations. Both ENU and similar chemicals are widely used as mutagens in genetic screens in all model organisms (Justice, 2000; Balling, 2001).

The mutagenic properties of ENU

N-ethyl-*N*-nitrosourea is an alkylating agent that acts without the need for metabolic activation. Its ethyl group can be transferred to nucleophilic sites on each of the four nucleotides of DNA. The resulting DNA adduct is resolved during the next round of replication, most frequently resulting in AT-to-TA transversions and AT-to-GC transitions (Balling, 2001).

As a germline mutagen in mice, ENU acts most efficiently on spermatogonial stem cells; mutation rates in post-spermatogonial cells and female germ cells are substantially lower. Optimal mutagenesis is achieved by fractionated dosage regimens between 200 and 400 mg/kg body weight, injected intraperitoneally. Different mouse strains show substantially different responses to ENU, requiring titration of the optimal dose (Justice *et al.*,

2000). Mutagenized males usually show transient sterility for up to 14 weeks, before the gonad is repopulated from the remaining stem cells. The observation of a threshold dose for mutagenesis points to the saturation of DNA repair systems at optimal dosage.

Because the mutagen acts predominantly on the level of spermatogenesis, each offspring (G1) of a mating between a mutagenized male (G0) and a wild-type female is heterozygous for a unique set of point mutations (Figure 9.2). These G1 animals can be used in three typical experimental settings: they can be screened directly for dominant traits; they can be set up for further breeding to screen for recessive mutation phenotypes; or they can be genotyped for heterozygous mutations in the genes of interest.

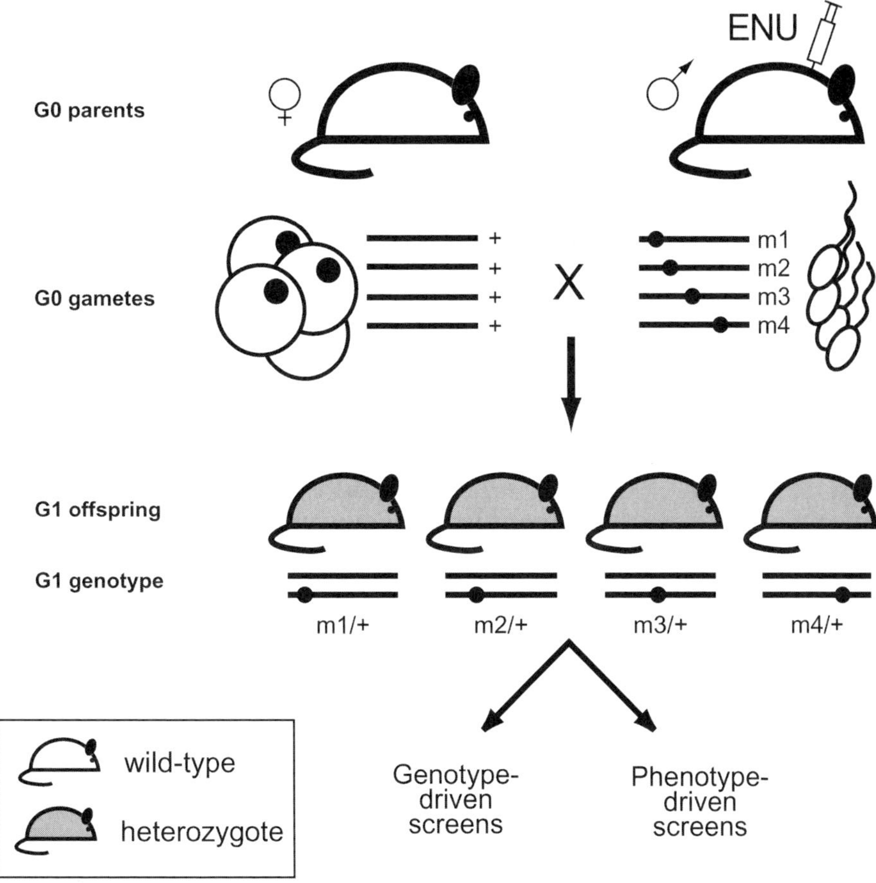

Figure 9.2 Schematic illustration of ENU mutagenesis. Wild-type males are injected with ENU and mated with wild-type females (G0 parents). Sperm from treated males carry individual sets of point mutations (G0 gametes, m1–m4); G1 offspring are heterozygous for ENU-induced mutations (G1 genotype, m1/+–m4/+)

Mutation load in ENU experiments

The mutation rate achieved by ENU treatment can be measured phenotypically as the rate of functionally relevant alterations, or at the DNA level as the rate of base pair exchanges, either silent or functionally relevant. Both parameters should be considered together in experimental design.

The classic assay to determine the rate of induced mutations is the specific locus test (SLT; Russell *et al.*, 1979). In the SLT, heterozygous carriers for the induced mutations (typically G1 animals) are mated with partners homozygous at tester loci for easily scorable recessive mutations, e.g. albino. The resulting offspring should be phenotypically normal, because they are heterozygous carriers of the mutated tester gene, inheriting the mutated allele only from the non-mutagenized parent. Only if the tester locus has been hit by a loss-of-function or strong hypomorphic mutation in the mutagenized parent will offspring mutant for the tester trait result. These offspring are compound heterozygotes for the tester allele and an ENU-induced mutation.

The average mutation rate reported in the literature for optimized ENU regimens is in the range of 1/1000 per locus, i.e. 1 in 1000 G1 animals is heterozygous for a functionally relevant mutation at a tester locus (Hitotsumachi *et al.*, 1985). Assuming 30 000–35 000 genes in the murine genome, this indicates the presence of 30–35 recessive mutations in each G1 animal. Currently, there are no data indicating that there is a strong site preference for ENU action that would result in mutagenesis 'hot spots', suggesting that the induced mutations are not linked in clusters.

Because technologies for high-throughput mutation detection became available only recently, the mutation load at DNA level is less well documented in the literature. Although older published data suggest a higher mutation load (Beier, 2000), the current consensus is a rate of 1 base pair exchange in 1–2.5 megabases (Mb) of genomic DNA. This includes silent as well as functionally relevant changes, at a total of 1000–2500 per haploid genome. The resulting assumption that 1%–3.5% of all point mutations lead to a functional change is generally consistent with the predicted size of the coding and regulatory regions of the genome and the redundancy of the genetic code.

According to the mutation loads described above, 1000 G1 animals would be sufficient to provide onefold statistical coverage of the whole genome with mutants; 5000 G1 animals would provide fivefold genome coverage, with a high likelihood of yielding at least one mutation in every gene and allelic series for many. It is this opportunity to obtain a very large number of mutations *in vivo* in one scalable, straightforward experiment that gives tremendous power to ENU mutagenesis.

The presence of multiple mutations in each G1 animal frequently leads to the concern that the downstream analysis of ENU-induced mutants might be confounded by the interaction of mutations. A closer look at the numbers described above indicates that this is extraordinarily unlikely; 30–35 mutations in a recombinational genome size of 1453 cM (Silver, 1995) amount to an average genetic distance between two functionally relevant mutations of 41.5–48.4 cM, indicating that adjacent mutations are almost certain to segregate in the next generation. The average distance of base pair exchanges of 1–2.5 per Mb is large enough so that for every functional mutation even the neighbouring silent can be segregated in a simple cross.

Because usually experimentation on a given mutant line will not be done on the founder animal, but in G2 and subsequent generations, the appropriate breeding strategy for the maintenance of a mutant will be enough to provide a clean genetic background. A similar routine backcrossing scheme is good practice also in embryonic stem (ES) cell-based experiments to eliminate unlinked spontaneous mutations that might have arisen during cell culture.

9.3 Reverse genetics by ENU mutagenesis

The application of ENU mutagenesis in reverse genetics, i.e. gene-driven strategies, is very straightforward: from a pool of carrier animals, mutations in a gene of interest can be identified rapidly and mouse lines carrying the desired mutations can be established (Coghill *et al.*, 2002) (Figure 9.3).

Compared with standard gene-driven mutagenesis approaches, such as gene targeting in ES cells, this strategy offers several advantages. An allelic series of point mutations, including hypomorph or domain specific changes, can be generated without extra effort. Because the mutant mouse line is established from frozen sperm samples (Figures 9.3B and 9.3C) rather than ES cells, both male and female carrier animals are available in the first generation, allowing direct intercrossing for the generation of homozygotes (Figure 9.3D). This cuts out the typical ES cell chimera stage and thus shortens the experimental schedule by one breeding generation, i.e. at least 3 months. Last, but not least, ENU mutagenesis is not restricted to certain genetic backgrounds, whereas ES cells are usually derived from the '129' family or from hybrid backgrounds.

On the downside, gene-driven ENU mutants do not allow the specific design of desired mutations to the extent possible in ES cells. Also, conditional mutants are not possible. It is most likely, therefore, that the ENU approach will complement but not supersede ES cell technology.

In practice, this approach requires a repository of G1 animals representing one or preferably several genome coverages. Rather than maintaining this repository as a constantly renewing pool of living animals, the genetic diversity is typically conserved by sperm freezing, with the establishment of a parallel repository of somatic DNA to be used in mutation screening

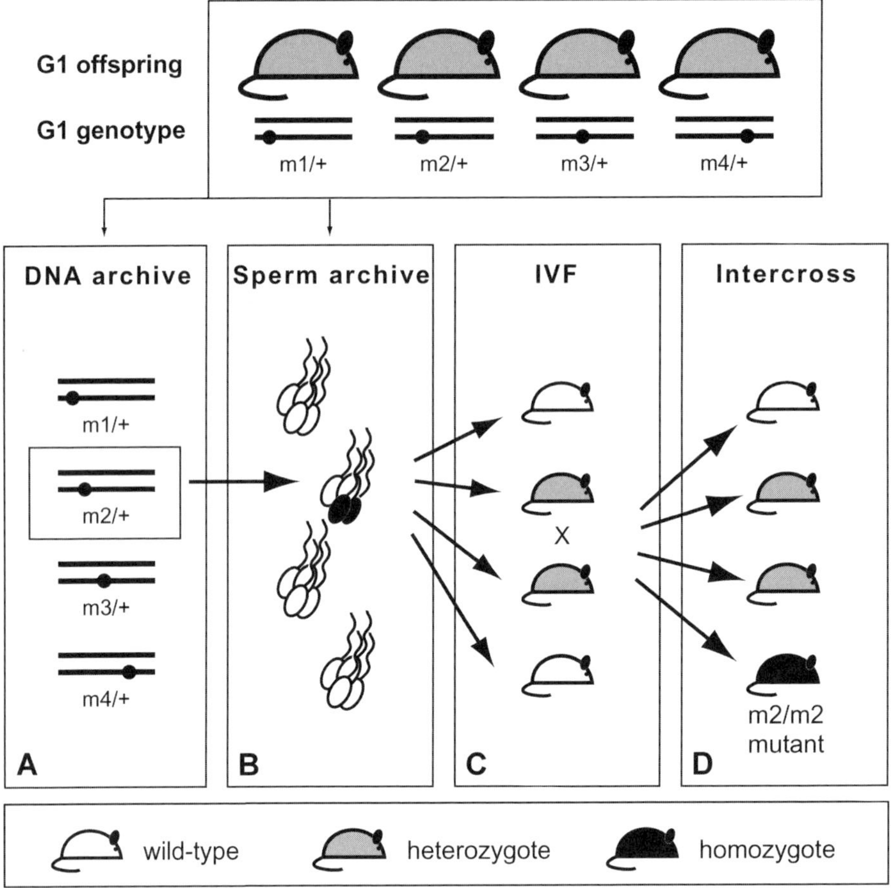

Figure 9.3 Schematic representation of reverse genetics by ENU mutagenesis. (A) A repository of DNA samples derived from G1 animals carrying heterozygous mutations is generated. The sample carrying the hypothetical mutation of interest is boxed. (B) A parallel repository of corresponding sperm samples derived from G1 animals is created. (C) The sperm sample carrying the mutation of interest is used to create heterozygous carriers by *in vitro* fertilization (IVF). (D) The mutation is bred to homozygosity and mutant animals are analyzed phenotypically

(Figure 9.3A). This parallel repository design provides a constant resource of assured quality.

Mutation detection in gene-driven ENU experiments

Mutations in the genes of interest are identified in the DNA repository using any one of the existing or emerging mutation detection technologies. Although single-strand conformation polymorphisms (SSCPs) and denaturing gradient HPLC (dHPLC; Coghill *et al.*, 2002) have been applied successfully, the most promising approach in terms of sensitivity and throughput is currently temperature gradient capillary electrophoresis (TGCE; Li *et al.*, 2002).

Using TGCE, the gene of interest is amplified from heterozygous DNA derived from G1 animals, typically one exon per polymerase chain reaction (PCR) fragment. If an ENU-induced mutation is present, the PCR product will contain heteroduplex molecules, showing melting curves in a temperature gradient different from homoduplexes (Figure 9.4A). Positive fragments are then sequenced to determine the exact nature of the mutation (silent, missense, nonsense or splice site; Figure 9.4B).

Recently introduced TGCE machines can sensitively and rapidly identify heteroduplexes in 2000 or more fragments per day. Assuming a typical gene with 10 exons, the mutation detection in a substantial G1 repository can be performed in 1 week or less. Extrapolating from the mutation loads outlined above, a repository of 1000 samples, equivalent to onefold genome coverage, would provide a 60% statistical chance of obtaining one loss-of-function allele in any given gene of interest. With a repository size of 5000, the likelihood of identifying one allele would be >95%, with a 70% chance to obtain two alleles (Coghill *et al.*, 2002). A repository size of 10 000 might be desirable and will be within the range of mutation detection technology.

Applications of gene-driven ENU mutagenesis

Reverse genetics by ENU is very scalable because it requires only minimal human input in comparison to the specific design and construction of recombinant vectors necessary for gene targeting. The availability of the mouse genomic sequence facilitates the semi-automatic design of primers for DNA amplification, and rapid mutation detection technologies such as TGCE are highly amenable to industrialization.

Its main application is in the physiological validation of candidate drug targets. The rapid and cost-effective availability of informative genetic variation in target genes in various genetic backgrounds lowers the threshold to the use of the mouse model as early as possible in a drug discovery program. In particular, the investigation of orphan druggable genes will profit from application of the technology, because targeted knock-outs frequently only uncover the first step in development where the mutated gene is absolutely required, whereas point mutations frequently provide additional information due to hypomorphic changes or partial loss of function.

Point mutations can mimic drug action more closely than gene deletions. Because drugs usually do not work by eliminating the target protein, but by inducing specific changes in function, removal of the protein in a null allele

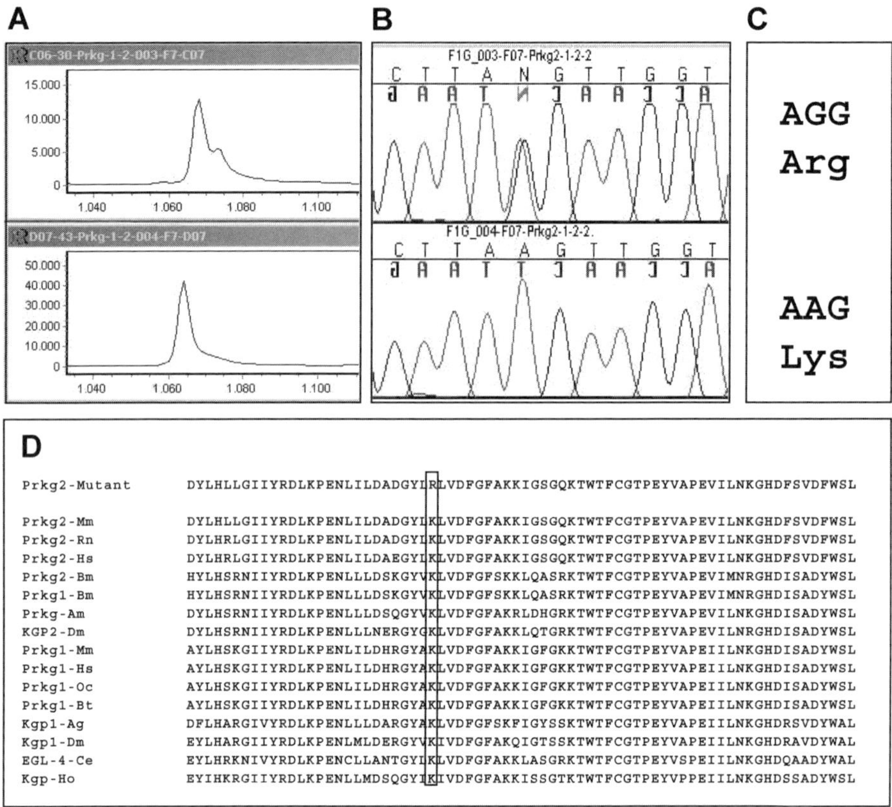

Figure 9.4 A typical experiment applying gene-driven ENU mutagenesis. The exon encoding the catalytic loop of Type II cGMP-dependent protein kinase Prgk2 was amplified by PCR and analysed for point mutations using TGCE. (A) The TGCE profile of the amplified fragment indicates the presence of a point mutation (upper panel; lower panel is wild-type control). (B) The point mutation is verified by direct DNA sequencing. (C) The base pair exchange identified leads to a missense mutation changing lysine (K) to Arginine (R). (D) The mutation in Prgk2 occurs at an evolutionary highly conserved amino acid position, as illustrated by the protein sequence alignment. Mm, *Mus musculus*; Hs, *Homo sapiens*; Rn, *Rattus norvegicus*; Dm, *Drosophila melanogaster*; Ce, *Caenorhabditis elegans*; Ag, *Anopheles gambiae*; Am, *Apis mellifera*; Bm, *Bombyx mori*; Oc, *Oryctolagus cuniculus*; Bt, *Bos taurus*; Ho, *Hydra oligactis*

does not model drug action very well. Rather, alteration of a specific domain should be the desired change.

Figure 9.4 shows a typical experiment implementing this strategy. In a search for mutations specifically affecting the enzymatic activity of orphan protein kinases, a missense mutation leading to a change in an evolutionary highly conserved amino acid in the catalytic domain of kinase Prkg2 was identified (Figure 9.4C). The associated sperm sample was used to revitalize the mutant mouse line, and homozygous offspring are currently under investigation to identify the mutant phenotype. Although a similar experiment using gene targeting might have taken more than 1 year (design of a specific point mutation, targeting in ES cells, generation of chimeras, germline transmission) and three generations of mice, the ENU-based approach took less than 6 months from the start of experimentation to the homozygous mutant animal.

9.4   Forward genetics in the discovery of new pathways

Although reverse genetics in the mouse model has its strengths in the physiological validation of drug targets identified by various approaches, the realm of forward genetics using phenotype-driven screens is the *de novo* identification of novel physiological pathways.

For decades, forward genetic screens have been invaluable for generating molecular maps of the pathways controlling some of the most fundamental functions of living systems. The only requirements to generate molecular entry points into new physiological pathways are assays to identify variations of the phenotype of interest. Two screens that opened new fields for molecular investigation were awarded Nobel Prizes: Lewis, Wieschaus and Nusslein-Volhard received the award in 1995 for the dissection of early pattern formation in animal development. (Nusslein-Volhard and Wieschaus, 1980), whereas Hartwell, Nurse and Hunt were honored in 2001 for the discovery of the cell cycle control machinery (Hartwell *et al.*, 1974). Both examples illustrate not only the power of genetic screens within the chosen model organism but also the tremendous impact on human biology and medicine due to the evolutionary conservation of the mechanisms identified.

The list of examples can be extended to cases where the discovery of basic biological principles by forward genetics has ultimately led to marketed drugs. The development of the statins – powerful cholesterol-lowering drugs representing one of the most important advances of modern pharmaco-therapy – can be traced back to the 'phenotype first' investigations of Goldstein and Brown into the molecular basis of familial hypercholesterol-emia (Brown and Goldstein, 1986).

These examples illustrate that forward genetics approaches in their application to drug discovery provide an opportunity for breaking new ground and opening up very innovative therapeutic approaches. This advantage comes at the price that the biochemical nature of the novel targets, and thus their amenability to classical medicinal chemistry, cannot be predicted. But as technologies are available now that go beyond traditional small-molecule pharmaceuticals and secreted protein approaches (monoclonal antibodies, RNA interference, antisense, gene therapy, cell therapy), even targets that are not considered typically drugable proteins can be used now to develop therapeutics.

The impact of mouse forward genetics on drug discovery

A number of classical spontaneous mouse mutants have identified new physiological pathways. Typical examples include the discovery of stem cell factor and its receptor, defined by the classical mutants *steel* and *kit*, and the discovery of the apoptosis-inducing *fas* receptor and its ligand encoded by the *lpr* locus.

The synergy between 'opening' new pathways by forward genetic studies of the molecular nature of classical mouse mutants, and using reverse genetics to refine their understanding towards clinical application, is very well illustrated by the progress that has been made in our understanding of obesity (Barsh *et al.*, 2000).

Although the endocrinology of obesity was not a mainstream field until the mid-1990s, a flurry of fruitful activity was catalyzed by the molecular cloning of classical mouse mutants displaying genetic obesity syndromes. The identification of leptin (Zhang *et al.*, 1994) and its receptor (Lee *et al.*, 1996), mutated in the mouse lines *ob* and *db*, established a previously unknown peptide hormone as a key player in the regulation of body weight and energy expenditure.

The molecular defect in yellow agouti – a classical mouse mutant first described in 1905 – implicated the melanocortin system as a second pathway controlling body composition (Miller *et al.*, 1993). Elegant studies were performed to investigate the epistatic interactions between leptin, the melanocortin system and other neuropeptides (Erickson *et al.*, 1996). Human genetics studies showed that the physiological activities of the leptin and melanocortin systems were highly conserved between rodent and humans (Montague *et al.*, 1997; Yeo *et al.*, 1998).

As these studies opened new routes for investigation, obvious questions arose that were addressed by reverse genetics. Agouti-related peptide *Agrp* was identified as the likely antagonist acting on melanocortin receptors in the brain (Shutter *et al.*, 1997), and its physiological function was confirmed by transgenic expression in mice (Ollmann *et al.*, 1997). Gene-targeting experiments identified the melanocortin receptor subtype 4 (*MC4-R*) as a strong drug target in the melanocortin pathway (Huszar *et al.*, 1997), and a function completely unrelated to obesity was identified for the closely related receptor *MC5-R* (Chen *et al.*, 1997).

Mutagenesis screens as a source of qualitative and quantitative genetic variation

Although classical mendelian mouse mutants still have some surprises to offer, their supply and the range of phenotypes that contribute to the understanding of human disease are limited. These limitations can be overcome with mutagenesis screens.

The majority of the genetic variation contributing to prevalent human diseases results from complex interactions of multiple loci, the rare exception being the fully penetrant disease genes causing qualitative Mendelian disease predisposition. Understanding the complex interactions of disease-modifying genes requires the identification of all players as well as quantitative analysis of their interactions. This requires investigations in model organisms because, by their very nature, such phenomena cannot be treated easily in reductionist *in vitro* experiments.

As part of genome projects, large-scale mutagenesis screens have been set up as national core resource centers in several countries, and are being supported by the major funding agencies (NIH, MRC, DFG, etc.). Also, spin-out biotech companies have been set up to apply the technology commercially. The currently running large-scale mutagenesis screens are summarized in Table 9.1.

With the choice of the appropriate experimental design, a phenotype driven mutagenesis screen provides the potential to dissect the molecular pathways contributing to any phenotype of interest. Qualitative, Mendelian phenotypes aid the identification of pathway components, while alleles modifying quantitative phenotypes and epistatic experiments can shed light on gene interactions. A number of genetic strategies for genome wide screens are available.

Table 9.1 Weblinks to the major academic and industrial ENU screening programs

ENU screen	Website
ANU, Canberra, Australia	http://jcsmr.anu.edu.au/group_pages/mgc/MedGenCen.html
Baylor College, Houston, USA	http://www.mouse-genome.bcm.tmc.edu
Genomics Institute of the Novartis Research Foundation, San Diego, USA	http://www.gnf.org/
GSF, Neuherberg, Germany	http://www.gsf.de/ieg/groups/enu-mouse.html
Ingenium Pharmaceuticals AG, Germany	http://www.ingenium-ag.com
Jackson Laboratory, Bar Harbor, USA	http://pga.jax.org//index.html
Jackson Laboratory, Bar Harbor, USA	http://www.jax.org/nmf/
MRC, Harwell, UK	http://www.mut.har.mrc.ac.uk
Oak Ridge Nat. Lab., Tennessee, USA	http://bio.lsd.ornl.gov/mouse
Phenomix, San Diego, USA	http://www.phenomixcorp.com
Tennessee Genome Consortium, Tennessee, USA	http://Tnmouse.org
RIKEN, Yokohama, Japan	http://www.gsc.riken.go.jp/Mouse/
University of Toronto, Toronto, Canada	http://www.cmhd.ca
Northwestern University, Chicago, USA	http://Genome.northwestern.edu

Genome-wide screens for dominant mutations

A very straightforward experimental design is the screening of G1 offspring derived from mutagenized G0 fathers (Figure 9.2). As discussed above, each individual G1 animal is heterozygous for a unique set of induced mutations. Dominant mutations thus can be identified in a very simple and efficient mating scheme and verified by further breeding. A limitation of this strategy is that phenotypes can be recovered only if they do not severely impair viability or fertility.

In the two major large-scale screens reported to date (Hrabe de Angelis *et al.*, 2000; Nolan *et al.*, 2000), 1%–2% of the G1 animals displayed a heritable alteration in the phenotypes investigated.

A substantial number of alleles affecting therapeutically relevant phenotypes have been isolated by these screens. A number of mutations have been characterized already at the molecular level, e.g. the mutations *Bth* (Beethoven, Vreugde *et al.*, 2002) and *Htu* (headturner, Kiernan *et al.*, 2001), causing defects in inner ear development and progressive hearing loss, respectively. Several mutations causing dominant cataracts also have been identified (Graw *et al.*, 1997, 1999, 2002a,b; Favor and Neuhauser-Klaus, 2000).

Genome-wide screens for recessive mutations

Recessive screens require more logistic effort but they extend the scope of the experiment to loss-of-function mutations and alterations that reduce viability and fertility. The mutated alleles are passed to the G2 generation in the heterozygous state (Figure 9.5A) and are bred to homozygosity in G3 by intercrossing of G2 siblings or backcrossing of G2 females to their G1 father (Figure 9.5B). In this way, recessive phenotypes manifest in G3 siblings and the appearance of multiple affected animals in Mendelian frequency provides a first level of confirmation that the phenotype is indeed genetically determined. This greatly facilitates the selection of mutant lines for further analysis (Figure 9.5C).

Because the mutations introduced by the G1 founder segregate freely, 30–35 mutations are analyzed in each pedigree. A phenotype that occurs in G3 with close to the expected mendelian frequency is very unlikely to be a compound effect of more than one mutation. The a priori likelihood that two genes acting in the same pathway are mutated in the same founder is very low, and the chances that two unlinked genes co-segregate in 25% of G3 animals are even lower. For three affected animals from a family of 12, the chances for two unlinked loci to cosegregate is in the range of 1.5%; for 4 out of 16 it is down to 0.4%.

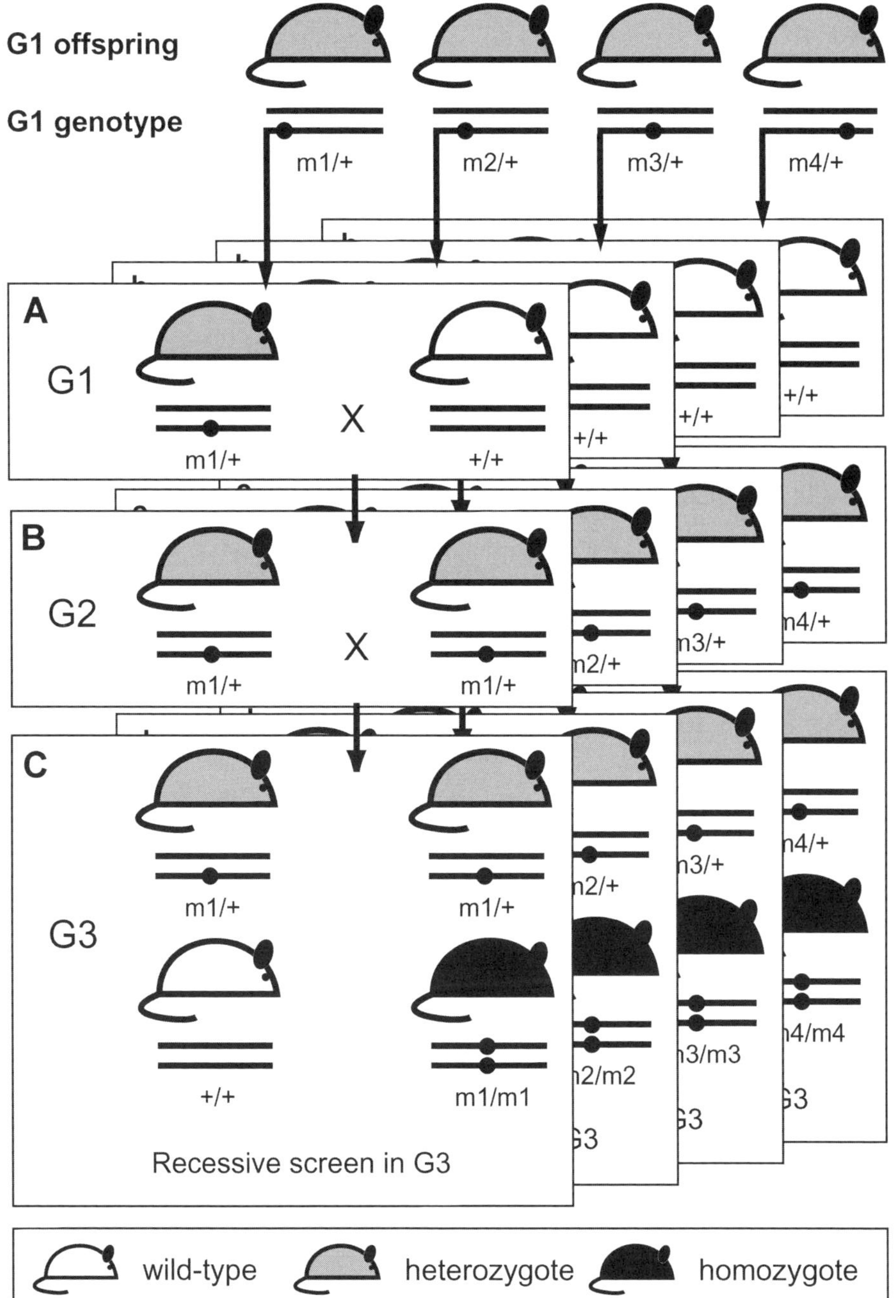

Figure 9.5 Schematic representation of a typical breeding protocol. In a recessive screen, mutant alleles are propagated in a heterozygous state from G1 to G2 (A, B), and homozygote animals (C) derived from G2 inter- or backcrosses are screened for variant phenotypes. Note that each G1 animal generates a separate micro-pedigree, in which mutations m1–m4 segregate

In theory, a recessive screen would require 1000 G3 pedigrees for onefold genome coverage, i.e. analysis of 30–35 000 mutations. Because the analysis of 10–15 animals per pedigree will only allow the recovery of about 50% of the segregating phenotypes, and a larger number of animals per pedigree would provide diminishing returns, the observation of 2000 pedigrees is required in practice. Although the phenotyping of 20–30 000 animals is a substantial effort, this scale can be achieved readily in an industrial setting. A typical physiological pathway can be estimated to have 5–10 components, so a recessive screen of only 0.3–0.5-fold coverage has a high likelihood to provide at least one hit in each pathway.

Although only preliminary data have been reported from recessive screens investigating postnatal phenotypes, a number of results from screens for recessive mutations affecting development (Kasarskis *et al.*, 1998; Hentges *et al.*, 1999; Zoltewicz *et al.*, 1999; Anderson, 2000; Herron *et al.*, 2002) have been published. Examples include the identification of a mutation in the FRAP/mTOR gene as the cause of the *flat-top* phenotype (Hentges *et al.*, 2001), and alterations of Rab23 (Eggenschwiler *et al.*, 2001) in the *open brain* mutation (Sporle *et al.*, 1996; Gunther *et al.*, 1997; Sporle and Schughart, 1998).

Screens for specific chromosomal regions

Region-specific screens use large deletions or inversions to facilitate the recovery of recessive mutations. By breeding ENU-treated males with females carrying specific chromosomal deletions, recessive mutations induced in the hemizygous region are uncovered and can be identified in G1 animals (Rinchik *et al.*, 1990; Rinchik and Carpenter, 1999). The advantages are the simple breeding scheme and the fact that mutations are immediately mapped by failing to complement the deletion. The application of this strategy is limited by the availability of chromosomal deletions.

A very elegant improvement of the region-specific screen design is the use of genetically engineered balancer chromosomes carrying large inversions. This scheme also requires a three-generation breeding scheme and has been reviewed in detail by Justice (Justice, 2000).

The role of mutagenesis in the analysis of quantitative traits

Substantial effort has been put into analysis of the molecular basis of quantitative trait loci (QTL) occurring spontaneously as differences between inbred strains of mice and rats (Stoll *et al.*, 2001). Analyzing a QTL in a model organism requires three major steps: genetic mapping of the chromosomal segment containing the QTL; identifying the molecular changes in the candidate region; and demonstrating that one or more of these changes are causing the observed phenotypic variation. Although low-resolution mapping of spontaneous QTL in rodents is comparatively straightforward and has been performed successfully many times, pinning down the molecular detail has proved to be extremely hard. Typically, each of the interstrain QTL only has a minor contribution to the phenotype, and each candidate QTL region is 'contaminated' by a substantial amount of irrelevant DNA sequence variation. The former poses a substantial problem to mapping strategies, whereas the latter makes it practically impossible to establish one of the many changes as the main culprit. A recent milestone paper (Steinmetz *et al.*, 2002) demonstrates the enormous complexity of the problem, even in a simple and genetically extremely well-tractable model organism such as *Saccharomyces cerevisiae*.

For these reasons it has been proposed that mutagenesis screens in rodents will be essential for uncovering the molecular basis of QTL (Nadeau and Frankel, 2000). Into physiological systems with a high degree of similarity to the human situation, genetic variation leading to qualitative as well as quantitative variation can be introduced into a 'clean' background. This provides the essential basis for the successful isolation and characterization of the trait.

As discussed above, the average spacing between base pair exchanges introduced by ENU mutagenesis is 1–2.5 Mb, a distance that allows genetic separation of polymorphisms with a manageable number of meioses. Once a candidate polymorphism has been isolated in this way, it can be investigated further by the armentarium of reverse genetics techniques. The fundamental difference between analysis of spontaneous and induced QTL is that the latter can be reintroduced in isolation into the parental background, thus eliminating the influence of background differences. Thus, quantitative traits arising from mutagenesis can be treated in a similar manner to qualitative Mendalian traits.

Sensitized screens

In addition to physiological challenges as part of the screening protocol, forward genetics also offers the possibility to use genetic challenges. In this case, the mutagenesis is performed on animals already carrying an alteration in a gene of interest, which are screened for changes of the primary mutant phenotype by a second, induced mutation.

This design offers important additional options. Firstly, it allows specific identification of new players in pathways of interest without prior molecular information. Many examples from invertebrate experiments illustrate the power of this approach, e.g. dissection of the *ras* signalling pathway (Gaul *et al.*, 1993). Secondly, a screen for medically relevant functions in mammals can be designed to improve a disease phenotype caused by the primary genetic defect, thus identifying 'health genes' rather than 'disease genes'. It can be assumed that players identified in such a design are prime candidates for pharmacological intervention.

An illustration of this approach is identification of the locus *Mom* (*modifier of min*), which influences the phenotype of the mutant line *Min* (Moser *et al.*, 1990); *Min* carries a mutation in the mouse homolog of the human familial polyposis gene (*Apc*) and suffers from multiple intestinal polyps; *Mom* strongly modifies the extent and progression of this polyposis. The *Mom* locus encodes the secretory phospholipase *Pla2g2aI*, which could be shown in transgenic rescue experiments to provide at least one component of the modifier function (Cormier *et al.*, 1997; 2000). The *Mom* locus is an allele that occurred spontaneously, but similar approaches are under investigation in several laboratories using ENU-induced modifiers.

9.5   The art of screen design: phenotyping

Whatever the genetic design of a screen, the right phenotyping protocol is a prerequisite for finding informative mutants that will lead to the identification of novel molecular pathways. The art of designing and implementing a successful screen lies in the choice of the appropriate target phenotype, combined with the establishment of scalable combinations of primary and secondary assays to detect this phenotype sensitively and specifically.

An excellent example is isolation of the mutant *clock*, which led to identification of the first gene affecting circadian rhythm in mammals (Vitaterna *et al.*, 1994). The assay employed – measurement of the circadian activity using a computer-monitored running wheel – is straightforward, scalable and very specific, although great care had to be taken to establish normal ranges and baselines. In contrast, the measurement of body weight would not be sufficient to identify specifically the lean animals with reduced body fat. Many animals initially would score positively, thus obscuring the desired mutants, because there are many reasons for a mouse to have lower body weight than normal, e.g. non-genetic runts and growth retardation secondary to many other genetic defects.

Thus, a typical screening protocol employs several levels of activities. The primary screen should employ simple parameters and assays that have a high sensitivity for rapid and efficient enrichment of candidate mutants with altered physiology in the areas of interest. Each 'hit' in these crude but sensitive primary assays has to be followed up with more elaborate assays of higher physiological specificity. These help to exclude false positives, confirm the relevance of true positives and place the mutant in the framework of known functions.

The primary phenotype might be related directly to the physiology of interest, e.g. behavioral alteration of the *clock* mutant (Vitaterna *et al.*, 1994), but also can be a surrogate marker, e.g. the anemia that led to identification of the *Min/apc* mutant showing multiple adenomas of the colon (Moser *et al.*, 1990).

The sensitivity and specificity of a screen can be improved by devising and implementing functional challenges that specifically test biological mechanisms in the disease area of interest. For example, neurological assays challenge locomotion, balance and muscular strength, thus exploring specific functions of the nervous system, such as mechano- and thermosensation, hearing, vision and motor coordination. Behavioral challenges are required to address the anxiety/exploration paradigm and sensorimotor gating, a central brain function disturbed in schizophrenia.

An example: screening for mutations affecting the immune system

The immune system of the mouse is particularly useful and valuable for identifying phenotypic mutants for several reasons. Firstly, it has been studied extensively and methods are well established that can be adapted easily for high-throughput screening. There are also numerous mouse models for human diseases of the immune system. Secondly, it is feasible to screen for a multitude of molecules and cell populations of the immune system in a small sample of blood, the taking of which does not require killing the animal. Thirdly, it is likely to be a rich source of potential drug targets because disorders of the immune system are responsible for a variety of pathological conditions in humans. Lack of an appropriate response to foreign molecules can result in infectious disease or cancer, whereas inappropriate stimulation of the immune system can lead to conditions such as autoimmune disease, asthma, allergy and transplant rejection.

The immune system is dynamic and changes upon responding to immunological challenge. In most animal facilities, the mice are housed in barriers where exposure to pathogens is low. Consequently, the immune system is in a relatively dormant state. Because many human immune disorders are only evident upon exposure to some foreign agent, it becomes necessary to stimulate the immune system to respond by treatment with some reagent or organism, e.g. immunization with a foreign protein or polysaccharide to elicit an immune response, or treatment with some non-foreign protein to induce autoimmune disease.

Two large-scale ongoing screens that did not involve any challenge to the immune system have already yielded interesting mutants. In both screens, blood was taken and examined for various lymphocyte subsets, as well as measuring the levels of the various immunoglobulin subclasses. Mutants were identified that had abnormal immunoglobulin levels, lack of T and/or B cells or various hemopoetic tumors (Flaswinkel *et al.*, 2000; Hrabe de Angelis *et al.*, 2000; Alessandrini *et al.*, 2001; Nelms and Goodnow, 2001).

The next generation of screens will involve challenging the immune system. Large-scale screens that require infection with live bacteria, viruses or parasites are difficult to implement due to the obvious problems in animal husbandry; screened mice would need to be housed separately in order to protect the rest of the colony. Additionally, if the dose induced illness this would generate problems and extra work in caring for the animals, as well as being ethically debatable. Such problems could be avoided by challenge with a dose that induces a measurable response (e.g. increase in antigen-specific antibody titers) but is not high enough to induce a clinical phenotype.

An added problem is that the stereotypical response in inbred strains proves to be highly variable. Even when inbred mice of the same strain, age, gender and housing are immunized with protein antigens, the antigen-specific antibody titers can vary by up to tenfold. For a disease screen such as late-onset autoimmunity, not every mouse will get the disease at the same time and to the same extent. In such cases, either a significant false-positive rate is accepted, requiring further screening to identify true mutant phenotypes, or only extreme phenotypes are selected for characterization. The problem with the latter strategy is that the screen could miss a lot of potentially interesting mutants. Thus, although the logistics and expense of challenge screens are greater than for passive screens, they have the potential to yield important new information.

9.6   Industrialized positional cloning

Identifying the causative mutation for an interesting phenotype by positional cloning has been a very time-consuming effort in the past. Most of this effort typically had been directed at the generation of a dense genetic and physical map of the candidate region, often limited by the cloning technologies and the limited availability of polymorphic genetic markers.

In the age of genome sequencing, most of these problems have simply disappeared. Dense maps of genetic markers, single-nucleotide polymorphisms and simple sequence-length polymorphisms are available covering virtually all of the mouse genome. State-of-the-art genotyping technology allows the mapping of a candidate region in a matter of days, once informative animals derived from a mapping cross are available (see below).

With high-quality genomic sequences available for humans, mice and rats, the gene content of a candidate region can be analyzed rapidly *in silico* and compared with syntenic regions of other species for detailed annotation of potential coding regions and conserved non-coding sequences. Candidate genes in the region can be prioritized by previously existing information about their function, as well as by information about the expression pattern derived from expressed sequence tag databases or microarray experiments. The final step – the testing of candidate genes for mutations – is limited only by the cost of DNA sequencing and alternative mutation detection technologies.

A typical positional cloning project can be finished in well under a year. For a fully penetrant recessive phenotype the generation of informative meiotic recombinations will be achieved in a simple outcross–intercross breeding strategy (Silver, 1995), that takes approximately 6 months. With modern genotyping technology, mapping the mutated locus to a resolution of 1–2 cM can be achieved in less than 1 month, once the F2 animals are phenotyped.

Thus, the limiting factor for positional cloning of ENU mutants is not the application of high-throughput genomics technologies, but the specific characteristics of the mutant under investigation. Phenotypes affecting breeding performance will delay the generation of informative meioses, and quantitative phenotypes modified by genetic background make it more challenging to classify affected and unaffected animals. For example, A/J mice are highly susceptible to allergen-induced airway hyperresponsiveness (AHR), an asthma-related phenotype, whereas C3H/HeJ and C57BL/6 are much more resistant (De Sanctis *et al.*, 1995; Ewart *et al.*, 2000; Karp *et al.*, 2000). Similarly, A/J mice respond much more than C57BL/6 in an immediate cutaneous hypersensitivity test, which is a test for atopy (Daser *et al.*, 2000). Such strain differences strongly affect the strategy for positional cloning of the mutation; also, the mutant phenotype needs to be expressed in the mixed background. Outcrossing a C3H/HeJ mutant that is resistant to allergen-induced AHR with A/J mice, which have a dominant allergen-resistant AHR phenotype (De Sanctis *et al.*, 1995), could lead to loss of the phenotype.

Finally, once the mutation hunt in the candidate region is ongoing, it is obvious that coding mutations in known genes already linked to the physiology of interest can be identified very rapidly, whereas mutations in novel 'orphan' genes or non-coding mutations require more downstream work to be proved beyond doubt as the cause of the mutant phenotype.

A recent report (Herron *et al.*, 2002) describes the rapid mapping of 7/15 ENU-induced recessive developmental mutations using interval haplotype analysis and the identification of the causative mutation in two of these lines. From the first phase of the ENU mutagenesis screen at Ingenium, 30 mutations were cloned in a time-frame of 15 months and 20 additional loci

Table 9.2 Summary of the molecular changes identified by positional cloning in the Ingenium ENU screen

Mutation type[1]	n^2	%
Missense	15	56%
Nonsense	3	11%
Splice site	7	26%
Non-coding	2	7%
AT/TA transversion	11	41%
AT/GC transition	8	30%
CG/AT transition	7	26%
GC/CG transversion	1	4%
Total	27	100%

[1]Type of change in gene structure and DNA sequence, respectively.
[2]Number of mutant mouse lines analyzed.

were mapped (manuscript in preparation), demonstrating that the positional cloning process can be established at an industrial scale. Table 9.2 shows the distribution of the molecular characteristics (missense, nonsense, splice site, non-coding) of the mutations characterized so far. No regional bias to specific chromosomes or chromosome segments was observed.

9.7   Conclusions and Prospects

The rediscovery and large-scale application of random mutagenesis using ENU is an important extension of the mouse genetics toolkit, which can be straightforwardly integrated into existing drug discovery strategies. When applied in gene-driven reverse genetics experiments, it has been established as a scalable alternative to ES-cell based mutagenesis technologies in target validation. As a tool for the discovery of new therapeutic principles, it provides the opportunity to investigate new physiological pathways by means of phenotype-driven forward genetics screens, which are unrestricted by existing knowledge of molecular entry points. The improved availability of mouse models will facilitate their systematic application in pharmaceutical research.

9.8   References

Alessandrini, F., Jakob, T., Wolf, A., Wolf, E., Balling, R., Hrabe, de Angolis, M. H., Ring, J., *et al.* (2001). Enu mouse mutagenesis: generation of mouse mutants with aberrant plasma IgE levels. *Int. Arch. Allergy Immunol.* **124**, 25–28.

Anderson, K. V. (2000). Finding the genes that direct mammalian development: ENU mutagenesis in the mouse. *Trends Genet.* **16**, 99–102.

Balling, R. (2001). ENU mutagenesis: analyzing gene function in mice. *Annu. Rev. Genom. Hum. Genet.* **2**, 463–492.

Barsh, G. S., Farooqi, I. S. and O'Rahilly, S. (2000). Genetics of body-weight regulation. *Nature* **404**, 644–651.

Beier, D. R. (2000). Sequence-based analysis of mutagenized mice. *Mamm. Genome* **11**, 594–597.

Black, J. (1989). Drugs from emasculated hormones: the principle of syntopic antagonism. *Science* **245**, 486–493.

Brown, M. S. and Goldstein, J. L. (1986). A receptor-mediated pathway for cholesterol homeostasis. *Science* **232**, 34–47.

Brown, S. D. and Balling, R. (2001). Systematic approaches to mouse mutagenesis. *Curr. Opin. Genet. Dev.* **11**, 268–273.

Chen, W., Kelly, M. A., Opitz-Araya, X., Thomas, R. E., Low, M. J. and Cone, R. D. (1997). Exocrine gland dysfunction in MC5-R-deficient mice: evidence for coordinated regulation of exocrine gland function by melanocortin peptides. *Cell* **91**, 789–798.

Coghill, E. L., Hugill, A., Parkinson, N., Davison, C., Glenister, P., Clements, S., Hunter, J., *et al.* (2002). A gene-driven approach to the identification of ENU mutants in the mouse. *Nat. Genet.* **30**, 255–256.

Cohen, P., Miyazaki, M., Socci, N. D., Hagge-Greenberg, A., Liedtke, W., Soukas, A. A., Sharma, R., *et al.* (2002). Role for stearoyl-CoA desaturase-1 in leptin-mediated weight loss. *Science* **297**, 240–243.

Cormier, R. T., Hong, K. H., Halberg, R. B., Hawkins, T. L., Richardson, P., Mulherkar, R., Dove, W. F., *et al.* (1997). Secretory phospholipase Pla2g2a confers resistance to intestinal tumorigenesis. *Nat. Genet.* **17**, 88–91.

Cormier, R. T., Bilger, A., Lillich, A. J., Halberg, R. B., Hong, K. H., Gould, K. A., Borenstein, N., *et al.* (2000). The Mom1AKR intestinal tumor resistance region consists of Pla2g2a and a locus distal to D4Mit64. *Oncogene* **19**, 3182–3192.

Daser, A., Koetz, K., Batjer, N., Jung, M., Ruschendorf, F., Goltz, M., Ellerbrok, H., *et al.* (2000). Genetics of atopy in a mouse model: polymorphism of the IL-5 receptor alpha chain. *Immunogenetics* **51**, 632–638.

De Sanctis, G. T., Merchant, M., Beier, D. R., Dredge, R. D., Grobholz, J. K., Martin, T. R., Lander, E. S., *et al.* (1995). Quantitative locus analysis of airway hyperresponsiveness in A/J and C57BL/6J mice. *Nat. Genet.* **11**, 150–154.

Eggenschwiler, J. T., Espinoza, E. and Anderson, K. V. (2001). Rab23 is an essential negative regulator of the mouse Sonic hedgehog signalling pathway. *Nature* **412**, 194–198.

Erickson, J. C., Hollopeter, G. and Palmiter, R. D. (1996). Attenuation of the obesity syndrome of ob/ob mice by the loss of neuropeptide Y. *Science* **274**, 1704–1707.

Ewart, S. L., Kuperman, D., Schadt, E., Tankersley, C., Grupe, A., Shubitowski, D. M., Peltz, G., *et al.* (2000). Quantitative trait loci controlling allergen-induced airway hyperresponsiveness in inbred mice. *Am. J. Respir. Cell Mol. Biol.* **23**, 537–545.

Favor, J. and Neuhauser-Klaus, A. (2000). Saturation mutagenesis for dominant eye morphological defects in the mouse *Mus musculus*. *Mamm. Genome* **11**, 520–525.

Flaswinkel, H., Alessandrini, F., Rathkolb, B., Decker, T., Kremmer, E., Servatius, A., Jakob, T., *et al.* (2000). Identification of immunological relevant phenotypes in ENU mutagenized mice. *Mamm. Genome* **11**, 526–527.

Gaul, U., Chang, H., Choi, T., Karim, F. and Rubin, G. M. (1993). Identification of ras targets using a genetic approach. *Ciba Found. Symp.* **176**, 85–92.

Graw, J., Neuhauser-Klaus, A. and Pretsch, W. (1997). Detection of a point mutation (A to G) in exon 5 of the murine Mgf gene defines a novel allele at the Steel locus with a weak phenotype. *Mutat. Res.* **382**, 75–78.

Graw, J., Jung, M., Loster, J., Klopp, N., Soewarto, D., Fella, C., Fuchs, H., *et al.* (1999). Mutation in the betaA3/A1-crystallin encoding gene Cryba1 causes a dominant cataract in the mouse. *Genomics* **62**, 67–73.

Graw, J., Klopp, N., Neuhauser-Klaus, A., Favor, J. and Loster, J. (2002a). Crygf(Rop): the first mutation in the Crygf gene causing a unique radial lens opacity. *Invest Ophthalmol. Vis. Sci.* **43**, 2998–3002.

Graw, J., Neuhauser-Klaus, A., Loster, J., Klopp, N. and Favor, J. (2002b). Ethylnitrosourea-induced base pair substitution affects splicing of the mouse gammaE-Crystallin encoding gene leading to the expression of a hybrid protein and to a cataract. *Genetics* **161**, 1633–1640.

Gunther, T., Sporle, R. and Schughart, K. (1997). The open brain (opb) mutation maps to mouse chromosome 1. *Mamm. Genome* **8**, 583–585.

Harris, S. (2001). Transgenic knockouts as part of high-throughput, evidence-based target selection and validation strategies. *Drug Discov. Today* **6**, 628–636.

Harris, S. and Foord, S. M. (2000). Transgenic gene knock-outs: functional genomics and therapeutic target selection. *Pharmacogenomics* **1**, 433–443.

Hartwell, L. H., Culotti, J., Pringle, J. R. and Reid, B. J. (1974). Genetic control of the cell division cycle in yeast. *Science* **183**, 46–51.

Hentges, K., Thompson, K. and Peterson, A. (1999). The flat-top gene is required for the expansion and regionalization of the telencephalic primordium. *Development* **126**, 1601–1609.

Hentges, K. E., Sirry, B., Gingeras, A. C., Sarbassov, D., Sonenberg, N., Sabatini, D. and Peterson, A. S. (2001). FRAP/mTOR is required for proliferation and patterning during embryonic development in the mouse. *Proc. Natl. Acad. Sci. USA* **98**, 13 796–13 801.

Herron, B. J., Lu, W., Rao, C., Liu, S., Peters, H., Bronson, R. T., Justice, M. J., *et al.* (2002). Efficient generation and mapping of recessive developmental mutations using ENU mutagenesis. *Nat. Genet.* **30**, 185–189.

Hitotsumachi, S., Carpenter, D. A. and Russell, W. L. (1985). Dose-repetition increases the mutagenic effectiveness of *N*-ethyl-*N*-nitrosourea in mouse spermatogonia. *Proc. Natl. Acad. Sci. USA* **82**, 6619–6621.

Hrabe de Angelis, M. H. and Balling, R. (1998). Large scale ENU screens in the mouse: genetics meets genomics. *Mutat. Res.* **400**, 25–32.

Hrabe de Angelis, M. H., Flaswinkel, H., Fuchs, H., Rathkolb, B., Soewarto, D., Marschall, S., Heffner, S., *et al.* (2000). Genome-wide, large-scale production of mutant mice by ENU mutagenesis. *Nat. Genet.* **25**, 444–447.

Huszar, D., Lynch, C. A., Fairchild-Huntress, V., Dunmore, J. H., Fang, Q., Berkemeier, L. R., Gu, W., *et al.* (1997). Targeted disruption of the melanocortin-4 receptor results in obesity in mice. *Cell* **88**, 131–141.

Justice, M. J. (2000). Capitalizing on large-scale mouse mutagenesis screens. *Nat. Rev. Genet.* **1**, 109–115.

Justice, M. J., Noveroske, J. K., Weber, J. S., Zheng, B. and Bradley, A. (1999). Mouse ENU mutagenesis. *Hum. Mol. Genet.* **8**, 1955–1963.

Justice, M. J., Carpenter, D. A., Favor, J., Neuhauser-Klaus, A., Hrabe de Angelis, M. H., Soewarto, D., Moser, A., *et al.* (2000). Effects of ENU dosage on mouse strains. *Mamm. Genome* **11**, 484–488.

Karp, C. L., Grupe, A., Schadt, E., Ewart, S. L., Keane-Moore, M., Cuomo, P. J., Kohl, J., *et al.* (2000). Identification of complement factor 5 as a susceptibility locus for experimental allergic asthma. *Nat. Immunol.* **1**, 221–226.

Kasarskis, A., Manova, K. and Anderson, K. V. (1998). A phenotype-based screen for embryonic lethal mutations in the mouse. *Proc. Natl. Acad. Sci. USA* **95**, 7485–7490.

Kiernan, A. E., Ahituv, N., Fuchs, H., Balling, R., Avraham, K. B., Steel, K. P. and Hrabe de Angelis, M. (2001). The Notch ligand Jagged1 is required for inner ear sensory development. *Proc. Natl. Acad. Sci. USA* **98**, 3873–3878.

Lander, E. S., Linton, L. M., Birren, B., Nusbaum, C., Zody, M. C., Baldwin, J., Devon, K., *et al.* (2001). Initial sequencing and analysis of the human genome. *Nature* **409**, 860–921.

Langenbach, R., Loftin, C., Lee, C. and Tiano, H. (1999). Cyclooxygenase knockout mice: models for elucidating isoform-specific functions. *Biochem. Pharmacol.* **58**, 1237–1246.

Lee, G. H., Proenca, R., Montez, J. M., Carroll, K. M., Darvishzadeh, J. G., Lee, J. I. and Friedman, J. M. (1996). Abnormal splicing of the leptin receptor in diabetic mice. *Nature* **379**, 632–635.

Li, Q., Liu, Z., Monroe, H. and Culiat, C. T. (2002). Integrated platform for detection of DNA sequence variants using capillary array electrophoresis. *Electrophoresis* **23**, 1499–1511.

McPherron, A. C. and Lee, S. J. (2002). Suppression of body fat accumulation in myostatin-deficient mice. *J. Clin. Invest.* **109**, 595–601.

Miller, M. W., Duhl, D. M., Vrieling, H., Cordes, S. P., Ollmann, M. M., Winkes, B. M. and Barsh, G. S. (1993). Cloning of the mouse agouti gene predicts a secreted protein ubiquitously expressed in mice carrying the lethal yellow mutation. *Genes Dev.* **7**, 454–467.

Montague, C. T., Farooqi, I. S., Whitehead, J. P., Soos, M. A., Rau, H., Wareham, N. J., Sewter, C. P., *et al.* (1997). Congenital leptin deficiency is associated with severe early-onset obesity in humans. *Nature* **387**, 903–908.

Morteau, O., Morham, S. G., Sellon, R., Dieleman, L. A., Langenbach, R., Smithies, O. and Sartor, R. B. (2000). Impaired mucosal defense to acute colonic injury in mice lacking cyclooxygenase-1 or cyclooxygenase-2. *J. Clin. Invest.* **105**, 469–478.

Moser, A. R., Pitot, H. C. and Dove, W. F. (1990). A dominant mutation that predisposes to multiple intestinal neoplasia in the mouse. *Science* **247**, 322–324.

Nadeau, J. H. and Frankel, W. N. (2000). The roads from phenotypic variation to gene discovery: mutagenesis versus QTLs. *Nat. Genet.* **25**, 381–384.

Nelms, K. A. and Goodnow, C. C. (2001). Genome-wide ENU mutagenesis to reveal immune regulators. *Immunity* **15**, 409–418.

Nolan, P. M., Peters, J., Strivens, M., Rogers, D., Hagan, J., Spurr, N., Gray, I. C., *et al.* (2000). A systematic, genome-wide, phenotype-driven mutagenesis programme for gene function studies in the mouse. *Nat Genet.* **25**, 440–443.

Nusslein-Volhard, C. and Wieschaus, E. (1980). Mutations affecting segment number and polarity in *Drosophila*. *Nature* **287**, 795–801.

Ollmann, M. M., Wilson, B. D., Yang, Y. K., Kerns, J. A., Chen, Y., Gantz, I. and Barsh, G. S. (1997). Antagonism of central melanocortin receptors *in vitro* and *in vivo* by agouti-related protein. *Science* **278**, 135–138.

Rinchik, E. M. and Carpenter, D. A. (1999). *N*-ethyl-*N*-nitrosourea mutagenesis of a 6- to 11-cM subregion of the Fah- Hbb interval of mouse chromosome 7: completed testing of 4557 gametes and deletion mapping and complementation analysis of 31 mutations. *Genetics* **152**, 373–383.

Rinchik, E. M., Carpenter, D. A. and Selby, P. B. (1990). A strategy for fine-structure functional analysis of a 6- to 11-centimorgan region of mouse chromosome 7 by high-efficiency mutagenesis. *Proc. Natl. Acad. Sci. USA* **87**, 896–900.

Russell, L. B., Kelly, L. B., Hunsicker, P. R., Bangham, J. W., Maddux, S. C. and Phipps, E. L. (1979). Specific-locus test shows ethylnitrosourea to be the most potent mutagen in the mouse. *Proc. Natl. Acad. Sci. USA* **76**, 5818–5819.

Sanseau, P. (2001). Impact of human genome sequencing for *in silico* target discovery. *Drug Discov. Today* **6**, 316–323.

Shutter, J. R., Graham, M., Kinsey, A. C., Scully, S., Luthy, R. and Stark, K. L. (1997). Hypothalamic expression of ART, a novel gene related to agouti, is up-regulated in obese and diabetic mutant mice. *Genes Dev.* **11**, 593–602.

Silver, L. M. (1995). *Mouse Genetics. Concepts and Applications.* Oxford: Oxford University Press.

Sporle, R. and Schughart, K. (1998). Paradox segmentation along inter- and intrasomitic borderlines is followed by dysmorphology of the axial skeleton in the open brain (opb) mouse mutant. *Dev. Genet.* **22**, 359–373.

Sporle, R., Gunther, T., Struwe, M. and Schughart, K. (1996). Severe defects in the formation of epaxial musculature in open brain (opb) mutant mouse embryos. *Development* **122**, 79–86.

Steinmetz, L. M., Sinha, H., Richards, D. R., Spiegelman, J. I., Oefner, P. J., McCusker, J. H. and Davis, R. W. (2002). Dissecting the architecture of a quantitative trait locus in yeast. *Nature* **416**, 326–330.

Stoll, M., Cowley, A. W., Jr., Tonellato, P. J., Greene, A. S., Kaldunski, M. L., Roman, R. J., Dumas, P., *et al.* (2001). A genomic-systems biology map for cardiovascular function. *Science* **294**, 1723–1726.

Venter, J. C., Adams, M. D., Myers, E. W., Li, P. W., Mural, R. J., Sutton, G. G., Smith, H. O., *et al.* (2001). The sequence of the human genome. *Science* **291**, 1304–1351.

Vitaterna, M. H., King, D. P., Chang, A. M., Kornhauser, J. M., Lowrey, P. L., McDonald, J. D., Dove, W. F., *et al.* (1994). Mutagenesis and mapping of a mouse gene, Clock, essential for circadian behavior. *Science* **264**, 719–725.

Vreugde, S., Erven, A., Kros, C. J., Marcotti, W., Fuchs, H., Kurima, K., Wilcox, E. R., *et al.* (2002). Beethoven, a mouse model for dominant, progressive hearing loss DFNA36. *Nat. Genet.* **30**, 257–258.

Yeo, G. S., Farooqi, I. S., Aminian, S., Halsall, D. J., Stanhope, R. G. and O'Rahilly, S. (1998). A frameshift mutation in MC4R associated with dominantly inherited human obesity. *Nat. Genet.* **20**, 111–112.

Zhang, Y., Proenca, R., Maffei, M., Barone, M., Leopold, L. and Friedman, J. M. (1994). Positional cloning of the mouse obese gene and its human homologue. *Nature* **372**, 425–432.

Zoltewicz, J. S., Plummer, N. W., Lin, M. I. and Peterson, A. S. (1999). oto is a homeotic locus with a role in anteroposterior development that is partially redundant with Lim1. *Development* **126**, 5085–5095.

10

Saturation Screening of the Druggable Mammalian Genome

Hector Beltrandelrio, Francis Kern, Thomas Lanthorn, Tamas Oravecz, James Piggott, David Powell, Ramiro Ramirez-Solis, Arthur T. Sands and Brian Zambrowicz

Functional annotation of the mammalian genome has become an important goal in the post-genome era. Genetic studies in model organisms provide an excellent approach for understanding gene function. The development of technologies for massive parallel production and analysis of mouse mutants is making it possible to screen through mutations in all druggable genes to identify targets with high value for drug discovery. By carrying out genetic screens in a mammalian model system, it is possible to screen directly for changes in physiology relevant to human disease treatment. Here we describe our biological screening strategy being carried out on 1000 mouse gene knockouts per year. This screen is focused on discovering the targets for the next generation of therapeutic products in the areas of metabolism, endocrinology, immunology, neurology, cardiology, ophthalmology, reproductive biology and oncology.

10.1 Introduction

Genetic screens can be used potentially to scan a genome for genes that play a role in any process of interest. Early genetic screens were carried out in invertebrate model organisms and included saturation screens of *Drosophila*

Model Organisms in Drug Discovery. Edited by Pamela M. Carroll and Kevin Fitzgerald.
© 2003 John Wiley & Sons, Ltd. ISBN 0 470 84893 6

to identify genes involved in organization of the body plan during development (Nusslein-Volhard and Wieschaus, 1980) and screens in *Caenorhabditis elegans* to identify genes involved in producing the invariant cell lineage pattern (Horvitz and Sulston, 1980; Chalfie *et al.*, 1981; Hedgecock *et al.*, 1983). These screens relied on saturation mutagenesis to interrogate the genome for the set of genes involved in these processes and led to the discovery of genes such as the homeobox genes and apoptosis regulators involved in development across all invertebrate and vertebrate species examined. Since these early screens, a tremendous number of additional genetic screens have been carried out in the fly and worm, further demonstrating the power of genetics for the dissection of pathways and processes. These screens require only a method for creating large numbers of tractable mutations in genes and a phenotype that can be measured.

More recently, genetic screening has been adapted for the vertebrate model organisms of zebrafish and mice. In zebrafish, both chemical mutagenesis (Mullins *et al.*, 1994; Haffter *et al.*, 1996) and gene trapping (Golling *et al.*, 2002) have been combined with phenotypic screens to identify mutations affecting development of the neural crest, pigmentation, jaw, branchial arches, visual system, heart and other internal organs, ear, retina, brain, midline, shape and movement (Brockerhoff *et al.*, 1995; Abdelilah *et al.*, 1996; Baier *et al.*, 1996; Brand *et al.*, 1996; Chen *et al.*, 1996; Granato *et al.*, 1996; Kelsh *et al.*, 1996; Malicki *et al.*, 1996a,b; Neuhauss *et al.*, 1996; Odenthal *et al.*, 1996; Piotrowski *et al.*, 1996; Schier *et al.*, 1996; Solnica-Krezel *et al.*, 1996; Stemple *et al.*, 1996). These screens take advantage of the large number of offspring, oviparous development and transparent nature of the zebrafish embryo that make it an excellent system for the study of vertebrate development. These studies undoubtedly will result in the identification of a large number of genes required for vertebrate development. For the purpose of drug discovery, effective genetic screens in mammals would allow one to dissect mammalian physiology to identify key genes with therapeutic relevance as potential drug targets. Some might consider genetic screens in mammals to be impossible logistically, but recent advances in mutagenesis and screening methods in mice are facilitating functional dissection of the mammalian genome. Advances in the scale and speed of gene targeting (Walke *et al.*, 2001; Abuin *et al.*, 2002) and the development of genome-wide gene trapping (Zambrowicz *et al.*, 1998; Wiles *et al.*, 2000; Leighton *et al.*, 2001; Mitchell *et al.*, 2001) in mouse embryonic stem cells have resulted in saturation of the mammalian genome with tractable mutations in large numbers of genes. This has been combined with the recent miniaturization of a broad array of medical technologies and the transfer of many disease challenge assays to the mouse model to enable detailed diagnostic analysis of mice. The mouse is a model organism that is ideal for studying many aspects of mammalian physiology with direct medical relevance. Screens are currently being used to identify genes involved in insulin sensitivity, hypertension, body fat deposition, energy expenditure, bone deposition and breakdown, angiogenesis and many other processes with significance for the treatment of human disease.

These advances have brought together the two requirements for genetic screens in mammals: the ability to produce large numbers of mutations and the ability to screen for phenotypes of interest. The development of mutagenesis strategies to mutate large numbers of mouse genes has been described elsewhere (Zambrowicz *et al.*, 1998; Wiles *et al.*, 2000; Leighton *et al.*, 2001; Mitchell *et al.*, 2001). Here we describe the development of phenotypic screens designed to identify genes that could be used as targets to ameliorate diseases in the areas of diabetes/metabolism, cardiology, neurology, ophthalmology, reproductive biology, oncology and immunology/inflammation. Mammalian genetics is now identifying the targets for future pharmaceutical development.

10.2   Saturating the druggable genome

One of the major advantages of doing genetic screens in the mouse model system is the ability to measure directly the physiological parameters relevant to human disease. These direct measures allow the identification of gene products that, when modulated by small-molecule drugs, may provide a therapeutic effect. This approach is supported by the excellent correlation between the knock-out phenotypes of the targets of marketed pharmaceutical drugs and the known efficacy and side-effects of those drugs (Zambrowicz and Sands, 2003). One excellent example is knock-out of the H^+/K^+ ATPase: the target of drugs such a Prilosec used to lower gastric acid secretion for the treatment of gastric ulcer disease. Knock-out of either the alpha or beta subunit of ATPase results in animals with neutral stomach pH – a phenotype that correlates exactly with the action of the pharmacological antagonists of ATPase (Scarff *et al.*, 1999; Spicer *et al.*, 2000). Similarly, mammalian screens can be set up to identify the genes that play a role in any specified therapeutic area. For instance, if one is interested in genes that may be important for the treatment of diabetes, it is possible to screen mutations in mice for direct effects on blood glucose and insulin levels, insulin sensitivity and other parameters such as obesity that play an important role in the diabetic process. There are clearly genes to be found that help to regulate glucose homeostasis, and one example being the insulin receptor, which when mutated in mice results in animals with severe insulin resistance and frank diabetes (Accili *et al.*, 1996; Joshi *et al.*, 1996). Likewise, if one is interested in genes important for the treatment of osteoporosis, one can screen for mutations that increase or decrease bone mineral density, as has been observed for mice with mutations of the cathepsin K (Saftig *et al.*, 1998) and osteoprotegerin genes

(Bucay *et al.*, 1998; Mizuno *et al.*, 1998), respectively. This genetic approach leaves little question as to the role of a gene within the mammalian organism and its likely medical relevance.

This ability to measure directly the parameters of mammalian physiology stands in stark contrast to attempts to identify genes with human disease relevance in lower model organisms such as *Drosophila*. The *Drosophila* system is excellent for defining genetic pathways because of the ability to perform saturation screens for genetic modifiers of phenotypes that have been established already. The problem is that these genetic screens often are designed artificially and are far removed from mammalian physiology. For instance, primary phenotypes to be used for modifier screens often are developed based upon overexpression, ectopic expression or expression of dominant or activated forms of a gene known to be involved in human disease in the *Drosophila* eye (Therrien *et al.*, 2000; Hirose *et al.*, 2001; LaJeunesse *et al.*, 2001; Schreiber *et al.*, 2002; Sullivan and Rubin, 2002). Screens then are used to identify modifier genes that ameliorate or exacerbate the eye phenotype originally produced. These screens are clearly able to elucidate genetic pathways and the types of genes that might play a role in a pathway of interest, but the corresponding mammalian genes still must be identified and tested for any relevance to the original human disease or physiology of interest.

Saturation screens for genetic modifiers in non-mammalian organisms can provide clues for finding genes that may play a role in a disease-relevant pathway in humans, but what if one could rapidly carry out genetic screens directly in mammals for those genes? The question is whether the ability to scan a genome using saturation mutagenesis in invertebrate organisms outweighs the ability to screen directly in a more focused manner for genes that modulate disease-relevant mammalian physiology. Two of the challenges of conducting genetic screens in the mouse mammalian model have centered on the issues of the speed at which tractable genetic mutations can be generated and the large number of genes that must be processed to provide broad genomic coverage. Although, in mice, saturation modifier screens remain a logistical challenge, it has become possible to create mutations in all members of the so-called druggable classes of genes through gene targeting and gene trapping. This creates an opportunity to saturate the druggable mammalian genome, which is an extremely important milestone in the evolution of drug discovery in the post-genome era. These druggable genes include secreted proteins that could be biotherapeutics themselves, potential targets for antibody-based therapeutics and small-molecule drug targets that belong to gene families that have proved themselves to be amenable to small-molecule modulation based upon marketed drugs (Hopkins and Groom, 2002). The druggable genes include GPCRs, ion channels, nuclear hormone receptors, key enzymes, kinases, proteases, secreted proteins and cell surface proteins. Indeed, one could argue that all human disease or disease treatment pathways of interest probably contain druggable genes, so that by mutating all the druggable genes in the genome one can interrogate all pathways for points of therapeutic intervention.

Demonstrating the scale at which mammalian genes can be mutated, we have industrialized gene knock-out technologies for saturation of the druggable genome within the next 4 years. We have implemented our genome 5000 program to knock out and analyze the resulting phenotypes for 5000 genes from the mammalian genome. The 5000 genes chosen are all members of the currently druggable gene families. Because others have suggested that the druggable genome may be as small as about 3000 genes (Hopkins and Groom, 2002), this scale is sufficient to saturate the mammalian druggable genome in order to identify those genes that have the greatest potential for human disease treatment.

10.3   Screening the genome effectively for novel drug targets

Given the possibility of generating knock-out mouse lines at a rate of 1000 per year, the next challenge is to implement a biological evaluation process that has a high probability of identifying potential drug targets, as assessed by the physiological consequences of gene disruption. We have developed a process that maximizes our potential to identify therapeutically significant genes.

This process represents the application of increasingly fine filters to genomic information. First, the genome is mined for members of druggable families. Second, knock-out mice are generated for selected genes at an average rate of 20 lines of mutant mice per week. A minimum cohort for initial evaluation is 16 animals; 8 homozygous nulls, 4 heterozygotes and 4 wild-type animals for each gene. This cohort size has produced reliable data from the primary screen upon which decisions for secondary screens can be made. Implementation of this plan has necessitated the integration of bioinformatics, mouse genetics, robotics and high-speed physiological evaluation in a unique and robust infrastructure that has demonstrated already the ability to operate at the required rate. The logistics of generating, maintaining, genotyping and characterizing the required number of animals have been satisfied.

Our first biological evaluation of the animals is a comprehensive clinical assessment of all the physiological parameters that we can measure effectively in high-throughput mode. Each test has direct relevance to one or more of our therapeutic areas and is designed to yield information that can be correlated directly with therapeutic intervention. This process includes an extensive battery of behavioral evaluations (neurology), blood pressure and heart rate measurements (cardiology) and a complete hematology survey supplemented with fluorescence-activated cell sorting (FACS) scans for immune function (immunology). The animals also are evaluated for body fat content, lean body mass (metabolism), bone mineral density, bone mineral content (endocrinology) and retinal integrity/vascularization (ophthalmology). Effects on cell proliferation and reproductive organ development are studied (oncology) and fertility (reproductive biology) is assessed. This screening phase of biological investigation is called Level 1 analysis.

This initial analysis of the physiological consequence of creating null mutations is designed to be unbiased with regard to potential outcome but to encompass phenotypes indicative of utility to our chosen therapeutic areas. All animals in all projects are submitted to the same tests in the same temporal sequence. This means that each test must be self-contained and have minimal impact on the outcome of subsequent tests. The aim of Level 1 analysis is to obtain a comprehensive understanding of gene function within the context of mammalian physiology. Variations from normal in any parameter are detected by comparison with internal cohort controls and, very importantly, with the pooled historical data for all controls. Historical control data are now based on over 2500 animals, giving us a precise quantitative measure of 'normal' for each test and the level of background variation. Most of the tests are of primary importance to one particular therapeutic area (e.g. blood pressure and cardiology), but the total picture gained from this type of analysis is critical in identifying possible side-effects of target modulation. This allows the identification of targets with a high potential for success, provided that specific modulators can be developed. Figure 10.1 is a schematic outline of our Level 1 protocol.

In addition to therapeutic area-specific tests, multiple general diagnostic tests are performed. Level I pathology examines 52 tissues for the female and 53 tissues for the male. A complete gross necropsy is performed, with collection of tissues and photography of any significant gross lesions. Tissues are immersion-fixed in 10% neutral buffered formalin for 24 h, trimmed, processed to paraffin, embedded, sectioned at 4–5 μm, and stained with hematoxylin and eosin for histopathological examination. A board-certified pathologist examined tissues from one male and one female homozygote for each project (heterozygotes are examined for homozygous lethal projects). The recent introduction of computer-assisted tomography (CAT) scanners, which operate effectively on mice, has enabled non-invasive evaluation of soft-tissue anatomy in addition to very refined skeletal analysis. Application of CAT (MicroCAT, ImTek Inc.) can be used to obtain important morphological information non-invasively. All lesions are recorded and compared with controls in order to facilitate interpretation of phenotypes.

The output from all Level 1 tests is reduced to digital data and ported to a relational database. Data acquisition is rapid to the point that no Level 1 test

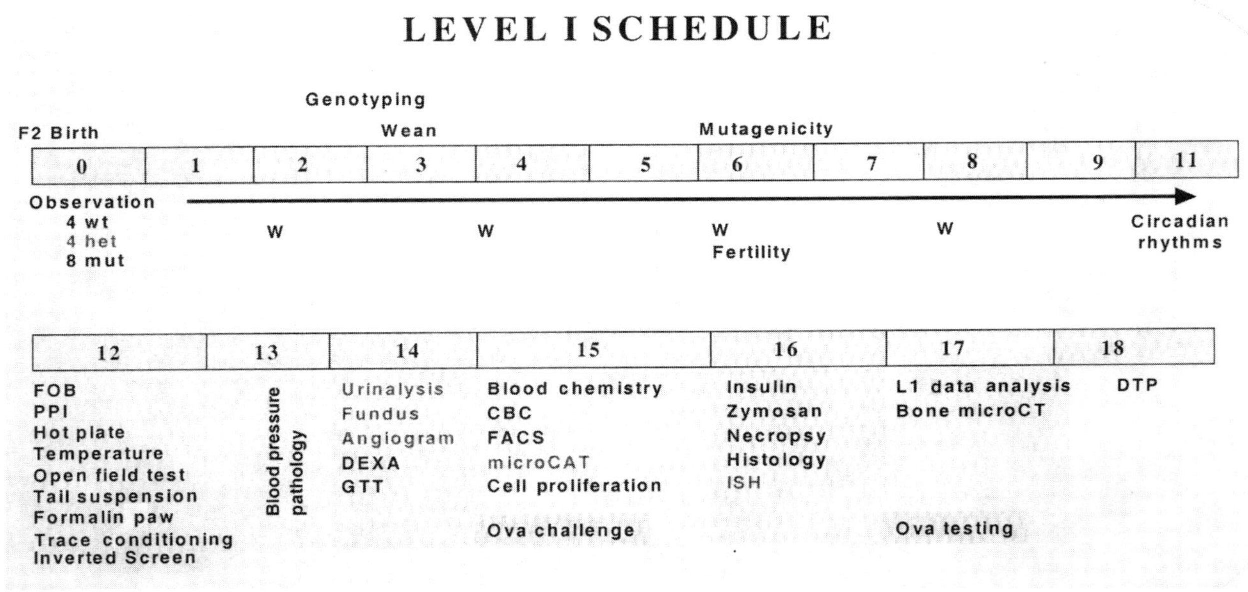

LEVEL I SCHEDULE

Figure 10.1 Outline of Level 1 testing protocol. Individual tests are described more fully in the most relevant therapeutic area description. Abbreviations in order of occurrence: FOB, functional observation battery; PPI, prepulse inhibition; DEXA, dual-energy emission X-ray absorptiometry; GTT, glucose tolerance test; CBC, complete blood cell count; FACS, fluorescence-activated cell sorting; CAT, computer-assisted tomography; OVA, ovalbumin; ISH, *in situ* hybridization; CT, computed tomography; DTP, drug target prioritization

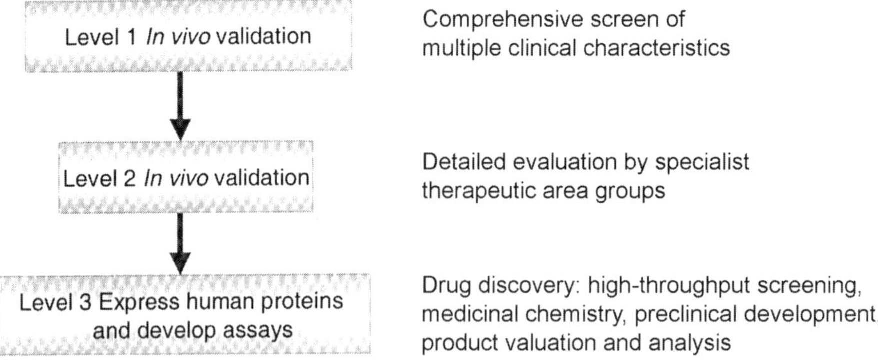

Figure 10.2 *In vivo* target validation and drug discovery

is rate-limiting for the overall process. Numerical data is represented graphically with appropriate statistical tools, images are annotated by project scientists and interpretation of pharmaceutical relevance is summarized. It is therefore possible to gain very quickly a comprehensive view of the physiological function of every gene that is studied. This view encompasses those features that are most indicative of therapeutic potential in specific disease areas. Level 1 analysis has been a rich source of targets for drug discovery programs. Level 2 analysis entails the confirmation of Level 1 observations using additional animals and the application of specialized tests in a given project in reaction to Level 1 observations. Level 2 includes numerous therapeutic area-specific tests and challenge assays that cannot be used in the screening phase. Level 2 analysis may be triggered also through a hypothesis-driven approach. Level 3 analysis is designed for in-depth biological study in order to determine the merits of each target for assay development and high-throughput screening (see Figure 10.2).

The decision to submit a given gene product to actual drug discovery is based on three major criteria: modulation of the target by a small molecule, antibody or therapeutic protein could provide significant therapeutic effect with minimal or no discernable on-target side-effects; the target represents a potential breakthrough for the treatment of disease with significant advantages over existing therapies; and the program addresses a major unmet medical need. These are strict criteria, and after having completed more than 500 full analyses to date we have committed 14 projects to drug discovery.

What follows is a brief description of the capabilities of the therapeutic area biology groups, including Level 1 and some Level 2 tests that are most directly relevant to them.

10.4 High-throughput biology: maximizing return from reverse genetics

Endocrinology/metabolism

Three of the most prevalent diseases of endocrinology/ metabolism are Type II diabetes, obesity and osteoporosis. We have implemented a comprehensive panel of physiological tests for each disease process that have proved to provide reliable clinical descriptions of disease-related symptoms. These tests include measures of body composition index, glucose homeostasis and bone mass.

Level 1 diabetes tests

Glucose tolerance test

The glucose tolerance test (GTT) is the standard for defining impaired glucose homeostasis in mammals. For example, intraperitoneal glucose tolerance tests showed improved glucose clearance and the serum glucose and insulin levels were significantly lower in protein tyrosine phosphatase-1B (PTP-1B) and SHIP2 knock-out mice (Klaman *et al.*, 2000; Clement *et al.*, 2001). These findings indicate improved insulin sensitivity, a possibility that was confirmed by hyperinsulinemic–euglycemic clamp studies in the PTP-1B knock-out mice (Klaman *et al.*, 2000). These results suggest that these two proteins are potential targets for new therapeutics aimed at Type II diabetes. In addition, the ability of retinoid X receptor agonists to lower serum glucose and insulin levels has been used as evidence that these agonists act as insulin sensitizers *in vivo* (Mukherjee *et al.*, 1997). These examples validate the effectiveness of GTT for the identification of potential targets for diabetes. Glucose tolerance tests are performed using a Lifescan glucometer. Animals are injected i.p. with 2 g/kg D-glucose, delivered as a 20% solution, and blood glucose levels are measured at 0, 30, 60 and 90 min after injection (Klaman *et al.*, 2000).

Urinalysis

Elevated glucose and/or ketone levels in urine are diagnostic markers for diabetes. We perform qualitative urinalysis using Chemstrip 10 UA reagent strips (Roche) for the detection of glucose, bilirubin, ketones, blood, pH, protein, urobilinogen, nitrites and leukocytes in urine. Results are recorded using a Chemstrip 101 urine analyser.

Serum insulin

Serum insulin levels are also diagnostic markers for diabetes. Insulin levels are assayed using a sensitive rat radioimmunoassay kit from Linco, which is sensitive to 0.02 ng/ml insulin in serum.

Level 2 diabetes tests

In Level 2, other tests are performed to verify and further define the role of targets in glucose homeostasis:

- Insulin tolerance test

- Insulin levels during GTT

- Insulin clearance (serum c-peptide/insulin ratio)

- Measurement of serum free fatty acids, glycerol, glucagon, leptin, corticosterone

- Insulin content of pancreatic islets (radioimmunoassay)

- Immunohistochemical analysis of pancreas for insulin, glucagon, somato-statin and pancreatic polypeptide

- Muscle and liver pathology, including glycogen and lipid content

- Pharmacological evaluation of liver slices, isolated soleus muscle and adipocytes

Level 1 obesity tests

Animal weight and percent body fat are measured in Level 1 to identify obesity phenotypes.

Body weight

All mice are weighed at 2, 4, 6, 8 and 16 weeks of age.

Dual-energy X-ray absorptiometry

Dual-energy X-ray absorptiometry (DEXA) has been used successfully to identify increased total body fat in melanocortin-3 receptor knock-out mice (Butler *et al.*, 2000) and decreased total body fat in melanin concentrating hormone 1 receptor knock-out mice; the latter observation was confirmed by direct analysis of fat pad weights (Marsh *et al.*, 2002). Such results suggest that these proteins may be targets for novel obesity therapies. In addition, DEXA was used to show that the small-molecule insulin mimetic cpd2 blocks the accumulation of body fat in mice fed a high fat diet, an observation that was confirmed by direct analysis of fat pad weights (Air *et al.*, 2002). A DEXA instrument (Lunar Piximus) is used to record bone mineral density, bone mineral content, percent body fat and total tissue mass (Nagy and Clair, 2000; Punyanitya *et al.*, 2000). Although primarily aimed at metabolic and osteoporotic conditions, DEXA is a sensitive measure of all-round well-being and often contributes to diagnosis in other therapeutic areas.

Level 2 obesity tests

In Level 2, obesity targets are analyzed to determine whether they regulate metabolism, feeding, appetite or food absorption. Level 2 obesity tests include:

- Metabolic cages to measure food intake, water intake and fat malabsorption

- Mini-Mitter telemetry for physical activity, core body temperature, drinking frequency and feeding frequency and duration

- Oxymax measurement of metabolic rate and physical activity

- Home cage diet studies, including high-fat-diet challenge, food intake measurement and pair-feeding studies

- Fat mass by DEXA or nuclear magnetic resonance (Bruker Minispec)

- Body composition analysis (analysis of carcass fat mass by Sohxlet; fat pad and organ weights)

- Crosses to ob/ob mice

- Pharmacological challenge with leptin, melanocortin II and neuropeptide Y

- Blood pressure

Level 1 osteoporosis tests

Bone microcomputed tomography

Osteoporosis is characterized by a decreased bone mineral density due to a deficiency in bone production or increased bone absorption resulting in brittle bones. Specialized microcomputed tomography (micro-CT) machines have been developed with the capacity to provide quantitative and imaging data on the three-dimensional structure of mouse bones. This technique has been used to demonstrate the efficacy of parathyroid hormone in a mouse model of osteoporosis (Alexander *et al.*, 2001) and to describe in three dimensions the changes in bone resulting from the osteopetrotic mutation, which leads to osteopetrosis (Abe *et al.*, 2000). We use a Scanco Medical μCT40 machine for measurements of bone mineral density. This machine permits visualization of trabecular bone structure, which is critical in evaluating overall bone quality. This is a much more sensitive analysis of bone than can be achieved using DEXA alone and is a specialized test for osteoporosis that we have implemented as part of our Level 1 analysis.

Level 2 osteoporosis tests

In Level 2, targets are analyzed to determine whether changes in bone mineral density are due to effects on bone deposition or bone resorption using the following tests:

- DEXA
- Micro-CT
- Undecalcified bone histomorphometry
- Bone histopathology
- Measurement of urinary helical peptide

Cardiology

The major disease areas of interest in cardiology are hypertension, thrombosis, atherosclerosis and heart failure.

Level 1 tests

Blood pressure

Blood pressure measurements allow us to find targets that, upon inhibition, lead to a reduction in blood pressure. Angiotensin-converting enzyme inhibitors and angiotensin receptor antagonists are very successful drugs in the treatment of hypertension. Both knock-outs have low blood pressure. Blood pressure is measured using a non-invasive computerized tail-cuff system, the Visitech Systems BP-2000. First described by Krege *et al.* (1995), this technique has been validated by several studies (Ito *et al.*, 1995; Oliver *et al.*, 1998; Sugiyama *et al.*, 2001). Ten measurements of blood pressure are made per day on each of 4 days for each animal evaluated. Results are recorded as the pooled average of 40 measurements.

Zymosan challenge assay

Peritoneal leukocyte recruitment assays are used to identify targets that may regulate the inflammatory component of atherosclerosis. These assays detect abnormalities in immune cell recruitment to a site of inflammation. It has been shown in mutant such as C-C chemokine receptor 2 (CCR2) knock-outs that a defect in immune cell recruitment in these assays correlates well with a significant reduction in the inflammatory component of atherosclerosis and the subsequent plaque formation (Boring *et al.*, 1997).

Blood lipids

High cholesterol and triglyceride levels are recognized risk factors in the development of cardiovascular disease. Measuring blood lipids allows us to find the biological switches that regulate blood lipid levels; inhibition of these switches should lead to a reduction in the risk for cardiovascular disease.

Optic fundus photography and angiography

Optic fundus photography is performed on conscious animals using a Kowa Genesis small-animal-fundus camera modified according to Hawes *et al.* (1999). Intraperitoneal injection of fluorescein permits the acquisition of direct light fundus images and fluorescent angiograms for each examination. In addition to direct ophthalmological changes, this test can detect retinal changes associated with systemic diseases such as diabetes and atherosclerosis.

Level 2 cardiology tests

- Platelet aggregation
- Vascular injury by carotid cuff
- Chemically induced thrombosis
- Poloxamer-induced atherosclerosis
- Aortic banding
- Permanent coronary occlusion
- Crosses with apolipoprotein E, low-density lipoprotein receptor and knock-outs

Immunology

Our focus indications include acute inflammation, inflammatory bowel disease, transplantation, asthma, allergy, multiple sclerosis, rheumatoid arthritis and blood coagulation. The process of hematopoietic cell development and the regulation of mature immune cell function share several key signaling pathways, which are the result of similar molecular or cellular interactions. As an example, activation events via the antigen-specific T-cell receptor and co-stimulatory molecules are indispensable for both normal T-cell development in the thymus and normal T-cell function during an immune response. Comprehensive phenotypic analysis of functionally relevant immune cell subpopulations in knock-out mice is essential for two reasons: it can reveal the role of a novel gene or expose the central role of a known gene in immune cell development and function; and at the same time it can provide the first hint about the potential mechanism that can lead to the observed immune deficiency.

Level 1 tests

Complete blood cell count

Routine evaluation of the cellular components of the immune system in knock-out mice and wild-type littermates is performed by automated determination of the absolute numbers of various cell types and ratios in the peripheral blood, i.e. complete blood cell count (CBC). This analysis is followed by a more detailed study using flow cytometry, which is designed to determine the relative proportions of $CD4^+$ and $CD8^+$ T cells, B cells, NK cells and monocytes in the mononuclear cell population. In the absence of a single molecular entity, disturbances in the proportion of any of the analyzed cell types could signal a key role for that molecule in governing the immune system, as exemplified in the following knock-out phenotypes.

The immunosuppressants cyclosporin A and FK506, which are used to prevent transplant rejection, inhibit the immune response by inhibiting the catalytic activity of one or both isoforms of calcineurin A (can) in lymphocytes. Mice deficient in the β-isoform of the enzyme have a significant reduction in peripheral T lymphocytes due to 75% and 65% reductions in $CD4^+$ and $CD8^+$ positive thymocytes, respectively (Bueno *et al.*, 2002). Mice deficient in expression of granulocyte colony-stimulating factor (G-CSF) exhibit chronic neutropenia with a 70–80% reduction in circulating neutrophils, whereas recombinant GCSF (Neupogen) stimulates neutrophil production and is used to treat neutropenia (Lieschke *et al.*, 1994).

This test requires 135 µl of whole blood and employs a Cell-Dyn 3500R hematology analyzer. It reports on white blood cell count, neutrophils, lymphocytes, monocytes, eosinophils, basophils, red blood cell count and other standard hematology markers.

Blood chemistry

A Cobas Integra 400 serum analyzer is used to measure a range of soluble serum components using approximately 85 µl of serum. We record serum levels of alkaline phosphatase, albumin, total cholesterol, triglycerides, blood urea nitrogen, glucose, alanine aminotransferase, bilirubin, phosphate, creatinine, calcium and uric acid.

Fluorescence-activated cell sorting (FACS)

Flow cytometry is designed to determine the relative proportions of CD4$^+$ and CD8$^+$ T cells, B cells, NK cells and monocytes in the mononuclear cell population. We use a Becton-Dickinson FACSCalibur 3-laser FACS machine to assess immune status. For Level 1 screening, this machine records CD4+/CD8−, CD8+/CD4−, NK, B cell and monocyte numbers, in addition to the CD4+/CD8+ ratio.

Ovalbumin challenge

Chicken ovalbumin (OVA) is a T-cell-dependent antigen commonly used as a model protein for studying antigen-specific immune responses in mice. It is non-toxic and inert and therefore will not cause harm to the animals even if no immune response is induced. The murine immune response to OVA has been well characterized, to the extent that the immunodominant peptides for eliciting T-cell responses have been identified. Anti-OVA antibodies are detectable 8–10 days after immunization using enzyme-linked immunosorbent assay, and determination of different isotypes of antibodies gives further information on the complex processes that may lead to a deficient response in genetically engineered mice.

The cyclosporin-mediated suppression of immune response once again demonstrates the similarity of phenotype using the suppressive agent or the genetic knock-out mice in this challenge model. Both cyclosporin-treated animals and mice knocked out for calcineurin A, in this case the α-isoform, show deficiency in T-cell-dependent antigen response (Puignero *et al.*, 1995; Zhang *et al.*, 1996).

Another example is the cytokine tumor necrosis factor α (TNF-α), whose important role in modulating inflammatory and antibody responses is well known. Two novel treatment options are currently available for patients with rheumatoid arthritis, a soluble receptor (Enbrel) and antibody (Remicade), both based on blocking the TNF-α activity. Underlining the effectiveness of drug therapy, mice deficient in TNF-α exhibit impaired humoral response to both T-cell dependent and T-cell-independent antigens (Pasparakis *et al.*, 1996).

It is important to note that, even without antigenic challenge, the make-up of the immunoglobulin repertoire in a knock-out mouse is highly informative, because isotype switching of immunoglobulins is dependent on the interaction between B and T lymphocytes. Examples of the type of receptors required for normal function of T and B cells are the so-called co-stimulatory molecules, including CD28 and CD40 receptors, both of which are targets of antibody-based therapy with ongoing clinical trials for the treatment of various autoimmune diseases. In this case, mice deficient in either of these receptors register an impairment in immunoglobulin class switching, which is detectable in the serum of the animals (Shahinian *et al.*, 1993, Kawabe *et al.*, 1994). Our protocol assesses the ability of mice to raise an antigen-specific immune response. Animals are injected i.p. with 50 µg of OVA emulsified in Complete Feund's Adjuvant; 8 days later the serum titer of anti-OVA antibodies (IgG1 and IgG2 subclasses) is measured.

Level 2 immunology tests

The following Level 2 tests are used to elucidate the most likely disease indication for a given target:

- T-Cell activation, CD3 monoclonal antibody (mAb) + CD28 mAb induced
- B-Cell activation, CD40 mAb + IL4 induced
- Mixed lymphocyte reaction provoked by irradiated BALB/C spleen cells
- Lipopolysaccharide challenge to evaluate acute phase response
- Oxazolone sensitization and challenge for contact hypersensitivity
- Ovalbumin vaccine model
- Bovine collagen-induced arthritis
- Dextran sulfate gavage: inflammatory bowel disease model
- Ovalbumin + alum immunization followed by aerosol delivery of ovalbumin as asthma model
- Allograft rejection
- Blood coagulation assays: prothrombin time and activated partial thromboplastin
- Platelet aggregation
- Bone marrow transplantation

Neurology

Neurology focuses on the identification of targets for anxiety, depression, schizophrenia, pain, sleep disorders, learning and memory disorders, neuromuscular disease and neurodegenerative disorders. The Level 1 assays have been based upon the behavioral phenotypes associated with knock-outs of known central nervous system targets as well as the actions of known drugs.

Level 1 tests

Open field test

Several targets of known drugs have exhibited phenotypes in the open field test. These include knock-outs of the serotonin transporter (unpublished data), the dopamine transporter (Giros *et al.*, 1996), and the GABA receptor (Homanics *et al.*, 1997). Our automated open-field assay has been customized to address changes related to affective state and exploratory patterns related to learning. First, the field (40×40 cm) is relatively large for a mouse, which is designed to pick up changes in locomotor activity associated with exploration. In addition, there are four holes in the floor to allow for nose-poking, an activity specifically related to exploration. Several factors have been designed to heighten the affective state associated with this test. The open-field test is the first experimental procedure in which the mice are tested, and the measurements taken are the subjects' first experience with the chamber. In addition, the open field is brightly lit. All these factors will heighten the natural anxiety associated with novel and open spaces. Thus, pattern and extent of exploratory activity, especially the center-to-total distance traveled ratio, may be able to discern changes related to susceptibility to anxiety or depression. A large arena (40 cm×40 cm, VersaMax animal activity monitoring system from AccuScan Instruments) with infrared beams at three different levels is used to record rearing, hole poke and locomotor activity. The animal is placed in the center and its activity is measured for 20 min. Data from this test are analyzed in five 4-min intervals. The total distance traveled (cm), vertical movement number (rearing), number of hole pokes and the center-to-total distance ratio are recorded.

Inverted screen

This test is used to measure motor strength/coordination. Untrained mice are placed individually on top of a square (7.5 cm×7.5 cm) wire screen that is mounted horizontally on a metal rod. The rod is rotated 180° so that the mice are on the bottom of the screens. The following behavioral responses are recorded over a 1-min testing session: fell off, did not climb and climbed up.

Functional observation battery

This is a modified SHIRPA (Rogers *et al.*, 2001) analysis in which the animals are scored systematically for 37 individual behavioral and physical characteristics, such as vision, response to touch, palpebral closure, etc. It is a formalization of the complete observation of the whole organism, which often gives the first hint as to phenotype.

Hot plate and formalin paw

The 55°C hot plate is a standard assay for measuring nociception in animals. Knock-out of either the μ-opioid receptor (Sora *et al.*, 1997) or COX 1 (Ballou *et al.*, 2000) (both targets of analgesic drugs) results in effects on response latency in the hot-plate assay. Analgesia, such as that produced by morphine and other strong analgesics, is also detected using this assay. The hot-plate test is carried out by placing each mouse on a small, enclosed 55°C hot plate (Hot Plate Analgesia Meter, Columbus instruments). Latency to a hindlimb response (lick, shake or jump) is recorded, with a maximum time on the hot plate of 30 s. Each animal is tested once.

The formalin paw assay has been recognized for a number of years as an assay for hyperalgesia, as well as initial acute nociception. Recently, this assay has been automated and thus has become available for use in high-throughput analysis. Drugs that address novel mechanisms of hyperalgesia, without the side-effects of potent non-steroidal antiinflammatory drugs, will be very useful new therapeutics.

Prepulse inhibition

Prepulse inhibition is a pre-attentive process that has been shown to be deficient in patients with schizophrenia. This reduced ability to filter out environmental stimuli may contribute to both positive and negative symptoms of the disease. Antipsychotics can ameliorate some deficits in prepulse inhibition, therefore genetic inhibition of a target that can increase prepulse inhibition may presage a small-molecule therapeutic that can help patients with their disorder. The prepulse inhibition of the startle response assay is an automated measure of the startle response both with and without various intensities of prepulses. Targets whose genetic inhibition produces changes in prepulse inhibition without changes in the startle response itself may be excellent for the discovery of new therapeutics.

This test employs a San Diego Instruments SR-lab startle response system. Prepulse inhibition of the acoustic startle reflex occurs when a loud 120 decibel (dB) startle-inducing tone is preceded by a softer (prepulse) tone. The prepulse inhibition paradigm consists of six different trial types (70 dB background noise, 120 dB alone, 74 + 120 dB at postpartum day 4, 78 + 120 dB at postpartum day 8, 82 + 120 dB at postpartum day 12, and 90 + 120 dB at postpartum day 20) each repeated in pseudorandom order six times for a total of 36 trials. The maximum response to the stimulus (V_{max}) is averaged for each trial type. The percentage inhibition of the animal's response to the startle stimulus is calculated for each prepulse intensity and then graphed. This test is being used increasingly as a model of human schizophrenia and a test for antipsychotic drugs.

Tail suspension

The tail-suspension and forced-swim assays are the two mainstay assays for the discovery and validation of novel antidepressants. The knock-out of the noradrenalin transporter, one target of the antidepressant Welbutrin, demonstrates an increased struggle time in the tail-suspension assay (Xu *et al.*, 2000). The tail-suspension assay has been automated, giving it added objectivity and making it appropriate for high-throughput analysis. Both of these assays measure the efforts of the subject to extricate itself from an inescapable situation, i.e. they measure a tendency toward 'giving up'. Compounds known to reduce depressive symptoms in patients reduce the immobility time in tail suspension, therefore gene knock-outs that result in decreased time spent being immobile, in the absence of any general increase in activity levels (as measured in assays such as the open field), point to excellent opportunities for the discovery of novel therapeutics for the treatment of depression. In this particular set-up (PHM-300 Tail Suspension Test Cubicle) a mouse is suspended by its tail for 6 min, and in response the mouse will struggle to escape from this position. Extended struggle is taken as antidepressive behavior, whereas curtailed struggle is interpreted as depressive.

Circadian rhythms

Changes in sleep patterns can be detected by examining activity continuously over a period of days and nights. We use an infrared beam system that monitors the horizontal locomotor activity of individual mice in their home cage environment for 3 days and nights. This allows us to obtain an accurate indication of their sleep–wake cycle as well as overall locomotor activity rates. Changes in the normal circadian rhythm or an increase or decrease in the periods of activity during the normal sleep cycle can indicate genes controlling sleep and can be supportive of therapeutic potential for other conditions, such as depression or schizophrenia, in which normal sleep patterns are disrupted.

Trace aversive conditioning

Cognition, especially the loss of cognitive abilities in dementias such as Alzheimer's disease, later-stage Parkinson's and Huntington's disease, as well as in schizophrenia, is a major focus for drug discovery. This area has been hampered particularly by the lack of rapid assays that specifically target the learning and memory losses associated with these diseases, i.e. learning and memory dependent on areas of the brain such as the hippocampus. Assays generally used, such as the eight-arm radial arm maze or delayed-non-matching-to-sample procedures require significant time and training. However, animals learn aversive conditioning very easily and it has been found that this can be combined with 'trace' conditioning, in which there is a time interval between the signal stimulus and the aversive stimulus itself, to provide a rapidly (3–5 trials) learned response that is dependent upon the function of the hippocampus. As with most of our other assays, this assay has been automated to increase objectivity and make it appropriate for high-throughput behavioral analysis. Gene knock-outs that affect learning and memory in this assay, without changes in basic sensory or motor function, will point to targets for the discovery of new treatments for cognitive disorders.

Level 2 neurology tests

- Neurochemical analysis of dopamine, norepinephrine, serotonin and their primary metabolites in urine, blood, cerebrospinal fluid (CSF) and brain tissue

- Levels of melatonin and homocysteine in urine, blood, CSF and brain tissue

- *In situ* hybridization/immunocytochemical analyses using Neo, LacZ or radioactivity

- Immunohistochemical analyses of markers of choice

- Pharmacological challenges *in vivo*

- Electroretinogram (vision)

- Auditory brainstem response (hearing)

- Detailed neuroanatomical/pathological analysis of brain, spinal cord, eye, ear and peripheral ganglia

- Field potential and whole-cell patch clamp in brain slices

- Whole-cell patch clamp of cultured neurons and other cells (HEK, etc.)

- Fluorescence imaging of brain slices and cells

- Olfactory discrimination test (olfaction and social recognition)

- Trace and delay aversive conditioning

- Social interaction and social recognition tests

- Zero maze (anxiety)

Oncology

The targets of current oncology therapeutics fall into three major categories: cytotoxic agents such as DNA damaging agents or inhibitors of tubulin or topoisomerase, tissue-specific growth regulators such as estrogen receptor blockers and leutinizing hormone blockers, and disease-specific antitumor agents such as Gleevec, Herceptin and Rituxan. The oncology Level 1 screen is based on the hypothesis that targets for the next generation of cancer drugs are likely to fall into the same categories operating through control points in mammalian cell cycle, apoptosis or response to DNA damage.

Level 1 tests

Embryonic lethality and reduced viability

Targets for future cytotoxic agents are likely to be identified first by embryonic lethality or reduced viability. These phenotypes are examined further to detemine effects on cell cycle, apoptosis and angiogensis.

Tissue-specific growth regulation

Targets affecting growth, differentiation and function of reproductive organs are examined through histopathologic survey of males, virgin females and lactating female mice.

Cell proliferation

Oncogene targets that have a direct effect on cell cycle, DNA repair or apoptosis can manifest their function through changes in adult skin fibroblast proliferation. Punch biopsies are taken of skin samples from the backs of mutant mice and cohort controls. These are developed into primary fibroblast cultures and the fibroblast proliferation rates are measured in a strictly controlled protocol. The ability of this assay to detect hyperproliferative and hypoproliferative phenotypes has been demonstrated with p53 and Ku80 (unpublished results).

Level 2 oncology tests

Targets identified from Level 1 are characterized further for their potential role in human tumorigenesis. Focus is placed on targets that are highly expressed in human tumor cell lines and capable of driving the tumor phenotype as demonstrated by gene knock-down studies or overexpression-driven tumorigenesis models in nude mice.

Quantitative polymerase chain reaction for analysis of expression in cancerous and normal cell lines and tissues

Quantitative polymerase chain reaction of candidate genes is done using cDNA prepared from 66 cancer and nine normal cell lines from ATCC, seven primary cell strains from Clonetics, about three cancer lines and matched adjacent normal tissue controls from Ambion, MCF-7 breast cancer cells $+/-17\beta$-estradiol and LNCaP prostate cancer cells $+/-$ dihydrotestosterone. This is done to identify targets that are overexpressed in cancerous cell lines relative to normal cell and tissue controls.

Gene knock-down studies with short interfering RNA

Cancer cell lines determined to be overexpressing a target of interest are co-transfected with 3–6 short hairpin RNA vectors and blasticidin resistance vectors or synthetic short interfering RNAs to knock down the expression of specific targets. Assessment is made of the effects of RNA interference on *in vitro* proliferation, anchorage-dependent and anchorage-independent colony formation and the ability of cell lines to form tumors in nude mice.

Overexpression studies for putative oncogenes

Potential oncology targets are tested to determine whether they can drive tumor formation. Full-length genes of interest are cloned into a mammalian expression vector and co-transfected into NIH3T3 and RK3E cells with a blasticidin-resistance vector. The resulting blasticidin-resistant polyclonal pools are tested *in vitro* for acquisition of anchorage independence, reduced serum dependence and increased focus-forming ability. Stably transfected cell lines expressing exogenous cDNAs of interest are then analyzed for their ability to form tumors in athymic nude mice.

10.5 Conclusions

We have described a new conceptual framework for the discovery of drugs with the mammalian genome as starting material. The framework requires genetic antagonism of the drug target combined with a comprehensive *in vivo* physiological characterization of target function before any chemical screens for pharmaceutical agents are launched. This process constitutes a powerful genetic screen for the targets that allow, ultimately, for maximizing therapeutic effects while minimizing side-effects resulting from therapies modulating the target. In addition, determination of the role of the target in mammalian physiology enables identification of the likely medical indications for the therapeutics to be developed. Although this may appear an obvious prerequisite, it is important to note that many screens are conducted today against molecular targets for which the medical utility is either completely unknown or hypothesized based on only biochemical, gene expression or lower model organism data.

The mammalian genetic screen that we have described has been engineered specifically to reveal those genes that encode control points in physiology that may be used to treat major disease processes. Although there are literally thousands of assays that could be incorporated into the screen, we have selected those key tests that measure important medical parameters of physiology that are associated with accepted points of therapeutic intervention and major unmet medical needs. Additionally, the tests must be robust in their application to thousands of animals.

Once established, the screen used to discover therapeutic targets can be applied again to demonstrate the efficacy and potential side-effects of candidate therapeutic agents. This broad phenotypic screen, guided by mammalian genetics, provides a new level of power to the preclinical testing of compounds that are developed to interact with chosen targets. The screen enables identification of the key biomarker indicators of efficacy that should be followed when a compound is at the first-time-in-mammal stage. The genetic tools available for preclinical studies include not only wild-type animals but also knock-outs and knock-ins containing the actual human gene targets. The knock-out animals provide guidance for determining the efficacy of novel therapeutic agents. Another powerful aspect of the preclinical testing capabilities includes the treatment of knock-out animals themselves with compounds specific for the target. In such a scenario any effects seen, outside those associated with the knock-out state, are, by definition, off-target side-effects attributable to the compound itself. Clearly, the ability to manipulate the mouse genome at will provides exciting new opportunities to define accurately the on-target versus off-target side-effects produced by a given agent. Such new approaches are being incorporated into medicinal chemistry strategies to guide lead optimization for the invention of superior therapeutic agents.

In the post-genome era, a systematic *in vivo* screen for targets is becoming a necessary precondition for any high-throughput screen to identify small-molecule therapeutics. The recent revelation of tens of thousands of genes does not necessarily translate to the existence of thousands of drug targets. Strict criteria must be applied to the druggable genome in order to identify the targets for the next-generation breakthrough treatments for human disease.

10.6 References

Abdelilah, S., Mountcastle-Shah, E., Harvey, M., Solnica-Krezel, L., Schier, A. F., Stemple, D. L., Malicki, J., *et al.* (1996). Mutations affecting neural survival in the zebrafish *Danio rerio. Development* **123**, 217–227.

Abe, S., Watanabe, H., Hirayama, A., Shibuya, E., Hashimoto, M. and Ide, Y. (2000). Morphological study of the femur in osteopetrotic (op/op) mice using microcomputed tomography. *Br. J. Radiol.* **73**, 1078–1082.

Abuin, A., Holt, K. H., Platt, K. A., Sands, A. T. and Zambrowicz, B. P. (2002). Full-speed mammalian genetics: *in vivo* target validation in the drug discovery process. *Trends Biotechnol.* **20**, 36–42.

Accili, D., Drago, J., Lee, E. J., Johnson, M. D., Cool, M. H., Salvatore, P., Asico, L. D., *et al.* (1996). Early neonatal death in mice homozygous for a null allele of the insulin receptor gene. *Nat. Genet.* **12**, 106–109.

Air, E. L., Strowski, M. Z., Benoit, S. C., Conarello, S. L., Salituro, G. M., Guan, X. M., Liu, K., *et al.* (2002). Small molecule insulin mimetics reduce food intake and body weight and prevent development of obesity. *Nat. Med.* **8**, 179–183.

Alexander, J. M., Bab, I., Fish, S., Muller, R., Uchiyama, T., Gronowicz, G., Nahounou, M., *et al.* (2001). Human parathyroid hormone 1-34 reverses bone loss in ovariectomized mice. *J Bone Miner. Res.* **16**, 1665–1673.

Baier, H., Klostermann, S., Trowe, T., Karlstrom, R. O., Nusslein-Volhard, C. and Bonhoeffer, F. (1996). Genetic dissection of the retinotectal projection. *Development* **123**, 415–425.

Ballou, L. R., Botting, R. M., Goorha, S., Zhang, J. and Vane, J. R. (2000). Nociception in cyclooxygenase isozyme-deficient mice. *Proc. Natl. Acad. Sci. USA* **97**, 10272–10276.

Boring, L., Gosling, J., Chensue, S. W., Kunkel, S. L., Farese, R. V., Jr., Broxmeyer, H. E. and Charo, I. F. (1997). Impaired monocyte migration and reduced type 1 (Th1) cytokine responses in C-C chemokine receptor 2 knockout mice. *J. Clin. Invest.* **100**, 2552–2561.

Brand, M., Heisenberg, C. P., Warga, R. M., Pelegri, F., Karlstrom, R. O., Beuchle, D., Picker, A., Jiang, Y. J., Furutani-Seiki, M., van Eeden, F. J., *et al.* (1996). Mutations affecting development of the midline and general body shape during zebrafish embryogenesis. *Development* **123**, 129–142.

Brockerhoff, S. E., Hurley, J. B., Janssen-Bienhold, U., Neuhauss, S. C., Driever, W. and Dowling, J. E. (1995). A behavioral screen for isolating zebrafish mutants with visual system defects. *Proc. Natl. Acad. Sci. USA* **92**, 10545–10549.

Bucay, N., Sarosi, I., Dunstan, C. R., Morony, S., Tarpley, J., Capparelli, C., Scully, S., *et al.* (1998). Osteoprotegerin-deficient mice develop early onset osteoporosis and arterial calcification. *Genes Dev.* **12**, 1260–1268.

Bueno, O. F., Brandt, E. B., Rothenberg, M. E. and Molkentin, J. D. (2002). Defective T cell development and function in calcineurin A beta-deficient mice. *Proc. Natl. Acad. Sci. USA* **99**, 9398–9403.

Butler, A. A., Kesterson, R. A., Khong, K., Cullen, M. J., Pelleymounter, M. A., *et al.*, (2000). A unique metabolic syndrome causes obesity in the melanocortin-3 receptor-deficient mouse. *Endocrinology* **141**, 3518–3521.

Chalfie, M., Horvitz, H. R. and Sulston, J. E. (1981). Mutations that lead to reiterations in the cell lineages of *C elegans*. *Cell* **24**, 59–69.

Chen, J. N., Haffter, P., Odenthal, J., Vogelsang, E., Brand, M., van Eeden, F. J., Furutani-Seiki, M., *et al.* (1996). Mutations affecting the cardiovascular system and other internal organs in zebrafish. *Development* **123**, 293–302.

Clement, S., Krause, U., Desmedt, F., Tanti, J. F., Behrends, J., Pesesse, X., Sasaki, T., *et al.* (2001). The lipid phosphatase SHIP2 controls insulin sensitivity. *Nature* **409**, 92–97.

Giros, B., Jaber, M., Jones, S. R., Wightman, R. M. and Caron, M. G. (1996). Hyperlocomotion and indifference to cocaine and amphetamine in mice lacking the dopamine transporter. *Nature* **379**, 606–612.

Golling, G., Amsterdam, A., Sun, Z., Antonelli, M., Maldonado, E., Chen, W., Burgess, S., *et al.* (2002). Insertional mutagenesis in zebrafish rapidly identifies genes essential for early vertebrate development. *Nat. Genet.* **31**, 135–140.

Granato, M., van Eeden, F. J., Schach, U., Trowe, T., Brand, M., Furutani-Seiki, M., Haffter, P., *et al.* (1996). Genes controlling and mediating locomotion behavior of the zebrafish embryo and larva. *Development* **123**, 399–413.

Haffter, P., Granato, M., Brand, M., Mullins, M. C., Hammerschmidt, M., Kane, D. A., Odenthal, J., *et al.* (1996). The identification of genes with unique and essential functions in the development of the zebrafish, *Danio rerio*. *Development* **123**, 1–36.

Hawes, N. L., Smith, R. S., Chang, B., Davisson, M., Heckenlively, J. R. and John, S. W. (1999). Mouse fundus photography and angiography: a catalogue of normal and mutant phenotypes. *Mol. Vis.* **5**, 22.

Hedgecock, E. M., Sulston, J. E. and Thomson, J. N. (1983). Mutations affecting programmed cell deaths in the nematode *Caenorhabditis elegans*. *Science* **220**, 1277–1279.

Hirose, F., Ohshima, N., Shiraki, M., Inoue, Y. H., Taguchi, O., Nishi, Y., Matsukage, A., *et al.* (2001). Ectopic expression of DREF induces DNA synthesis, apoptosis, and unusual morphogenesis in the *Drosophila* eye imaginal disc: possible interaction with Polycomb and trithorax group proteins. *Mol. Cell Biol.* **21**, 7231–7242.

Homanics, G. E., DeLorey, T. M., Firestone, L. L., Quinlan, J. J., Handforth, A., Harrison, N. L., Krasowski, M. D., *et al.* (1997). Mice devoid of gamma-aminobutyrate type A receptor beta3 subunit have epilepsy, cleft palate, and hypersensitive behavior. *Proc. Natl. Acad. Sci. USA* **94**, 4143–4148.

Hopkins, A. L. and Groom, C. R. (2002). The druggable genome. *Nat. Rev. Drug Discov.* **1**, 727–730.

Horvitz, H. R. and Sulston, J. E. (1980). Isolation and genetic characterization of cell-lineage mutants of the nematode *Caenorhabditis elegans*. *Genetics* **96**, 435–454.

Ito, M., Oliverio, M. I., Mannon, P. J., Best, C. F., Maeda, N., Smithies, O. and Coffman, T. M. (1995). Regulation of blood pressure by the type 1A angiotensin II receptor gene. *Proc. Natl. Acad. Sci. USA* **92**, 3521–3525.

Joshi, R. L., Lamothe, B., Cordonnier, N., Mesbah, K., Monthioux, E., Jami, J. and Bucchini, D. (1996). Targeted disruption of the insulin receptor gene in the mouse results in neonatal lethality. *EMBO J.* **15**, 1542–1547.

Kawabe, T., Naka, T., Yoshida, K., Tanaka, T., Fujiwara, H., Suematsu, S., Yoshida, N., *et al.* (1994). The immune responses in CD40-deficient mice: impaired immunoglobulin class switching and germinal center formation. *Immunity* **1**, 167–178.

Kelsh, R. N., Brand, M., Jiang, Y. J., Heisenberg, C. P., Lin, S., Haffter, P., Odenthal, J., *et al.* (1996). Zebrafish pigmentation mutations and the processes of neural crest development. *Development* **123**, 369–389.

Klaman, L. D., Boss, O., Peroni, O. D., Kim, J. K., Martino, J. L., Zabolotny, J. M., Moghal, N., *et al.* (2000). Increased energy expenditure, decreased adiposity, and tissue-specific insulin sensitivity in protein-tyrosine phosphatase 1B-deficient mice. *Mol. Cell Biol.* **20**, 5479–5489.

Krege, J. H., Hodgin, J. B., Hagaman, J. R. and Smithies, O. (1995). A noninvasive computerized tail-cuff system for measuring blood pressure in mice. *Hypertension* **25**, 1111–1115.

LaJeunesse, D. R., McCartney, B. M. and Fehon, R. G. (2001). A systematic screen for dominant second-site modifiers of Merlin/NF2 phenotypes reveals an interaction with blistered/DSRF and scribbler. *Genetics* **158**, 667–679.

Leighton, P. A., Mitchell, K. J., Goodrich, L. V., Lu, X., Pinson, K., Scherz, P., Skarnes, W. C., *et al.* (2001). Defining brain wiring patterns and mechanisms through gene trapping in mice. *Nature* **410**, 174–179.

Lieschke, G. J., Grail, D., Hodgson, G., Metcalf, D., Stanley, E., Cheers, C., Fowler, K. J., *et al.* (1994). Mice lacking granulocyte colony-stimulating factor have chronic neutropenia, granulocyte and macrophage progenitor cell deficiency, and impaired neutrophil mobilization. *Blood* **84**, 1737–1746.

Malicki, J., Neuhauss, S. C., Schier, A. F., Solnica-Krezel, L., Stemple, D. L., Stainier, D. Y., Abdelilah, S., *et al.* (1996a). Mutations affecting development of the zebrafish retina. *Development* **123**, 263–273.

Malicki, J., Schier, A. F., Solnica-Krezel, L., Stemple, D. L., Neuhauss, S. C., Stainier, D. Y., Abdelilah, S., *et al.* (1996b). Mutations affecting development of the zebrafish ear. *Development* **123**, 275–283.

Marsh, D. J., Weingarth, D. T., Novi, D. E., Chen, H. Y., Trumbauer, M. E., Chen, A. S., Guan, X. M., *et al.* (2002). Melanin-concentrating hormone 1 receptor-deficient mice are lean, hyperactive, and hyperphagic and have altered metabolism. *Proc. Natl. Acad. Sci. USA* **99**, 3240–3245.

Mitchell, K. J., Pinson, K. I., Kelly, O. G., Brennan, J., Zupicich, J., Scherz, P., Leighton, P. A., *et al.* (2001). Functional analysis of secreted and transmembrane proteins critical to mouse development. *Nat. Genet.* **28**, 241–249.

Mizuno, A., Amizuka, N., Irie, K., Murakami, A., Fujise, N., Kanno, T., Sato, Y., *et al.* (1998). Severe osteoporosis in mice lacking osteoclastogenesis inhibitory factor/ osteoprotegerin. *Biochem. Biophys. Res. Commun.* **247**, 610–615.

Mukherjee, R., Davies, P. J., Crombie, D. L., Bischoff, E. D., Cesario, R. M., Jow, L., Hamann, L. G., *et al.* (1997). Sensitization of diabetic and obese mice to insulin by retinoid X receptor agonists. *Nature* **386**, 407–410.

Mullins, M. C., Hammerschmidt, M., Haffter, P. and Nusslein-Volhard, C. (1994). Large-scale mutagenesis in the zebrafish: in search of genes controlling development in a vertebrate. *Curr. Biol.* **4**, 189–202.

Nagy, T. R. and Clair, A. L. (2000). Precision and accuracy of dual-energy X-ray absorptiometry for determining *in vivo* body composition of mice. *Obes. Res.* **8**, 392–398.

Neuhauss, S. C., Solnica-Krezel, L., Schier, A. F., Zwartkruis, F., Stemple, D. L., Malicki, J., Abdelilah, S., *et al.* (1996). Mutations affecting craniofacial development in zebrafish. *Development* **123**, 357–367.

Nusslein-Volhard, C. and Wieschaus, E. (1980). Mutations affecting segment number and polarity in *Drosophila*. *Nature* **287**, 795–801.

Odenthal, J., Haffter, P., Vogelsang, E., Brand, M., van Eeden, F. J., Furutani-Seiki, M., Granato, M., *et al.* (1996). Mutations affecting the formation of the notochord in the zebrafish, *Danio rerio*. *Development* **123**, 103–115.

Oliver, P. M., John, S. W., Purdy, K. E., Kim, R., Maeda, N., Goy, M. F. and Smithies, O. (1998). Natriuretic peptide receptor 1 expression influences blood pressures of mice in a dose-dependent manner. *Proc. Natl. Acad. Sci. USA* **95**, 2547–2551.

Pasparakis, M., Alexopoulou, L., Episkopou, V. and Kollias, G. (1996). Immune and inflammatory responses in TNF alpha-deficient mice: a critical requirement for TNF alpha in the formation of primary B cell follicles, follicular dendritic cell networks and germinal centers, and in the maturation of the humoral immune response. *J. Exp. Med.* **184**, 1397–1411.

Piotrowski, T., Schilling, T. F., Brand, M., Jiang, Y. J., Heisenberg, C. P., Beuchle, D., Grandel, H., *et al.* (1996). Jaw and branchial arch mutants in zebrafish II: anterior arches and cartilage differentiation. *Development* **123**, 345–356.

Puignero, V., Salgado, J. and Queralt, J. (1995). Effects of cyclosporine and dexamethasone on IgE antibody response in mice, and on passive cutaneous anaphylaxis in the rat. *Int. Arch. Allergy Immunol.* **108**, 142–147.

Punyanitya, M., Leibel, R. L., Heymsfield, S. B. and Boozer, C. N. (2000). Evaluation of a new dual-energy x-ray absorptiometry technique for *in vivo* body composition measurements in mice. *FASEB J.* **14**, 497.

Rogers, D., Peters, J., Martin, J. E., Ball, S., Nicholson, S. J., Witherden, A. S., Hafezparast, M., Latcham, J., Robinson, T. L., Quilter, C. A. and Fisher, E. M. (2001). SHIRPA, a protocol for behavioral assessment: validation for longitudinal study of neurological dysfunction in mice. *Neurosci. Lett.* **306**, 89–92.

Saftig, P., Hunziker, E., Wehmeyer, O., Jones, S., Boyde, A., Rommerskirch, W., Moritz, J. D., *et al.* (1998). Impaired osteoclastic bone resorption leads to osteopetrosis in cathepsin-K-deficient mice. *Proc. Natl. Acad. Sci. USA* **95**, 13453–13458.

Scarff, K. L., Judd, L. M., Toh, B. H., Gleeson, P. A. and Van Driel, I. R. (1999). Gastric H(+),K(+)-adenosine triphosphatase beta subunit is required for normal function, development, and membrane structure of mouse parietal cells. *Gastroenterology* **117**, 605–618.

Schier, A. F., Neuhauss, S. C., Harvey, M., Malicki, J., Solnica-Krezel, L., Stainier, D. Y., Zwartkruis, F., *et al.* (1996). Mutations affecting the development of the embryonic zebrafish brain. *Development* **123**, 165–178.

Schreiber, S. L., Preiss, A., Nagel, A. C., Wech, I. and Maier, D. (2002). Genetic screen for modifiers of the rough eye phenotype resulting from overexpression of the notch antagonist hairless in *Drosophila*. *Genesis* **33**, 141–152.

Shahinian, A., Pfeffer, K., Lee, K. P., Kundig, T. M., Kishihara, K., Wakeham, A., Kawai, K., *et al.* (1993). Differential T cell costimulatory requirements in CD28-deficient mice. *Science* **261**, 609–612.

Solnica-Krezel, L., Stemple, D. L., Mountcastle-Shah, E., Rangini, Z., Neuhauss, S. C., Malicki, J., Schier, A. F., *et al.* (1996). Mutations affecting cell fates and cellular rearrangements during gastrulation in zebrafish. *Development* **123**, 67–80.

Sora, I., Takahashi, N., Funada, M., Ujike, H., Revay, R. S., Donovan, D. M., Miner, L. L., *et al.* (1997). Opiate receptor knockout mice define mu receptor roles in endogenous nociceptive responses and morphine-induced analgesia. *Proc. Natl. Acad. Sci. USA* **94**, 1544–1549.

Spicer, Z., Miller, M. L., Andringa, A., Riddle, T. M., Duffy, J. J., Doetschman, T. and Shull, G. E. (2000). Stomachs of mice lacking the gastric H,K-ATPase alpha -subunit have achlorhydria, abnormal parietal cells, and ciliated metaplasia. *J. Biol. Chem.* **275**, 21555–21565.

Stemple, D. L., Solnica-Krezel, L., Zwartkruis, F., Neuhauss, S. C., Schier, A. F., Malicki, J., Stainier, D. Y., *et al.* (1996). Mutations affecting development of the notochord in zebrafish. *Development* **123**, 117–128.

Sugiyama, F., Churchill, G. A., Higgins, D. C., Johns, C., Makaritsis, K. P., Gavras, H. and Paigen, B. (2001). Concordance of murine quantitative trait loci for salt-induced hypertension with rat and human loci. *Genomics* **71**, 70–77.

Sullivan, K. M. and Rubin, G. M. (2002). The Ca(2 +)-calmodulin-activated protein phosphatase calcineurin negatively regulates EGF receptor signaling in *Drosophila* development. *Genetics* **161**, 183–193.

Therrien, M., Morrison, D. K., Wong, A. M. and Rubin, G. M. (2000). A genetic screen for modifiers of a kinase suppressor of Ras-dependent rough eye phenotype in *Drosophila*. *Genetics* **156**, 1231–1242.

Walke, D. W., Han, C., Shaw, J., Wann, E., Zambrowicz, B. and Sands, A. (2001). *In vivo* drug target discovery: identifying the best targets from the genome. *Curr. Opin. Biotechnol.* **12**, 626–631.

Wiles, M. V., Vauti, F., Otte, J., Fuchtbauer, E. M., Ruiz, P., Fuchtbauer, A., Arnold, H. H., *et al.* (2000). Establishment of a gene-trap sequence tag library to generate mutant mice from embryonic stem cells. *Nat. Genet.* **24**, 13–14.

Xu, F., Gainetdinov, R. R., Wetsel, W. C., Jones, S. R., Bohn, L. M., Miller, G. W., Wang, Y. M., *et al.* (2000). Mice lacking the norepinephrine transporter are supersensitive to psychostimulants. *Nat. Neurosci.* **3**, 465–471.

Zambrowicz, B. P. and Sands, A. T. (2003). Knockouts model the 100 best-selling drugs – will they model the next 100? *Nat. Rev. Drug Discov.* **2**, 38–51.

Zambrowicz, B. P., Friedrich, G. A., Buxton, E. C., Lilleberg, S. L., Person, C. and Sands, A. T. (1998). Disruption and sequence identification of 2,000 genes in mouse embryonic stem cells. *Nature* **392**, 608–611.

Zhang, B. W., Zimmer, G., Chen, J., Ladd, D., Li, E., Alt, F. W., Wiederrecht, G., *et al.* (1996). T cell responses in calcineurin A alpha-deficient mice. *J. Exp. Med.* **183**, 413–420.

Index

Page numbers in *italic* indicate tables.